SUPERSYMMETRY, SUPERGRAVITY AND RELATED TOPICS

Edited by

F del Aguila
J A de Azcárraga
L E Ibáñez

SUPERSYMMETRY, SUPERGRAVITY AND RELATED TOPICS

Proceedings of the
XVth GIFT International Seminar on Theoretical Physics

4-9 June 1984
Sant Feliu de Guixols, Girona, Spain

World Scientific

Published by

World Scientific Publishing Co. Pte. Ltd.
P. O. Box 128, Farrer Road, Singapore 9128

SUPERSYMMETRY, SUPERGRAVITY AND RELATED TOPICS

ISBN 9971-966-79-4
9971-966-92-1

Printed in Singapore by Kim Hup Lee Printing Co Pte Ltd.

FOREWORD

The XVth GIFT International Seminar in Theoretical Physics was held in Sant Feliu de Guixols (Girona), Spain, from the 4th to the 9th of June and was devoted to "Supersymmetry, Supergravity and Related Topics". Its aim was to provide the audience with an introduction to the subject and some of its recent developments.

The Seminar was sponsored by the *Instituto de Estudios Nucleares,* the *Consejo Superior de Investigaciones Científicas,* the *Dirección General de Política Científica,* the *Generalitats de Catalunya i València,* the *British Council,* the *Institutos de Ciencias de la Educación de las Universidades Autónomas de Barcelona y de Madrid y de las Universidades de Barcelona y de Valencia,* the *Diputació de Girona* and the *Ajuntament de Sant Feliu de Guixols.*

These proceedings contain the written versions of the main series of lectures and the seminars. We are grateful to all lecturers and participants for their collaboration in the success (in spite of the weather) of the XVth GIFT Seminar. Finally, we wish to thank the secretary, Miss Mercè Pascual, for her efficiency.

F. del Aguila *(Barcelona)*
J. A. de Azcárraga *(Valencia)*
L. E. Ibáñez *(Madrid)*

CONTENTS

SUPERSYMMETRY, SUPERGRAVITY AND RELATED TOPICS

THE ELEMENTS OF SUPERSYMMETRY

K.S. Stelle[1]
Institute for Theoretical Physics
University of California
Santa Barbara, California 93106

ABSTRACT

An introduction to supersymmetry is presented, starting from the representations of the supersymmetry algebra. These are derived in Chapter 1 using Wigner's method of induced representations. The representations of the supersymmetry algebra on fields are introduced in Chapter 2. The natural formalism for supersymmetric field multiplets is superspace, which is applied to the Wess-Zumino model. In Chapter 3, supersymmetric Yang-Mills theories are discussed and superspace methods applied to their quantization, including coupling to supersymmetric matter. The structure of the quantum supermultiplet containing the stress tensor, *i.e.*, the supercurrent, is presented, in both conformally invariant and nonconformally invariant theories. In Chapter 4, an overview of the striking ultraviolet cancellations in supersymmetric field theories is presented, based upon the background field method. In order to test the completeness of our understanding of these cancellations, comparison is made with explicit calculations of two-loop divergences in the maximal six- and seven-dimensional super Yang-Mills theories. The onset of divergences at two loops in the seven-dimensional theory confirms the completeness of this analysis.

1. Supersymmetry Algebras and their Representations

Supersymmetry is the only known way to have a nontrivial unification of space-time and internal symmetries of the S-matrix in a four-dimensional relativistic particle theory. In the context of ordinary groups of symmetries for a

[1]Permanent address: Blackett Laboratory, Imperial College, London SW7 2BZ England

relativistic nontrivial S-matrix, the theorem of Coleman and Mandula[1] showed that the only allowed groups were locally isomorphic to the direct product of an internal symmetry group and the Poincaré group, subject to some general assumptions on analyticity and finiteness of the number of particle types. This direct product structure earned these results the name 'no-go theorem', because an internal symmetry can change neither spin nor mass.

The way to avoid the strictures of the no-go theorem proved to be the generalization from groups of symmetries to graded groups. Graded Lie groups are characterized by Graded Lie algebras, whose composition rules contain both commutators and anticommutators. The graded Poincaré algebra was first considered by Gol'fand and Likhtman[2] in 1971. A four-dimensional field theory with nonlinearly realized supersymmetry was constructed by Volkov and Akulov,[3] while the first four-dimensional theory with linearly realized supersymmetry was constructed by Wess and Zumino,[4] generalizing the supergauge transformations of dual models.

Another way to avoid the strictures of the no-go theorem is to allow an infinite number of particle types. This is realized in the Kaluza-Klein program, based upon higher dimensions of space-time with spontaneous compactification to four dimensions.

The difficulties with internal symmetry currents carrying nontrivial Lorentz representations are codified by the Coleman-Mandula theorem.[1] The basic idea is that new conserved quantities with nontrivial Lorentz representations would force the scattering matrix to be unity except when certain kinematical conditions are met. For example, conservation of momentum P_μ and angular momentum $M_{\mu\nu}$ in a 2-body collision leaves only the scattering angle unknown. Additional conservation laws would allow only a discrete set of scattering angles, but then the analyticity of the S-matrix would rule out scattering at all angles.

Suppose there were a conserved symmetric traceless tensor charge $Q_{\alpha\beta}$. Lorentz invariance would then require that the matrix element $\langle p|Q_{\alpha\beta}|p\rangle$ take the form $(p^2 + m^2 = 0)$

$$\langle p|Q_{\alpha\beta}|p\rangle = (p_\alpha p_\beta + \frac{1}{4}\eta_{\alpha\beta}m^2)\, f(m^2). \tag{1.1}$$

Conservation of this quantity in a two-particle interaction with momenta $p_1{}^\mu$, $p_2{}^\mu$ scattering to $q_1{}^\mu$, $q_2{}^\mu$ would require

$$p_{1\alpha}p_{1\beta} + p_{2\alpha}p_{2\beta} = q_{1\alpha}q_{1\beta} + q_{2\alpha}q_{2\beta}, \tag{1.2}$$

but this can happen only for zero scattering angle. Then, by analyticity, the scattering would have to be zero for all angles.

The above difficulty occurs with all bosonic symmetries carrying nontrivial Lorentz indices, beyond those already present in the Poincaré algebra. Supersymmetry escapes the requirements of the Coleman-Mandula theorem because the generators are *fermionic*. The simplest example is the Wess-Zumino model.[4)] Starting with the free theory of a massless complex scalar and a Majorana spinor field ($\bar{\psi} = \psi^T C^\dagger$)

$$I^{free} = \int d^4x (\partial_\mu \phi^* \partial^\mu \phi + i\bar{\psi} \not{\partial} \psi), \tag{1.3}$$

we find that there is a conserved spinorial current

$$Q^{free}_{\mu\alpha} - (\partial_\nu \phi^* \gamma^\nu \gamma_\mu \psi)_\alpha, \tag{1.4}$$

as may be checked using the free field equations:

$$\begin{aligned} \partial^\mu Q^{free}_{\mu\alpha} &= \partial^\mu \partial_\nu \phi^* \gamma^\nu \gamma_\mu \psi + \partial_\nu \phi^* \gamma^\nu \gamma^\mu \partial_\mu \psi \\ &= \frac{1}{2} \partial_\mu \partial_\nu \phi^* [\gamma^\mu \gamma^\nu + \gamma^\nu \gamma^\mu] \psi + \not{\partial} \phi^* \not{\partial} \psi \\ &= \Box \phi^* \psi + \not{\partial} \phi^* \not{\partial} \psi \\ &= 0. \end{aligned} \tag{1.5}$$

Of course, it is not surprising to find such a conserved current in a free theory — in noninteracting purely bosonic theories there are always an infinite number of conserved 'Zilch' currents. What is truly remarkable about the Wess-Zumino model, however, is that it is possible to find a conserved supercharge if the interactions take a certain form:

$$I_{WZ} = \int d^4x \left[\partial_\mu \phi^* \partial^\mu \phi + i\bar{\psi} \not{\partial} \psi - g^2 (\phi^* \phi)^2 - g \left(\phi \bar{\psi} \left(\frac{1 + i\gamma_5}{2} \right) \psi + h.c. \right) \right]. \tag{1.6}$$

In this interacting theory, the full current

$$Q^{WZ}_{\mu\alpha} = Q^{free}_{\mu\alpha} + g\gamma_\mu (\phi^*)^2 \left(\frac{1 - i\gamma_5}{2} \right) \psi \tag{1.7}$$

is conserved by virtue of the interacting field equations. The supercharge is given by the Majorana spinor

$$Q_\alpha = \int d^3x\, Q_{o\alpha}. \tag{1.8}$$

Since $Q_{o\alpha}$ is linear in ψ, it will anticommute with itself at spacelike separations. More generally, we will need to know the algebra of anticommutators of the fermionic charge Q_α. Since Q_α is conserved, the anticommutator

$$\{Q_\alpha, \bar{Q}_\beta\} \tag{1.9}$$

must be conserved as well. But this is a bosonic operator with nontrivial Lorentz structure. In order to be consistent with the Coleman-Mandula theorem, it must be a Poincaré generator, so we obtain the (flat space) supersymmetry algebra

$$\{Q_\alpha, \bar{Q}_\beta\} = 2\gamma^\mu_{\alpha\beta} P_\mu. \tag{1.10}$$

Since the bosonic part of this algebra satisfies the Coleman-Mandula theorem, Q_α can be conserved even in an interacting theory. The Poincaré algebra together with its extension (1.10) is known as a graded Lie algebra.

The most general grading of the Poincaré algebra involves the addition of N spin 1/2 fermionic generators $Q_\alpha{}^i (i = 1, \ldots, N)$ to the bosonic, or even part of the algebra. These fermionic generators are required to be irreducible under the Lorentz group in four dimensions. Accordingly, we impose upon them the Majorana constraint $C(\bar{Q}_\alpha{}^i)^T = Q_\alpha{}^i$, or equivalently we could use Weyl spinors. The bosonic part of the algebra is restricted by the Coleman-Mandula theorem to be a direct product $P \otimes (T \otimes Z)$, where P is the Poincaré algebra, T is a semisimple internal symmetry Lie group acting on the indices i, j, and Z is an Abelian Lie group, the center of the algebra. The structure constants of the full algebra are then restricted via the Jacobi identities by the required structure of the bosonic part of the algebra. It is convenient to use Van der Waerden notation and the Weyl representation for the spinors, splitting them up into complex two component spinors and their complex conjugates, $Q^i = (Q_\alpha{}^i, \bar{Q}^{\dot{\alpha}}{}_i)$, where $\bar{Q}^{\dot{\alpha}}{}_i = \epsilon^{\alpha\beta}(Q_\beta{}^i)*$ and $\epsilon^{\alpha\beta} = \epsilon^{\dot{\alpha}\dot{\beta}} = -\epsilon_{\alpha\beta} (\epsilon^{12} = 1)$. The resulting most general grading is given by[5)]

$$\begin{aligned}
\{Q_\alpha{}^i, \bar{Q}^{\dot{\beta} j}\} &= 2\,\sigma^\mu{}_{\alpha\dot{\beta}}\, \delta^i{}_j P_\mu \\
\{Q_\alpha{}^i, Q_\beta{}^j\} &= 2\,\epsilon_{\alpha\beta} (\Omega^e)^{ij} Z_e \\
[Q_\alpha{}^i, T_a] &= (f_a)^i{}_j Q_\alpha{}^j \\
[Q_\alpha{}^i, M_{\mu\nu}] &= (\sigma_{\mu\nu})_\alpha{}^\beta Q_\beta{}^i \\
[Q_\alpha{}^i, P_\mu] = [Q_\alpha{}^i, Z_e] &= [Z_e, P_\mu] = [Z_e, Z_k] = 0.
\end{aligned} \tag{1.11}$$

The spinorial generators $Q_\alpha{}^i$ that form the odd part of the algebra transform according to the spin 1/2 representation matrices f^i_{aj} of the internal symmetry group, whose generators are the T_a. This internal symmetry group is an outer automorphism of the algebra of Q's, P's, and Z's, since the T_a generators do not occur as the result of the (anti)commutation of these other generators. The Z_e do occur in the $\{Q, Q\}$ anticommutator, and through the analysis of the Jacobi identities it can be shown[5)] that they must commute with all of the generators of the algebra, *i.e.*, they form the center of the algebra. The structure constants $(\Omega^e)^{ij}$ must be antisymmetric numerically invariant matrices under the action of the internal symmetry group.

It is clear from the algebra (1.11) that in the case $Z_e = 0$ the algebra has a $U(N)$ outer automorphism, and for this reason we have written the internal symmetry index in the lower position on $\bar{Q}_{\dot{\alpha}i}$. $U(N)$ may or may not be a symmetry of a theory that is invariant under the Q's and the Poincaré generators. If $U(N)$ is a symmetry of a given theory, the Q's and spinor fields must be taken as Weyl spinors, for a Majorana spinor mixes upper and lower i, j indices: $Q^i = (Q_\alpha{}^i, \bar{Q}^{\dot{\alpha}}{}_i)$. If the spinors are taken to be Majorana, then the maximal automorphism group is $SO(N)$.

When the central charges are present, the requirement that the $(\Omega^e)^{ij}$ be numerically invariant under the action of the internal symmetry group, *i.e.*, $(f_a)^i{}_j(\Omega^e)^{jk} + (f_a)^k{}_l(\Omega^e)^{il} = 0$, reduced the internal symmetry group to some subgroup of $U(N)$. In the case of just one central charge the maximal internal symmetry group is $USP(N)$, for N even, in which case $(\Omega)^{ij}$ is the antisymmetric numerically invariant metric for that group. With more than one central charge, the internal symmetry group is further reduced.

We begin with the massive one-particle states in the absence of central changes ($Z_e = 0$). Since $P^\mu P_\mu$ is still a Casimir operator for the supersymmetry algebra, all the states in an irreducible representation must have the same mass M. We choose the Lorentz rest frame so that $P^\mu = (M, 0)$. The little group is now $SU(2)$, under which the α and $\dot{\alpha}$ indices transform in the same way.

The anticommutators of the Q's in the rest frame take the form

$$\begin{aligned}\{Q_\alpha{}^i, \bar{Q}_{\dot{\beta}j}\} &= M\delta_{\alpha\dot{\beta}}\delta^i{}_j \\ \{Q_\alpha{}^i, Q_{\beta\, j}\}\} &= \{\bar{Q}_{\dot{\alpha}i}, \bar{Q}_{\dot{\beta}j}\} = 0.\end{aligned} \tag{1.12}$$

This is the algebra of $2N$ fermionic creation and annihilation operators. It has a maximal automorphism group $SO(4N)$, for which it is, in fact, the Clifford algebra,[7] as may be more easily seen by going over to a Majorana representation for the four component spinors, in which case the stability sub-algebra becomes

$$\{Q_\gamma{}^i, Q_\lambda{}^j\} = M\delta_{\gamma\lambda}\delta^{ij}; \qquad \gamma, \lambda = 1\ldots 4; \quad i,j = 1\ldots N. \tag{1.13}$$

The unique irreducible representation of the Clifford algebra (1.12) has dimension 2^{2N}, and can be derived by starting from a singlet Clifford vacuum Ξ satisfying

$$Q_\alpha{}^i \Xi = 0 \qquad \forall \alpha, i \tag{1.14}$$

and building up the 2^{2N} states

$$\begin{array}{ccccccc} \Xi, \bar{Q}_{\dot\alpha i}\Xi, & \bar{Q}_{\dot\alpha i_1}\bar{Q}_{\dot\alpha i_2}\Xi, & \ldots, & \bar{Q}_{\dot\alpha}\bar{Q}_{\dot\alpha} & \ldots & \bar{Q}_{\dot\alpha}\Xi, \ldots \\ 1+2N & + \cdots + & & \frac{2N(2N-1)}{2} & = & (1+1)^{2N} \end{array} \tag{1.15}$$

These states span the spinorial representation of $SO(4N)$, which can be split into the + and - eigenstates of γ^{4N+1}, each of which have dimension 2^{2N-1}, containing the bosons and fermions, respectively.

The $SO(4N)$ symmetry that classifies the one-particle states is not a symmetry of a supersymmetric Lagrangian, for it refers only to the rest-frame stability sub-algebra. Another classification symmetry applies to the states of a given spin. If we define the $2N$ component spinors[7]

$$\begin{aligned} Q_\alpha{}^a &= Q_\alpha{}^i, \qquad & a = i = 1, 2, \ldots, N \\ &= \bar{Q}^{\dot\alpha}{}_i = \epsilon^{\dot\alpha\dot\beta}\bar{Q}_{\dot\beta i}, \qquad & a = N + i = N+1, \ldots, 2N \end{aligned} \tag{1.16}$$

we find that they satisfy the reality condition

$$(Q_\alpha{}^a)^* = \epsilon^{\alpha\beta}\Omega_{ab}Q_\beta{}^b \tag{1.17}$$

with

$$\Omega_{ab} = -\Omega_{ba} = \begin{pmatrix} 0 & \mathbf{1} \\ -\mathbf{1} & 0 \end{pmatrix}_{ab}. \tag{1.18}$$

In this notation, the anticommutator of the Q's becomes, for $Z_e = 0$,

$$\{Q_\alpha{}^a, Q_\beta{}^b\} = M\epsilon_{\alpha\beta}\Omega^{ab}, \qquad \Omega^{ac}\Omega_{cb} = \delta^a{}_b. \tag{1.19}$$

In this form, the stability sub-algebra has manifest $SU(2)\times USP(2N)$ symmetry, under which the spinor charges Q still transform irreducibly in the vector representation: $4N \to (2, 2N)$. Thus, we see that the states of a given spin are classified by representations of $USP(2N)$, and the 2^{2N} states of the whole supermultiplet break up into $SU(2)$ and $USP(2N)$ representations as

$$2^{2N} = (N+1, 1) + (N, 2N) + \ldots (N+1-k, [2N]_k) + \cdots + (1, [2N]_N), \tag{1.20}$$

where the first label is the dimension of the $SU(2)$ representations ($J = (N-k)/2$) and $[2N]_k$ indicates the k index totally antisymmetric and traceless (with Ω_{ab}) representation of $USP(2N)$.

In the above we have assumed that the Clifford vacuum carries no spinor internal symmetry representation. The general irreducible massive representations of supersymmetry on one particle states are obtained by allowing Ω to carry spin and some representation of the internal symmetry group for the theory of interest, in general $U(N)$. For the spin states, we just multiply the Clifford vacuum spin into the $SU(2)$ representations given in (1.20). If the Clifford vacuum is not a singlet under the internal symmetry then the $USP(N)$ representations given in (1.20) must be reduced into internal symmetry representations, and then the product taken with the representation of the Clifford vacuum Ξ. The total dimension of the supermultiplet is then

$$D = 2^{2N} \times d_\Xi, \tag{1.21}$$

where d_Ξ is the dimension of the spin $\otimes$ variable symmetry representation carried by the Clifford vacuum.

For example, we have the following irreducible massive supermultiplets, with Clifford vacuums of spin J_Ξ (all examples are singlets under internal symmetry):

TABLE 1

N \ J	2	3/2	1	1/2	0	J_Ξ
1				1	2	0
			1	2	1	1/2
		1	2	1		1
	1	2	1			3/2
2			1	4	5	0
		1	4	5⊕1	4	1/2
	1	4	5⊕1	4	1	1
3		1	6	14	14	0
	1	6	14⊕1	14⊕6	14	1/2
4	1	8	27	48	42	0

The quantity J_Ξ is known as the superspin of the multiplet. As can be seen, the lowest dimensional massive representation has spin states from 0 to $N/2$. In general, the range is $\max(0, J_\Xi - \frac{N}{2})$ to $J_\Xi + \frac{N}{2}$.

Irreducible supersymmetry representations on massless states can be derived in the same fashion as above, only now choosing the standard four-momentum $P^\mu = (k, o, o, k)$. In this case, the stability sub-algebra is

$$\begin{aligned} \{Q_\alpha{}^i, \bar{Q}_{\dot{\beta} j}\} &= k(\mathbf{1} + \sigma_3)_{\alpha\dot{\beta}}\delta^i{}_j \\ \{Q_\alpha{}^i, Q_\beta{}^j\} &= 0, \end{aligned} \tag{1.22}$$

from which it can be seen that $Q_2{}^i$, $\bar{Q}_{2j}$ are no longer creation and annihilation operators, so we set $Q_2{}^i = 0$ and have a Clifford algebra for only N creation and annihilation operators, whose irreducible representation contains 2^N states. If we rescale $Q_1{}^i = \frac{1}{\sqrt{2}} Q^i$, we have

$$\begin{aligned} \{Q^i, \bar{Q}_j\} &= \delta^i{}_j \\ \{Q^i, Q^j\} &= \{\bar{Q}_i, \bar{Q}_j\} = 0. \end{aligned} \tag{1.23}$$

We can then define

$$Q^a = \left[\frac{Q^i + \bar{Q}_i}{\sqrt{2}}, i\frac{(Q^i - \bar{Q}_i)}{\sqrt{2}}\right] \qquad q = 1 \ldots 2N \tag{1.24}$$

and the anticommutation relations for the Q^a become the Clifford algebra of $SO(2N)$. The 2^N states in the fundamental massless multiplet span the spinor representation of the classification symmetry $SO(2N)$, and can again be separated into bosonic and fermionic states using the $\pm$ eigenvalues of $\gamma(2N+1)$.

In the massless case, the states of the multiplet can be classified by helicity, with the classification symmetry for states of the same helicity now reduced to $U(N)$; as can be seen since with ${Q_2}^i = 0$, the anticommutators (1.19) just reduce to (1.23), with manifest $U(1)_{helicity} \otimes U(N)$ symmetry. We are thus interested in the decomposition of the 2^N states in the spinor representation of $SO(2N)$ into $U(1)_{helicity} \otimes U(N)$:

$$2^N = (\lambda, 1) + (\lambda - \frac{1}{2}, \bar{N}) + \cdots + (\lambda - \frac{k}{2}[\bar{N}]_k) + \cdots + (\lambda - \frac{N}{2}, 1) \tag{1.25}$$

where λ is the helicity of the Clifford vacuum and $[\bar{N}]_k$ is here the totally antisymmetric representation of $U(N)$ with k indices.

In a field theory, we must have a PCT conjugate state for every helicity, and these are not in general contained in (1.25). Thus we generally have to double the multiplet (1.25) by adding the PCT conjugate multiplet $(\lambda, 1), \ldots, (-\lambda + \frac{N}{2}, 1)$. If, in addition, the clifford vacuum carries a representation of some internal symmetry group $\subset U(N)$, then the PCT conjugate multiplet must transform in the conjugate representation of the internal symmetry group.

The only case where the addition of PCT conjugate states is not necessary is when $\lambda = \frac{N}{4}$, when the supersymmetry multiplet (1.25) already contains the PCT conjugates. This also gives the minimum helicity range for a massless multiplet: $\lambda = 0$ to $\lambda = \pm\frac{N}{4}$.

Just as we may decompose the massive spin state representations of the Poincaré group into massless helicity states, in supersymmetry we may decompose the massive multiplets into massless ones. The pattern is given by the decomposition of the 2^{2N} states of the fundamental massive representation (1.15) into massless multiplets. The massless multiplets can be denoted $\{\lambda, [N]_k\}$ where λ is the maximum helicity in the massless multiplet, and this state (the Clifford vacuum of the massless multiplet) also carries the representation $[N]_k$ of $U(N)$, *i.e.*,

$$\{\lambda, [N]_k\} = (\lambda, [N]_k) + (\lambda - \frac{1}{2}, [N]_k \otimes \bar{N}) + \cdots + (\lambda - \frac{\ell}{2}, [N]_k \otimes [\bar{N}]_k) + \cdots + (\lambda - \frac{N}{2}, [N]_k). \tag{1.26}$$

The decomposition of the fundamental massive multiplet is then

$$2^{2N} = \{\frac{N}{2}, 1\} + \{\frac{N-1}{2}, N\} + \cdots + \{\frac{N-k}{2}, [N]_k\} + \cdots + \{0, 1\}. \qquad (1.27)$$

In the massive supermultiplets discussed above, we have set the central charges to zero. If the central changes are active, the multiplet structure may still be analyzed by the method of induced representations.[8)] As we have pointed out above, the presence of central charges also affects the internal symmetry structure that can be given to the supermultiplets. The simplest case is that of a single central charge occurring as

$$\{Q_\alpha{}^i, Q_\beta{}^j\} = 2\epsilon_{\alpha\beta}\Omega^{ij} Z \qquad (1.28)$$

with Ω^{ij} the numerically invariant antisymmetric metric of $USP(N)$, for N even. In this case, the maximal internal symmetry group is $USP(N)$, and the massive particle states are also classified by this symmetry.

The most striking feature of supermultiplets with central charges is the reduction in the range of spins. In the simplest case, as given above, the lowest dimensional multiplet has maximum spin $J = N/4$, half that of the massive multiplet without central charges. The structure of the multiplets is just that of the massive multiplets without central charges for $N/2$ extended supersymmetry (again, for N even), but doubled. The central charge just rotates every state into its double.

2. Superactions in Superspace

In order to formulate supersymmeric field theories, we need to know the representations of the supersymmetry algebra on fields. In fact, we can already discover much about these representations from the results of the last chapter, even before we discuss the details of the supersymmetry transformations. As with the representations of the Poincaré group, we can learn most of the structure of the irreducible representations on fields from the irreducible representations on massive particle states.

For example, a massive spin zero particle corresponds to a scalar field ϕ satisfying

$$(\Box - m^2)\phi = 0. \tag{2.1}$$

A more revealing example is that of a massive spin one particle, which can be represented by a vector field A_μ satisfying the Proca field equation

$$\partial_\mu F^{\mu\nu}(A) - m^2 A^\nu = 0. \tag{2.2}$$

Taking the divergence of this equation, we obtain

$$m^2 \partial_\nu A^\nu = 0, \tag{2.3}$$

so $A^\nu = A^{\nu T}$, *i.e.*, it is a Poincaré irreducible transverse vector field. The remaining content of (2.2) is then just

$$(\Box - m^2)A^{\nu T} = 0. \tag{2.4}$$

Thus, a massive spin one field is described by a vector satisfying the Klein-Gordon equation (2.4) together with the auxiliary condition (2.3) which ensures Poincaré irreducibility. This pattern persists for all massive fields: Poincaré irreducible fields which satisfy the massive Klein-Gordon equation describe particle states with spin determined by the Lorentz representation of the field.

The above correspondence between massive particles and Poincaré irreducible fields is always one-to-one in terms of the number of degrees of freedom of the particles and the number of components of the fields. Spinor representations of the Lorentz group have numbers of components equal to integral multiples of four. Spinor particles, however, have numbers of degrees of freedom equal to multiples of two. Thus, a Majorana or Weyl spinor field must correspond to two massive spin 1/2 particles when we impose the massive Klein-Gordon equation, and similarly for higher spins. In the case of spin 1/2 particles, we may separate the Klein-Gordon equation for a Majorana field

$$(\Box - m^2)\psi = 0 \tag{2.5}$$

into two first order equations,

$$\begin{aligned}(\lambda - \not\partial\psi) &= 0 \\ (\not\partial\lambda - m^2\psi) &= 0.\end{aligned} \tag{2.6}$$

If we now redefine

$$\begin{aligned}\lambda &= \chi + \phi \\ m\psi &= \chi - \phi\end{aligned} \tag{2.7}$$

we obtain the system of equations

$$\begin{aligned}(\not\partial + m)\phi &= 0 \\ (\not\partial - m)\chi &= 0\end{aligned} \tag{2.8}$$

and if one wishes to change the sign on the mass term in the second of these, it is sufficient to redefine $\chi = \gamma_5\chi'$. Thus we see explicitly that the massive Klein-Gordon equation for a Majorana spinor field describes two massive spin 1/2 particles.

The supersymmetry algebra without central charges is a direct extension of the Poincaré algebra, and we find that the above pattern extends to cover full irreducible representations of supersymmetry on fields. For example, take the simplest massive particle representation given in Table 1, containing one spin 1/2 particle and two spin 0 particles. Because of the 2-1 correspondence between spinor particles and fields explained above, we must double the particle representation in order to have the same number of spinor components as a Majorana (or Weyl) spinor field. The correspondence then gives us the simplest representation on fields,

$$\Big(A(x), B(x), \psi_\alpha(x)\left(\bar\psi_{\dot\alpha}(x)\right), F(x), G(x)\Big), \qquad J_\Xi = 0 \tag{2.9}$$

i.e., four scalars and one complex two-component spinor field. This multiplet is known as the chiral multiplet of $N = 1$ supersymmetry. The relative dimensions of the component fields are not fixed by the correspondence to massive particle states, and we shall return to this point later. The parity of the above fields is also not determined.

The next particle representation in Table 1 can be taken over without doubling, as there are already four spin 1/2 degrees of freedom:

$$\Big(C(x), \lambda_\alpha(x)\big(\bar{\lambda}_{\dot{\alpha}}(x)\big), V_\mu(x)\Big), \qquad J_\Xi = \frac{1}{2} \tag{2.10}$$

where the vector field V_μ must be transverse to ensure Poincaré irreducibility:

$$\partial^\mu V_\mu = 0. \tag{2.11}$$

The need for the condition (2.11) in the irreducible supermultiplet (2.10) is to ensure that there are the same number of effective bosonic as there are fermionic field components, as we always must have in representations of supersymmetry, either on fields or on particles. Thus, the vector V_μ counts for only three bosonic components, and in total in (2.10) we have four fermionic plus four bosonic components. The multiplet (2.10) is known as the $N = 1$ linear multiplet.

We may, of course, wish to consider reducible representations of supersymmetry as well, particularly in order to avoid having to impose differential subsidiary conditions like (2.11) on some component fields. Just as with ordinary fields we may combine a scalar and a transverse vector to obtain an unconstrained 4-vector, $V_\mu = V_\mu^T + \partial_\mu \phi$, so in supersymmetry the multiplets (2.9) and (2.10) can be combined to form the $N = 1$ general scalar multiplet:

$$\big(C(x), \lambda_\alpha(x), H(x), K(x), V_\mu(x), \chi_\alpha(x), D(x)\big). \tag{2.12}$$

The irreducible submultiplets of the general scalar multiplet are the linear multiplet $(C, \lambda_\alpha, V_\mu^T)$ and the chiral multiplet $(H, K, \lambda_\alpha, D, \partial^\mu V_\mu)$ containing the longitudinal part of V^μ.

Now we consider the detailed form of the supersymmetry transformations. The basic structure can be established by dimensional considerations. From the supersymmetry algebra (1.10) we see that the supersymmetry generator has dimensions of $(\text{mass})^{1/2}$, so the parameter ϵ^α of supersymmetry must have dimensions of $(\text{length})^{1/2}$. This is in accord with the canonical dimensions of scalar and spinor fields, so that in the chiral four component Majorana spinor $\psi = (\psi_\alpha, \bar{\psi}^{\dot{\alpha}})$ of canonical dimensions (respectively 1, 3/2), we have

$$\begin{aligned} \delta A &= i\bar{\epsilon}\psi \\ \delta B &= i\bar{\epsilon}\gamma_5\psi. \end{aligned} \tag{2.13}$$

In order to be dimensionally consistent, the spinor ψ in turn must transform into fields of dimension two or into derivatives of fields of dimension one. It does both:

$$\delta\psi = \partial_\mu(A - \gamma_5 B)\gamma^\mu \epsilon + (F + \gamma_5 G)\epsilon. \tag{2.14}$$

Since we have now exhausted the fields in the multiplet (2.9), the dimension two fields can only transform into derivatives of the spinor ψ:

$$\begin{aligned} \delta F &= i\bar{\epsilon}\gamma^\mu \partial_\mu \psi \\ \delta G &= i\bar{\epsilon}\gamma_5 \gamma^\mu \partial_\mu \psi. \end{aligned} \tag{2.15}$$

Due to the presence of the derivatives in the transformations (2.14) and (2.15), the commutator of two supersymmetry transformations with parameters ϵ_1 and ϵ_2 on any field gives a translation (generator $P_\mu = i\partial_\mu$) with parameter $(\bar{\epsilon}_2\gamma_\mu\epsilon_1 - \bar{\epsilon}_1\gamma_\mu\epsilon_2)$. In order for this not to vanish, the Majorana spinors ϵ_1, ϵ_2 must be taken to be anticommuting objects. This is in accord with the classical limit of quantized spinor fields, which must also be taken to be anticommuting. Before continuing with the structure of the supersymmetry transformations, we present a discussion due to Valuyev[9] on the nature of anticommuting $C-$ numbers.

In order to realize the Grassman algebra of anticommuting $C-$ numbers concretely, we take the Clifford algebra for $SO(2N)$ where N denotes some very large integer. The γ-matrices for this group satisfy

$$\{\gamma_i, \gamma_j\} = 2\delta_{ij}. \tag{2.16}$$

We can define the basis for the Grassman algebra by

$$\theta_n = \frac{1}{\sqrt{2}}(\gamma_{2n} + i\gamma_{2n+1}) \tag{2.17}$$

$$d\theta_n = \frac{1}{\sqrt{2}}(\gamma_{2n} - i\gamma_{2n+1}), \tag{2.18}$$

where the θ_n anticommute among themselves and the $d\theta_n$ anticommute among themselves, but give δ_{mn} when mixed anticommutators are taken:

$$\{\theta_m, \theta_n\} = \{d\theta_m, d\theta_n\} = 0 \tag{2.19}$$

$$\{d\theta_m, \theta_n\} = 2\delta_{mn}. \tag{2.20}$$

Thus, we can define the $SO(2N)$ invariant integral

$$\int d\theta\, f(\theta) = 2^{-N}\mathrm{tr}[d\theta\, f(\theta)], \tag{2.21}$$

where $d\theta$ and θ are now general elements of the Grassman algebra formed by taking linear combinations of the basis (2.17),(2.18). In particular for the integral over a single θ the tracelessness of the γ matrices implies

$$\int d\theta = 0 \tag{2.22}$$

while the integral

$$\int d\theta\, \theta = 1 \tag{2.23}$$

follows from (2.20) and a construction of $d\theta$ from the basis (2.18) conjugate to that of θ from the basis (2.17). The rules (2.22) and (2.23) are just Berezin's rules for integrating over anticommuting variables.[10]

Requiring that the spinor fields and the spinor parameter ϵ of the supersymmetry transformations be anticommuting elements of a Grassman algebra as above, we can now see that the transformations (2.13)-(2.15) do indeed realize the supersymmetry algebra. This is trivially seen in the transformations on the scalar A and pseudoscalar B. More interesting is the check of the algebra's closure on the spinor ψ. In order to check the algebra, we take the commutator of two supersymmetry transformations with parameters ϵ_1 and ϵ_2, since $[\delta\epsilon_2, \delta\epsilon_1]\psi = -\bar{\epsilon}_2{}^{\alpha}\epsilon_1{}^{\beta}\{Q_\alpha, \bar{Q}_\beta\}\psi$. Thus,

$$\delta_{\epsilon_1}\psi = \partial_\mu(A - \gamma_5 B)\gamma^\mu\epsilon_1 + (F + \gamma_5 G)\epsilon_1 \tag{2.24}$$

so

$$\begin{aligned}[\delta_{\epsilon_2}, \delta_{\epsilon_1}]\psi = {} & i\partial^\mu(\bar{\epsilon}_2\psi - \bar{\epsilon}_2\gamma_5\psi\gamma_5)\gamma^\mu\epsilon_1 \\ & + i(\bar{\epsilon}_2\gamma^\mu\partial_\mu\psi + \bar{\epsilon}_2\gamma_5\gamma^\mu\partial_\mu\psi\gamma_5)\epsilon_1 - (1 \leftrightarrow 2),\end{aligned}$$

and in order to bring the two ϵ's together, we must make a Fierz transformation:

$$\begin{aligned}
[\delta_{\epsilon_2}, \delta_{\epsilon_1}]\psi &= -\frac{i}{4}\bar{\epsilon}_2\gamma_\nu\epsilon_1(\gamma^\mu\gamma^\nu\partial_\mu\psi - \gamma_5\gamma^\mu\gamma^\nu\gamma_5\partial_\mu\psi) \\
&\quad + \frac{i}{2}\bar{\epsilon}_2\sigma_{\rho\tau}\epsilon_1(\gamma^\mu\sigma^{\rho\tau}\partial_\mu\psi - \gamma_5\gamma^\mu\sigma^{\rho\tau}\gamma_5\partial_\mu\psi) \\
&\quad - \frac{i}{4}\bar{\epsilon}_2\gamma_\nu\epsilon_1(\gamma^\nu\gamma^\mu\partial_\mu\psi + \gamma_5\gamma^\nu\gamma_5\gamma^\mu\partial_\mu\psi) \\
&\quad + \frac{i}{2}\bar{\epsilon}_2\sigma_{\rho\tau}\epsilon_1(\sigma^{\rho\tau}\gamma^\mu\partial_\mu\psi + \gamma_5\sigma^{\rho\tau}\gamma_5\gamma^\mu\partial_\mu\psi) \\
&= -i\bar{\epsilon}_2\gamma_\nu\epsilon_1(\gamma^\mu\gamma^\nu\partial_\mu\psi + \gamma^\nu\gamma^\mu\partial_\mu\psi) \\
&= -2i\bar{\epsilon}_2\gamma_\nu\epsilon_1\partial^\nu\psi
\end{aligned} \tag{2.25}$$

as required.

Thus, all the properties of anticommuting Majorana spinors come into play when verifying the supersymmetry algebra. If we were to repeat the above derivation using the two-component Van der Waerden notation, the Fierz transformation step would correspond to symmetizing and antisymmetizing on the two component indices to be joined. Note also that there is a consistent restriction on the chiral multiplet which sets F and G to zero and requires the fields A, B, and ψ to satisfy free field equations. In that case, the restricted multiplet (A, B, ψ) still forms a realization of the supersymmetry algebra, but the closure calculation (2.25) is then valid only subject to the imposition of these field equations. The dimension two fields F and G, while they do not contribute physical degrees of freedom, are necessary in order to have a complete linear representation of supersymmetry without imposing the field equations. These nondynamical fields are called auxiliary fields.

The transformations (2.13)-(2.15) cause the following combinations of fields to transform by total derivatives:

$$\mathcal{L}_{kinetic} = -\frac{1}{2}(\partial_\mu A)^2 - \frac{1}{2}(\partial_\mu B)^2 - \frac{i}{2}\bar{\psi}\not{\partial}\psi + \frac{1}{2}F^2 + \frac{1}{2}G^2 \tag{2.26}$$

$$\mathcal{L}_{mass} = m(FA + GB - \frac{i}{2}\bar{\psi}\psi) \tag{2.27}$$

$$\mathcal{L}_{interaction} = g(FA^2 - FB^2 + 2GAB - i\bar{\psi}(A - \gamma_5 B)\psi). \tag{2.28}$$

The integrals of these three terms over $\int d^4x$ are thus separately supersymmetrically invariant. The action for the interacting Wess-Zumino model[4)] is the sum of these three integrals. The auxiliary fields F and G then have algebraic equations of motion:

$$F + mA + g(A^2 - B^2) = 0 \tag{2.29}$$
$$G + mB + 2gAB = 0. \tag{2.30}$$

Since these equations of motion are algebraic, the dynamical consequences of the action are unchanged if we substitute for F and G back into the rest of the action. For $m = 0$, the result is the same as we had in *e.g.*, (1.6) for $\phi = A + iB/\sqrt{2}$. With $m \neq 0$, the mass terms are accompanied by trilinear scalar interaction terms: after elimination of F and G,

$$I_{WZ} = \int d^4x \Big(-\frac{1}{2}(\partial_\mu A)^2 - \frac{1}{2}(\partial_\mu B)^2 - \frac{i}{2}\bar{\psi}\partial\!\!\!/\psi - \frac{1}{2}m^2A^2 - \frac{1}{2}m^2B^2$$
$$-\frac{i}{2}m\bar{\psi}\psi - gmA(A^2 + B^2) - \frac{1}{2}g^2(A^2 + B^2)^2 - ig\bar{\psi}(A - \gamma_5 B\psi)\Big). \tag{2.31}$$

The supersymmetry invariance of the Wess-Zumino model gives rise to the existence of a conserved vector-spinor current

$$\sqrt{2}Q^\mu = \gamma^\lambda \partial_\lambda (A - \gamma_5 B)\gamma^\mu \psi - (F + \gamma_5 G)\gamma^\mu \psi. \tag{2.32}$$

Upon elimination of the auxiliary fields F and G using their equations of motion, this current coincides with the one given in Eq. (1.7).

The potential for the scalar fields in the supersymmetric model (2.31) is positive definite, as can clearly be seen since it is just the sum of the squares of the auxiliary fields,

$$V = \frac{1}{2}(F^2 + G^2). \tag{2.33}$$

In addition to allowing the supersymmetry algebra to close without use of the equations of motion, the auxiliary fields are necessary in order to have a linear representation of supersymmetry: the action (2.31) is still supersymmetric, but under nonlinear transformations obtained from (2.13)-(2.15) by substituting (2.29)-(2.30) for F and G.

In order to exploit the linear realization of supersymmetry afforded by the chiral representation and others to be discussed, we need a convenient notation that makes the linear realization manifest. Such a notation is provided by following the analogy of constructing the induced representations of the Poincaré group. Just as Minkowski space can be viewed as the space of left cosets of the Lorentz group H within the Poincaré group P, so one can construct superspace[3,6] as GP/H, where GP is the graded Poincaré group, whose algebra is the supersymmetry algebra given in (1.10).

The coordinates of superspace label the above cosets. From the supersymmetry algebra, we can see that these coordinates will carry the same Lorentz indices as the generators P_μ and Q_α, $\bar{Q}_{\dot{\alpha}}$. Thus we have a space of vectorial bosonic coordinates x^μ and spinorial fermionic coordinates θ_α, $\bar{\theta}_{\dot{\alpha}}$ (or equivalently, a four-component Majorana spinor). In order to define the action of the supersymmetry group on these coordinates, we start with a corresponding group element

$$G(x, \theta_\alpha, \bar{\theta}\dot{\alpha}) = e^{i[-x^\mu P_\mu + \theta^\alpha Q_\alpha + \bar{\theta}_{\dot{\alpha}} \bar{Q}^{\dot{\alpha}}]} \tag{2.34}$$

In writing the group element this way, we are treating the Q's as if they were Lie algebra generators, but with anticommuting parameters. Thus the θ_α, $\bar{\theta}_{\dot{\alpha}}$ must be taken to be anticommuting elements of a Grassman algebra, as discussed above.

Two group elements like (2.34) can be multipled together using Hausdorff's formula.

$$e^A e^B = e^{A+B+\frac{1}{2}[A,B]+\dots} \tag{2.35}$$

The higher order commutators in this formula are not needed, since they vanish for the supersymmetry algebra. The action of a translation $e^{-i(\delta x^\mu P_\mu)}$ works as in the Poincaré group: $x^\mu \to x^\mu + \delta x^\mu$. The supersymmetry transformation of x^μ, θ_α, $\bar{\theta}_{\dot{\alpha}}$ is given by computing

$$G(0, \epsilon_\alpha, \bar{\epsilon}_{\dot{\alpha}}) G(x^\mu, \theta_\alpha, \bar{\theta}_{\dot{\alpha}}) = G(x^\mu + i\theta\sigma^\mu\bar{\epsilon} - i\epsilon\sigma^\mu\bar{\theta}, \theta + \epsilon, \bar{\theta} + \bar{\epsilon}). \tag{2.36}$$

Thus, group multiplication induces a motion in the coset parameter space

$$(x^\mu, \theta_\alpha, \bar{\theta}_{\dot{\alpha}}) \to (x^\mu + i\theta\sigma^\mu\bar{\epsilon} - i\epsilon\sigma^\mu\bar{\theta}, \theta_\alpha + \epsilon_\alpha, \bar{\theta}_{\dot{\alpha}} + \bar{\epsilon}_{\dot{\alpha}}). \tag{2.37}$$

Just as the Poincaré shift in x^μ is generated by $i\frac{\partial}{\partial x^\mu}$, so the shifts induced by a supersymmetry transformation are generated by

$$Q_\alpha = \frac{\partial}{\partial\theta^\alpha} - i\sigma^\mu_{\alpha\dot\alpha}\bar\theta^{\dot\alpha}\partial_\mu \tag{2.38}$$

$$\bar Q^{\dot\alpha} = \frac{\partial}{\partial\bar\theta_{\dot\alpha}} + i\theta^\alpha\sigma^\mu_{\alpha\dot\beta}\epsilon^{\dot\alpha\dot\beta}\partial_\mu \tag{2.39}$$

which give the required algebra

$$\{Q_\alpha, \bar Q_{\dot\alpha}\} = 2i\sigma^\mu_{\alpha\dot\alpha}\partial_\mu \tag{2.40}$$

$$\{Q_\alpha, Q_\beta\} = \{\bar Q^{\dot\alpha}, \bar Q^{\dot\beta}\} = 0. \tag{2.41}$$

The representations of supersymmetry that follows from the above construction are superfields $F(x,\theta,\bar\theta)$, a priori general functions of the commuting and anticommuting coordinates of superspace. In order to relate them to the fields used in our ordinary formulations of physical theories, we may expand a superfield in a Taylor series in $\theta,\bar\theta$. Due to the anticommuting property of $\theta,\bar\theta$ this series will terminate since the highest possible product is $\theta^\alpha\theta_\alpha\bar\theta_{\dot\alpha}\bar\theta^{\dot\alpha}$. Thus we obtain a set of coefficient functions of x^μ only, carrying various Lorentz representations:

$$\begin{aligned} F(x,\theta,\bar\theta) = {} & f(x) + \theta^\alpha\chi_\alpha(x) + \bar\theta_{\dot\alpha}\bar\phi^{\dot\alpha}(x) + \theta\theta m(x) + \bar\theta\bar\theta n(x) \\ & + \theta\sigma^\mu\bar\theta V_\mu(x) + \theta\theta\bar\theta^{\dot\alpha}\bar\lambda_{\dot\alpha}(x) + \bar\theta\bar\theta\theta_\alpha\psi^\alpha(x) + \theta\theta\bar\theta\bar\theta d(x). \end{aligned} \tag{2.42}$$

While indicating correctly the number and types of "component" ordinary fields contained within a general complex superfield, the expansion (2.42) is not, in fact, the most convenient one. Just as we define the coefficients in a Taylor expansion by taking repeated derivatives of the function to be expanded and evaluating the result at the zero value of the expansion coordinate, we would like to do the same for superfields in terms of repeated fermionic derivatives. However, we run into the problem that the ordinary partial derivative $\partial/\partial\theta^\alpha$ is not covariant:

$$\left\{\frac{\partial}{\partial\theta^\alpha}, \bar Q_{\dot\beta}\right\} = i\sigma^\mu_{\alpha\dot\beta}\partial_\mu. \tag{2.43}$$

Because of this non-covariance, the derivative $\partial/\partial\theta^\alpha$ of a superfield does not transform according to the same rule as a superfield, *i.e.*, by operation with Q_α, $\bar{Q}_{\dot{\beta}}$. In order to cure this problem, we introduce fermionic covariant derivatives:

$$D_\alpha = \frac{\partial}{\partial\theta^\alpha} + i\sigma^\mu_{\alpha\dot{\alpha}}\bar{\theta}^{\dot{\alpha}}\partial_\mu \tag{2.44}$$

$$\bar{D}^{\dot{\alpha}} = -\frac{\partial}{\partial\bar{\theta}^{\dot{\alpha}}} - i\theta^\alpha\sigma^\mu_{\alpha\dot{\alpha}}\partial_\mu. \tag{2.45}$$

These covariant derivatives preserve the superfield character of whatever they operate on, since

$$\{D_\alpha, Q_\beta\} = \{D_\alpha, \bar{Q}_{\dot{\beta}}\} = \{\bar{D}_{\dot{\alpha}}, Q_\beta\} = \{\bar{D}_{\dot{\alpha}}, \bar{Q}_{\dot{\beta}}\} = 0. \tag{2.46}$$

Among themselves, they satisfy an algebra similar to that of the Q's,

$$\{D_\alpha, \bar{D}_\alpha\} = -2i\sigma^\mu_{\alpha\dot{\alpha}}\partial_\mu \tag{2.47}$$

$$\{D_\alpha, D_\beta\} = \{\bar{D}_{\dot{\alpha}}, \bar{D}_{\dot{\beta}}\} = 0. \tag{2.48}$$

In fact, it may be checked that D_α and $\bar{D}_{\dot{\alpha}}$ can also be viewed as the generators of parameter shifts induced by *right* multiplication, analogously to (2.36). Their definitions (2.44)-(2.55) and anticommutation relations (2.47) differ from those of the Q, $\bar{Q}$ only by the signs of the x_μ shifting term.

Using the covariant derivatives, a convenient expansion of a superfield is to take the $\theta = \bar{\theta} = 0$ components of a sequence of superfields obtained from the original one by differentiation with D_α, $\bar{D}_{\dot{\alpha}}$. Thus, we may define

$$\begin{aligned} f(x) &\equiv F|_{\theta=\bar{\theta}=0} \\ \chi_\alpha(x) &\equiv (D_\alpha F)|_{\theta=\bar{\theta}=0}\,, \; etc. \end{aligned} \tag{2.49}$$

The reason that this expansion into components is more convenient than the straight expansion in θ, $\bar{\theta}$ of (2.42) is that the supersymmetry transformations of the components may be obtained by a simple rule. The supersymmetry transformation of any superfield $F(x, \theta\bar{\theta})$ is given by

$$\delta F = (\epsilon^\alpha Q_\alpha + \bar{\epsilon}_{\dot{\alpha}}\bar{Q}^{\dot{\alpha}})F \tag{2.50}$$

so, in particular, the transformation of the $\theta = \bar{\theta} = 0$ component of F is

$$\delta f = [(\epsilon^\alpha Q_\alpha + \bar{\epsilon}_{\dot{\alpha}} \bar{Q}^{\dot{\alpha}}) F]\big|_{\theta=\bar{\theta}=0} \,. \tag{2.51}$$

Moreover, since the expression on the right is evaluated at $\theta = \bar{\theta} = 0$, we obtain the same result if we operate with D_α and $\bar{D}_{\dot{\alpha}}$ instead of the Q_α, $\bar{Q}_{\dot{\alpha}}$ since the difference is set to zero with $\theta = \bar{\theta} = 0$:

$$\begin{aligned} \delta f &= [(\epsilon^\alpha D_\alpha + \bar{\epsilon}_{\dot{\alpha}} \bar{D}^{\dot{\alpha}}) F]\big|_{\theta=\bar{\theta}=0} \\ &= \epsilon^\alpha \chi_\alpha + \bar{\epsilon}_{\dot{\alpha}} \bar{\phi}^{\dot{\alpha}}. \end{aligned} \tag{2.52}$$

This same rule applies to the transformations of all the higher components of the multiplet, since each of them is obtained from the $\theta = \bar{\theta} = 0$ component of some superfield. Thus, for example,

$$\delta \chi_\alpha = [(\epsilon^\alpha D_\alpha + \bar{\epsilon}_{\dot{\alpha}} \bar{D}^{\dot{\alpha}}) D_\alpha F]\big|_{\theta=\bar{\theta}=0} \,, \tag{2.53}$$

and so on: one needs to use only the algebra of the D's and $\bar{D}$'s together with the definitions of the various components.

So far, we have considered our basic superfield F to be a general function of superspace, with an expansion into components given as in (2.42), or the more convenient expansion in terms of D's and $\bar{D}$'s given above. However, not every scalar superfield need contain as many components as given in (2.42). For example, we could impose a reality condition $F = F^*$. In imposing such a condition, it is conventional to invert the order of strings of θ's and $\bar{\theta}$'s:

$$(\theta_\alpha \bar{\theta}_{\dot{\beta}})^* = \theta_\beta \bar{\theta}_{\dot{\alpha}}. \tag{2.54}$$

Supersymmetry allows another type of restriction, however, due to the fact that the superspace coordinates $(x^\mu, \theta_\alpha, \bar{\theta}_{\dot{\alpha}})$ fall into a reducible representation of the Lorentz group. As a consequence, it is possible to impose a fermionic differential constraint without constraining the entire superspace dependence of a superfield. The fundamental example is a chiral superfield $\phi(x, \theta, \bar{\theta})$ satisfying

$$\bar{D}_{\dot{\alpha}} \phi = 0. \tag{2.55}$$

Using the constraint (2.55), we find that the independent component fields of ϕ are just

$$a = \frac{A + iB}{\sqrt{2}} \equiv \phi|_{\theta=\bar{\theta}=0} \tag{2.56}$$

$$\chi_\alpha = (D_\alpha \phi)|_{\theta=\bar{\theta}=0} \tag{2.57}$$

$$h = \frac{F - iG}{\sqrt{2}} \equiv (D^2\phi)|_{\theta=\bar{\theta}=0}, \quad D^2\phi = D^\alpha D_\alpha \phi. \tag{2.58}$$

Note that a chiral superfield cannot be made real without constraining it to be a constant. Evaluating the supersymmetry transformations of the multiplet acording to the rule given above, we find

$$\begin{aligned} \delta a &= [(\epsilon^\alpha D_\alpha + \bar{\epsilon}_{\dot{\alpha}} \bar{D}^{\dot{\alpha}})\phi]|_{\theta=\bar{\theta}=0} \\ &= \epsilon^\alpha \chi_\alpha \end{aligned} \tag{2.59}$$

$$\begin{aligned} \delta \chi_\alpha &= [(\epsilon^\beta D_\beta + \bar{\epsilon}_{\dot{\beta}} \bar{D}^{\dot{\beta}}) D_\alpha \phi]\Big|_{\theta=\bar{\theta}=0} \\ &= [-\frac{1}{2}\epsilon_\alpha D^2\phi + 2i\sigma^\mu{}_{\alpha\dot{\beta}}\bar{\epsilon}^{\dot{\beta}}\partial_\mu\phi]\Big|_{\theta=\bar{\theta}=0} \\ &= -\frac{1}{2}h\epsilon_\alpha + 2i\partial_\mu a \sigma^\mu{}_{\alpha\dot{\beta}}\bar{\epsilon}^{\dot{\beta}} \end{aligned} \tag{2.60}$$

$$\begin{aligned} \delta h &= [(\epsilon^\beta D_\beta + \bar{\epsilon}_{\dot{\beta}} \bar{D}^{\dot{\beta}}) D^2 \phi]\Big|_{\theta=\bar{\theta}=0} \\ &= -2i\bar{\epsilon}_{\dot{\beta}}\sigma^{\dot{\beta}\alpha}_\mu \partial^\mu \chi_\alpha, \end{aligned} \tag{2.61}$$

where the constraint on ϕ has been used as well as the property

$$D_\alpha D_\beta D_\gamma = 0 \tag{2.62}$$

following from the anticommutation relations of the 2-component D's.

Due to the dimensionality (length)$^{1/2}$ of the supersymmetry parameter ϵ, the component of highest dimension of any supermultiplet must transform into a derivative of some component with one half unit of dimension less than the highest. This means that the integral over all of spacetime of such a component is a supersymmetric invariant. The existence of both chiral and general scalar supermultiplets thus implies the existence of two types of supersymmetric invariant. The multiplet whose highest component is to be taken as a supersymmetric density is generally a product of the basic field multiplets and their derivatives in a given model.

The superspace formalism provides a very compact notation for supersymmetric actions when account is taken of the Berezin superspace integration rules (2.22)-(2.23). Because of these rules, it is apparent that integration in superspace is equivalent to differentiation

$$\int d\theta = \frac{\partial}{\partial\theta}. \tag{2.63}$$

Furthermore, since in an action we also integrate over $\int d^4x$, we may replace an overall $\partial/\partial\theta^\alpha$ by D_α, which differs from it only by an overall spacetime derivative. Thus, the supersymmetric invariant action formula for a superfield Lagrangian F of the general scalar type, such as the superfield in (2.42), is given by the full superspace integral

$$I_{GS} = \int d^4x\, d^4\theta\, F(x,\theta,\bar\theta) \tag{2.64}$$

where $d^4\theta = d\theta^\alpha\, d\theta_\alpha\, d\bar\theta_{\dot\beta}\, d\bar\theta^{\dot\beta}$. The full superspace integral just picks out the highest component of the multiplet. Since we may add total spacetime derivatives at will under the spacetime integral, this action is equivalent to

$$I_{GS} = \int d^4x\, D^2\bar D^2 F. \tag{2.65}$$

We can check the invariance by evaluating the supersymmetry transformation of the integral as discussd above,

$$\delta(D^2\bar D^2 F) = [(\epsilon^\alpha D_\alpha + \bar\epsilon_{\dot\alpha}\bar D^{\dot\alpha})D^2\bar D^2 F]\big|_{\theta=\bar\theta=0}\,. \tag{2.66}$$

In fact, due to the fact that we are integrating over all spacetime, the restriction to $\theta = \bar\theta = 0$ is unnecessary here since the θ, $\bar\theta$ dependent terms in the derivatives give total spacetime derivatives. Thus,

$$\begin{aligned} D_\alpha(D^2\bar D^2 F) &= 0 \\ \bar D^{\dot\alpha}(D^2\bar D^2 F) &= [\bar D^{\dot\alpha}, D^2]\bar D^2 F + D^2\bar D_{\dot\alpha}\bar D^2 F \\ &= 4i\partial_\mu\sigma^\mu_{\alpha\dot\alpha}D^\alpha\bar D^2 F, \end{aligned} \tag{2.67}$$

so the integral in (2.65) varies by a total spacetime derivative, as required for supersymmetric invariance.

The chiral multiplet (2.59)-(2.61) gives rise to an action formula with integration only over $d^2\theta$:

$$I_{ch} = \int d^4x\, d^2\theta \mathcal{L}. \tag{2.68}$$

This is invariant subject to the chiral constraint $\bar{D}_{\dot{\alpha}}\mathcal{L} = 0$ on the chiral superspace density $\mathcal{L}$, as may again be checked

$$I_{ch} = \int d^4x\, D^2 \mathcal{L} \tag{2.69}$$

and

$$D_\alpha D^2 \mathcal{L} = 0 \tag{2.70}$$

$$\begin{aligned} \bar{D}_{\dot{\alpha}} D^2 \mathcal{L} &= [\bar{D}_{\dot{\alpha}}, D^2]\mathcal{L} + D^2 \bar{D}_{\dot{\alpha}}\mathcal{L} \\ &= 4i\partial_\mu \sigma_{\alpha\dot{\alpha}} D^\alpha \mathcal{L} \end{aligned} \tag{2.71}$$

where it has been necessary to use the chiral constraint to obtain (2.71).

Examples of the two types of action formulas are provided by the three terms (2.26)-(2.28) which make up the action for the Wess-Zumino model. The basic fields of this model are contained in a chiral superfield ϕ. The kinetic term may be written using either of the two action formulas:

$$I_{kinetic} = \int d^4x\, d^4\theta \bar{\phi}\phi = \int d^4x\, d^4\theta \phi \bar{D}^2 \bar{\phi}, \tag{2.72}$$

where the second form of the action uses the fact that $\bar{D}_{\dot{\beta}} \bar{D}^2 \bar{\phi} = 0$, so $\phi \bar{D}^2 \bar{\phi}$ is chiral. Although it might appear that the second form is complex and one should take the real part, this is, in fact, unnecessary since the imaginary part is the integral of a total spacetime derivative. We can evaluate this kinetic action as follows:

$$I_{kinetic} = \int d^4x\, D^2(\phi \bar{D}^2 \bar{\phi})$$

$$= \int d^4x\, D^2\phi \bar{D}^2\bar{\phi} + 2\int d^4x\, D^\alpha\phi(D_\alpha \bar{D}^2\bar{\phi}) + \int d^4x\, \phi D^2\bar{D}^2\bar{\phi}$$
$$= \int d^4x [hh^* + 4i\chi^\alpha \partial_\mu \sigma^\mu_{\alpha\dot{\alpha}} \bar{\chi}^{\dot{\alpha}} + 8a\Box a^*] \tag{2.73}$$

which, after some rescalings, just gives (2.26).

The mass term of the Wess-Zumino model can only be written using the chiral action formula (2.68):

$$I_{mass} = \mathrm{Re}[m \int d^4x\, d^2\theta \phi^2]. \tag{2.74}$$

This can also be evaluated using the rules given above,

$$m \int d^4x\, d^2\theta \phi^2 = m \int d^4x\, D^2\phi^2$$
$$= 2m \int d^4x (D^2\phi)\phi + 2m \int d^4x\, D^\alpha\phi\, D_\alpha\phi$$
$$= 2m \int d^4x [ha + \chi^\alpha \chi_\alpha], \tag{2.75}$$

which gives (2.27) after taking the real part.

Finally, the interaction term (2.28) must also be written as a chiral superspace integral:

$$I_{interaction} = g \int d^4x\, d^2\theta \phi^3. \tag{2.76}$$

If we generalize the Wess-Zumino model to include a number of chiral superfields ϕ^a, then the most general renormalizable Lagrangian is

$$I_{matter} = \int d^4x\, d^4\theta \bar{\phi}^a \phi^a + \mathrm{Re} \int d^4x\, d^2\theta\, f(\phi^a), \tag{2.77}$$

where the function f, known as the superpotential, must be at most trilinear for renormalizability:

$$f(\phi^a) = \eta_a \phi^a + m_{ab}\phi^a\phi^b + g_{abc}\phi^a\phi^b\phi^c. \tag{2.78}$$

In the case that this matter model possesses some rigid symmetry in which the ϕ^a transform according to the representation matrices $T_k{}^a{}_b$, then the superpotential f must satisfy

$$f_{,a}(\phi)T_k{}^a{}_b\phi^b = 0 \tag{2.79}$$

where $f_{,a} = \partial f/\partial\phi^a$. Evaluation of the interaction for the general matter model (2.77) gives

$$\mathrm{Re}\int d^4x(h^a f_{,a} + \chi^{\alpha a}\chi_\alpha{}^b f_{,ab}). \tag{2.80}$$

Taking this together with the $h^a\bar{h}^a$ term from the kinetic term, h^a can be eliminated to give a general form to the potential

$$V = \sum_a |f_{,a}|^2 \tag{2.81}$$

whose manifest positivity is an essential feature of supersymmetric theories.

It is because of the manifest positivity of (2.81) that the possibilities of spontaneous breaking of supersymmetry are strongly constrained. In any supersymmetric state, the vacuum expectation value of the supersymmetric variation of any operator must vanish. In particular, $\langle h^a\rangle$ must vanish, so in a supersymmetric state the potential (2.81) vanishes. Thus one cannot break supersymmetry merely by having a non-supersymmetric state with lowest energy. Flat-space supersymmetry can only be broken by choosing a theory where no state is supersymmetric.

3. Quantum Supersymmetric Gauge Theories

In order to describe a massless supersymmetric gauge theory,[11)] we must start from the irreducible representation of supersymmetry containing a vector V_μ and its spin 1/2 partner λ_α, *i.e.*, the superspin 1/2 multplet given in (2.10). The condition $\partial^\mu V_\mu = 0$, which is necessary for the irreducibility of this multiplet, indicates that in order to find an action for the multiplet in terms of unconstrained fields, it is necessary to add another supermultiplet containing the divergence of the vector field. In the final action, this extra multiplet must be eliminable by gauge transformations, so we can see that a supersymmetric gauge theory will have a whole multiplet's worth of gauge invariances.

The only other irreducible multiplet that can be used as the multiplet of gauge components is the $J_\Xi = 0$ chiral multiplet (2.9). The combination of the "physical" and the "gauge" multiplets may be described by a real general scalar superfield $V(x,\theta,\bar{\theta}) = V^*(x,\theta,\bar{\theta})$. Since the gauge variation of this superfield must itself be real, we conclude that for a chiral superfield of gauge transformation parameters Λ, in an Abelian theory we must have

$$\delta V = i(\Lambda - \Lambda^*), \qquad \bar{D}_{\dot{\alpha}}\Lambda = 0 \tag{3.1}$$

where the choice of the imaginary part of Λ is necessary for the vector V^μ to be a parity-even object. If the components of Λ are denoted $(A + iB, \varsigma, E - iF)$, then the components of the real multiplet (3.1) are $(B, \varsigma, E, F, \partial_\mu A, 0, 0)$, as can be worked out using the rules given in the last chapter. If the components of the superfield V are denoted $(C, \chi, H, K, V_\mu, \lambda, D)$, then the transformation (3.1) allows us to gauge away C, χ, H, K, and $\partial^\mu V_\mu$. The remaining gauge invariant quantities are described by a multiplet whose lowest component is the Abelian gauge invariant spinor λ_α. In accordance with our definition of the components of a superfield given in the last chapter, we see that the gauge invariant superfield is given by

$$W_\alpha = \bar{D}^2 D_\alpha V. \tag{3.2}$$

It is easy to check the gauge invariance of W_α:

$$\delta W_\alpha = i\bar{D}^2 D_\alpha(\Lambda - \Lambda^*) = i\bar{D}^2 D_\alpha \Lambda = -2i\partial_\mu \sigma^\mu_{\alpha\dot{\beta}} \bar{D}^{\dot{\beta}}\Lambda = 0. \tag{3.3}$$

Due to the derivative factor $\bar{D}^2$, W_α is a chiral spinor superfield,

$$\bar{D}_{\dot{\beta}} W_\alpha = \bar{D}_{\dot{\beta}} \bar{D}^2 D_\alpha V = 0, \tag{3.4}$$

and, in addition, the reality of V implies the reality of the scalar component of W_α.

$$D \equiv (D^\alpha W_\alpha)|_{\theta=\bar{\theta}=0} = (D^\alpha \bar{D}^2 D_\alpha V)\big|_{\theta=\bar{\theta}=0} = (\bar{D}_{\dot{\alpha}} D^2 \bar{D}^{\dot{\alpha}} V)\big|_{\theta=\bar{\theta}=0} = D^*. \tag{3.5}$$

Conditions (3.4) and (3.5) define an irreducible superfield, whose components are $(\lambda_\alpha, \partial_\mu V_\gamma - \partial_\nu V_\mu, D, \sigma^\mu_{\alpha\dot{\beta}} \lambda^{\dot{\beta}} \partial_\mu \lambda^{\dot{\beta}})$. This multiplet is known as the supersymmetric field strength multiplet, since it contains the field strength $F_{\mu\nu} = \partial_\mu V_\nu - \partial_\nu V_\mu$ of the vector field. It possesses in a local expression the same content as the superspin 1/2 multiplet $(\frac{1}{\Box}D, \frac{\not\partial}{\Box}\lambda, V_\mu{}^T)$.

Since the superfield W_α is chiral, we may form the gauge invariant action

$$I = \mathrm{Re}\int d^4x\, d^2\theta W^\alpha W_\alpha. \tag{3.6}$$

Expressing this in components, we have, after some rescalings, the action for the vector V_μ, spinor λ, and auxiliary field D:

$$I = \int d^4x\left(-\frac{1}{4}F_{\mu\nu}F^{\mu\nu} + \frac{i}{2}\bar{\lambda}\not\partial\lambda + \frac{1}{2}D^2\right). \tag{3.7}$$

If we had taken the imaginary part of the integral in (3.6) instead of the real part, we would have obtained a total divergence whose spin 1 part is the Pontryagin index $\int d^4x\, F^*_{\mu\nu}F^{\mu\nu}$.

In order to describe non-Abelian supersymmetric gauge theories, we must generalize the above discussion. The clue that tells us how to do this is provided by the gauge transformation of some doublet of chiral matter superfields ϕ^i covariantly coupled to the gauge superfield V. The natural generalization of the ordinary gauge transformation is

$$\phi^i \to [e^{-i\Lambda T}]^{ij}\phi^j, \qquad T = \begin{bmatrix} 0 & 1 \\ -1 & 0 \end{bmatrix}. \tag{3.8}$$

The chirality of the gauge parameter superfield Λ is essential for the consistency of (3.8). If we let $\delta V = i(2g)^{-1}(V - V^*)$, we can write the gauge invariant kinetic term for the ϕ^i as

$$I_{kinetic} = \int d^4x\, d^4\theta \bar{\phi}^i [e^{2gVT}]^{ij}\phi^j. \tag{3.9}$$

This may now be expanded into components using the rules given in the last chapter. Unlike the kinetic term for the gauge multiplet (3.6), however, the gauge coupled matter kinetic term depends upon all of the components of the gauge multplet, not just the gauge invariant parts contributing to W_α. In order to simplify the component expression, it is convenient to make a non-supersymmetric gauge choice known as the Wess-Zumino gauge, setting the low dimension components of V to zero: $C = \chi = H = K = 0$. The only remaining unfixed gauge transformation is then the ordinary Maxwell gauge transformation for the vector field V_μ. In this gauge, the action (3.9) becomes

$$I_{kinetic} = \int d^4x \Big[-8(\nabla^\mu a^*)^i(\nabla_\mu a)^i + 4i\chi^{\alpha i}\sigma^\mu_{\alpha\dot\beta}(\nabla_\mu\bar\chi^{\dot\beta})^i + h^{*i}h^i + \mathrm{Re}(4g\lambda^\alpha\chi^i_\alpha T^{ij}a^{*j}) + 2ga^iT^{ij}a^{*j}D\Big] \quad (3.10)$$

where $\nabla^\mu = \partial^\mu - igV^\mu T$ is the usual covariant derivative.

The generalization to non-Abelian matter couplings is now straightforward, for matter fields ϕ belonging to some gauge group representation with generator matrices T_r must now transform as

$$\phi \to e^{-i\Lambda}\phi, \qquad \begin{matrix}\Lambda = \Lambda^r \mathrm{T}_r \\ \bar D^{\dot\alpha}\Lambda = 0\end{matrix}. \quad (3.11)$$

which still preserves the chirality of ϕ, $\bar D_{\dot\alpha}\phi = 0$. In the matter kinetic term, we need to cancel a factor $e^{i\Lambda^*}$ coming from $\bar\phi$ and a factor $e^{-i\Lambda}$ coming from ϕ. Thus, if we again write the gauge coupled kinetic action as

$$I_{kinetic} = \int d^4x\, d^4\theta \bar\phi e^{2gV}\phi, \quad (3.12)$$

then the V^r must undergo nonlinear gauge transformations such that

$$e^{2gV} \to e^{-i\Lambda^*} e^{2gV} e^{i\Lambda}. \quad (3.13)$$

Taking the Abelian limit of this expression, we find that it agrees with (3.1).

In the non-Abelian case, there is a generalization of the field strength superfield (3.2):

$$W_\alpha = \frac{1}{2g}\bar D^2(e^{-2gV}D_\alpha e^{2gV}). \quad (3.14)$$

This is still a chiral superfield: $\bar{D}_{\dot\beta} W_\alpha = 0$, and it transforms covariantly in the adjoint representation of the gauge group:

$$\begin{aligned} W_\alpha \to & \frac{1}{2g}\bar{D}^2\left[e^{-i\Lambda}e^{-2gV}e^{i\Lambda^*}D_\alpha(e^{-i\Lambda^*}e^{2gV}e^{i\Lambda})\right] \\ &= \frac{1}{2g}e^{-i\Lambda}(\bar{D}^2e^{-2gV}D_\alpha e^{2gV})e^{i\Lambda} + \frac{1}{2g}e^{-i\Lambda}\bar{D}^2 D_\alpha e^{i\Lambda} \\ &= e^{-i\Lambda}W_\alpha e^{i\Lambda}. \end{aligned} \tag{3.15}$$

In order to expand the field strength multiplet into its components, we take repeated fermionic covariant derivatives as before. In this case, however, we want to preserve the gauge covariance as well as the supersymmetric covariance. Thus, we need a gauge covariantized fermionic derivative ∇_α. As one can see from the calculation (3.15), such a derivative is provided by

$$\nabla_\alpha \equiv e^{-2gV}D_\alpha e^{2gV} \equiv D_\alpha - igA^r_\alpha \mathrm{T}_r, \tag{3.16}$$

where the D_α is taken to act upon everything to the right. The superfield A^r_α is known as the superspace connection. Note that $\bar{D}_{\dot\alpha}$ is still covariant without a connection since Λ is chiral. Expanding W_α using ∇_α, we find the set of gauge-covariant components

$$W_\alpha \leftrightarrow (\lambda_\alpha, F_{\mu\nu} = \partial_\mu V_\nu - \partial_\nu V_\mu + ig[V_\mu, V_\nu], D, \sigma^\mu_{\alpha\dot\alpha}\nabla_\mu \bar\lambda^{\dot\alpha}), \tag{3.17}$$

where ∇_μ is the ordinary spacetime covariant derivative. Every term in (3.17) is actually a contraction of a set of gauge component fields into the Hermitian generators T_r; thus $D = D^r T_r$ is a Hermitian matrix. Note that in terms of the gauge covariantized ∇_α and $\nabla_{\dot\alpha} = \bar{D}_{\dot\alpha}$, the algebra of covariant derivatives becomes

$$\{\nabla_\alpha, \bar\nabla_{\dot\alpha}\} = -2i\sigma^\mu_{\alpha\dot\alpha}\nabla_\mu \tag{3.18}$$

$$\{\nabla_\alpha, \nabla_\beta\} = \{\bar\nabla_{\dot\alpha}, \bar\nabla_{\dot\beta}\} = 0 \tag{3.19}$$

$$[\nabla_\mu, \bar\nabla_{\dot\beta}] = i\sigma_\mu{}^\beta{}_{\dot\beta}W_\beta \tag{3.20}$$

where W_α appears as a superspace field strength.

The action for the supersymmetric Yang-Mills theory may be written down as before, since W_α is chiral as before:

$$I_{sym} = \mathrm{Re} \int d^4x\, d^2\theta\, \mathrm{tr}\, W^\alpha W_\alpha, \tag{3.21}$$

where the trace is over the gauge group indices. Expanding (3.21) into components, we find after some rescalings

$$I_{sym} = \mathrm{tr} \int d^4x \left(-\frac{1}{4} F_{\mu\nu} F^{\mu\nu} - \frac{i}{2} \bar{\lambda} \not{\nabla} \lambda + \frac{1}{2} D^2 \right). \tag{3.22}$$

In writing down the form (3.22), it is important that we expanded W^α into components using ∇_α. If we had performed a non-gauge-covariant expansion with D_α, we would have obtained the form (3.22) only after going into a Wess-Zumino gauge. As in the Abelian case, if we had taken the imaginary part of the integral in (3.21), we would have obtained a total divergence, with spin one part now the Pontryagin index for the Yang-Mills field.

The algebra of Yang-Mills gauge covariantized superspace derivatives (3.18)-(3.20) does not reveal the asymmetry between ∇_α and $\nabla_{\dot{\alpha}} = \bar{D}_{\dot{\alpha}}$. This suggests that there should be a more symmetrical way to include the Yang-Mills fields. Indeed, we may also start in direct analogy with ordinary non-supersymmetric Yang-Mills, and introduce connections for all the derivatives:

$$\begin{aligned} \nabla_\alpha &= D_\alpha - igA_\alpha^k T_k \\ \bar{\nabla}_{\dot{\alpha}} &= \bar{D}_{\dot{\alpha}} - ig\bar{A}_{\dot{\alpha}}{}^k T_k \\ \nabla_\mu &= \partial_\mu - igA_\mu^k T_k. \end{aligned} \tag{3.23}$$

Matter fields ϕ can be taken to transform with a general scalar superfield parameter

$$\phi \to e^{-iK^r T_r} \phi = e^{-iK} \phi \tag{3.24}$$

where K^r is a real general scalar. To preserve covariance, we require that the covariant derivatives ∇_A transform as

$$\nabla_A \to e^{-iK}\, \nabla_A\, e^{iK}. \tag{3.25}$$

Chiral matter fields transforming as in (3.24) must satisfy *covariant* constraints

$$\bar{\nabla}_{\dot{\alpha}}\phi = 0. \tag{3.26}$$

The covariant constraint (3.26) is an overdetermined equation. By applying another covariant derivative and symmetrizing, we derive the integrability condition

$$\{\bar{\nabla}_{\dot{\alpha}}, \bar{\nabla}_{\dot{\beta}}\}\phi = 0 \tag{3.27}$$

requiring

$$F_{\dot{\alpha}\dot{\beta}} = 0 \tag{3.28}$$

where $F_{\dot{\alpha}\dot{\beta}}$ is one of the field strength superfields

$$F_{AB} = D_A A_B \pm D_B A_A - ig\{A_A, A_B] - 2i\sigma^C_{AB} A_C \tag{3.29}$$

in which the plus sign occurs for fermi-fermi components and the only nonvanishing components of σ^C_{AB} are $\sigma^c_{\alpha\dot{\beta}}$, $\bar{\sigma}^c_{\dot{\alpha}\beta}$.

The constraint (3.28) and its conjugate

$$F_{\alpha\beta} = 0 \tag{3.30}$$

are thus required in order to allow consistent coupling of chiral superfields to the Yang-Mills multiplet. These constraints are also needed to give the correct field content to the Yang-Mills supermultiplet itself. Since they set fermi-fermi field strengths to zero, they can easily be solved:

$$\begin{aligned} A_\alpha &= \frac{i}{g} e^{-gS} D_\alpha e^{gS} \\ A_{\dot{\alpha}} &= \frac{i}{g} e^{g\bar{S}} \bar{D}_{\dot{\alpha}} e^{-g\bar{S}} \end{aligned} \tag{3.31}$$

where S is a complex general scalar prepotential. Under the K-gauge transformations (3.25), we require

$$\begin{aligned} e^{gS} &\to e^{gS} e^{iK} \\ e^{g\bar{S}} &\to e^{-iK} e^{g\bar{S}}. \end{aligned} \tag{3.32}$$

There is now an additional invariance, however, that only arises upon solving the constraints as in (3.31), since

$$e^{gS} \rightarrow e^{i\bar{\lambda}} e^{gS}, \qquad D_\alpha \bar{\lambda} = 0 \tag{3.33}$$

leaves A_α invariant. The complete gauge transformation of e^{gS}, $e^{g\bar{S}}$ is thus

$$\begin{aligned} e^{gS} &\rightarrow e^{i\bar{\Lambda}} e^{gS} e^{iK}, & D_\alpha \bar{\Lambda} = 0 \\ e^{g\bar{S}} &\rightarrow e^{-iK} e^{g\bar{S}} e^{-i\Lambda}, & \bar{D}_{\dot{\alpha}} \Lambda = 0 \end{aligned} \tag{3.34}$$

where K is a general scalar superfield, and Λ is chiral (both contracted with Hermitian group generators).

The K-gauge transformations can be used to gauge away the imaginary part of the complex scalar prepotential S. Alternatively, one may form the K-gauge invariant combination

$$e^{2gV} = e^{gS} e^{g\bar{S}}, \tag{3.35}$$

which transforms only under the Λ-transformations,

$$e^{2gV} \rightarrow e^{i\bar{\Lambda}} e^{2gV} e^{-i\Lambda}. \tag{3.36}$$

The complex object $e^{-g\bar{S}}$ can be used to tranform a K-gauge transforming object such as ϕ in (3.24) into an object transforming under the chiral Λ-transformation:

$$\phi_{(\Lambda)} = e^{-g\bar{S}} \phi_{(K)}. \tag{3.37}$$

Moreover, this field is chiral with an ordinary superspace derivative,

$$\bar{D}_{\dot{\alpha}} \phi_{(\Lambda)} = 0, \tag{3.38}$$

and, in general, when acting on objects like $\phi_{(\Lambda)}$, the covariant derivatives must be rewritten in the new Λ-basis,

$$\nabla_\alpha = e^{-2gV} D_\alpha e^{2gV} \tag{3.39}$$

$$\nabla_{\dot{\alpha}} = \bar{D}_{\dot{\alpha}}, \tag{3.40}$$

which is just what we had in (3.16).

There is one remaining constraint that can be imposed on the fermi-fermi field strengths. If we set

$$F_{\alpha\dot{\beta}} = 0, \tag{3.41}$$

then, as can be seen from (3.29), it is possible to solve for A_μ:

$$A_\mu = -\frac{i}{4}\bar{\sigma}_\mu^{\dot{\beta}\alpha}(D_\alpha \bar{A}_{\dot{\beta}} + \bar{D}_{\dot{\beta}} A_\alpha - ig\{A_\alpha, \bar{A}_{\dot{\beta}}\}). \tag{3.42}$$

This constraint is called the conventional constraint. It is analogous to the vanishing of the torsion in general relativity. From (3.28), (3.30), and (3.41) we recover the algebra (3.18)-(3.20).

Combining the superymmetric Yang-Mills theory with a general gauge coupled matter sector, the general supersymmetric matter plus Yang-Mills theory can be written

$$I = \int d^4x\, d^2\theta \bar{\phi} e^{2gV} \phi + \mathrm{Re} \int d^4x\, d^2\theta (\mathrm{tr} W^\alpha W_\alpha + f(\phi)), \tag{3.43}$$

where $f(\phi)$ is a group invariant superpotential in the sense of Eq. (2.79). If the gauge group contains some Abelian $U(1)$ factors, then it is possible to add also the Fayet-Iliopoulos terms

$$I_{FI} = \int d^4x\, d^4\theta\, C^{\hat{k}} V_{\hat{k}} \tag{3.44}$$

where $C^{\hat{k}}$ are a set of coefficients, with $\hat{k}$ running over the number of Abelian $U(1)$ factors. These terms play an important role in the study of spontaneous gauge symmetry and supersymmetry breaking.

So far, we have described supersymmetric field theories at the level of classical physics. In order to develop such theories into full quantum field theories, one could, of course, just explicitly write a given theory out in components and then proceed *via* the standard methods of perturbative quantum field theory. However, many of the surprising features of supersymmetry have to do with its control over the higher order quantum corrections and renormalizations, and in order to discuss these, we need to keep the supersymmetry manifest at the quantum level. For this reason, we need to be able to quantize directly in the superfield formalism. It is not within the purview of this article to give an exhaustive treatment of superfield quantization techniques, for which we refer the reader to the series of articles by Grisaru, Siegel, and Roček.[12,13] Here we shall indicate only the main features.

We return to the Wess-Zumino model for a single chiral multiplet with a general renormalizable superpotential and add a source term in order to develop propagators and vertices:

$$I_{WZ} = \int d^4x\, d^4\theta \bar{\phi}\phi - \mathrm{Re}\int d^4x\, d^2\theta (m\phi^2 + \frac{1}{3}g\phi^3) + 2\mathrm{Re}\int d^4x\, d^2\theta J\phi \quad (3.45)$$

where the source J is also a chiral superfield.

The rules for developing propagators and vertices in this theory are derived directly from those for ordinary field theory. If one uses the functional integral formalism, an integral over a superfield is defined to be the integral over its component fields.

Following the standard procedure, we are led to invert the free part of (3.45). Since in (3.45) we have chiral as well as full superspace integrals, it is convenient to rewrite the free part as a full superspace integral, so that we may do fermionic integration by parts. This can be done if we allow nonlocal expressions to appear, and since the propagator we are deriving will itself be nonlocal, there is no reason not to do so at this stage. The starting action (3.45) is, of course, local regardless of how we rewrite it. In order to rewrite (3.45), we select, for example, the source term and rewrite it as

$$\int d^4x\, d^2\theta J\phi = -\frac{1}{4}\int d^4x\, d^4\theta \frac{D^2}{\Box}\phi. \quad (3.46)$$

This equality can be checked by rewriting the right-hand side again as a chiral integral with an extra $-\frac{1}{4}D^2$:

$$-\frac{1}{4}\int d^4x\, d^4\theta J\frac{D^2}{\Box}\phi = \frac{1}{16}\int d^4x\, d^2\theta \bar{D}^2\left(J\frac{D^2}{\Box}\phi\right)$$

$$= \frac{1}{16}\int d^4x\, d^2\theta J \frac{\bar{D}^2 D^2}{\Box}\phi$$

$$= \int d^4x\, d^2\theta J\phi. \tag{3.47}$$

The relative factor of $-1/4$ that enters in the first line is a conventional one introduced to make $\int d^2\theta\theta^2 = 1$. Doing the same for the other bilinear chiral integrals, we can write the free part of the action as

$$I_{(2)} = \int d^4x\, d^4\theta \left[\bar{\phi}\phi - \mathrm{Re}\left(m\phi\left(-\frac{D^2}{4\Box}\right)\phi - 2J\left(-\frac{D^2}{4\Box}\right)\phi\right)\right]. \tag{3.48}$$

Performing the Euclidean functional integral, we obtain

$$Z_0(J) = \int [d\phi][d\bar{\phi}]e^{I_{(2)}} = \exp[-\frac{1}{2}\int d^4x\, d^4\theta B^T A^{-1} B] \tag{3.49}$$

where

$$B = \begin{pmatrix} -\frac{1}{4}(D^2/\Box)J \\ -\frac{1}{4}(\bar{D}^2/\Box)\bar{J} \end{pmatrix} \tag{3.50}$$

and the propagators are given by

$$A^{-1} = \begin{pmatrix} -\frac{1}{4}\frac{m\bar{D}^2}{\Box - m^2} & 1 + \frac{m^2\bar{D}^2D^2}{16\Box(\Box - m^2)} \\ \mathrm{i} + \frac{m^2D^2\bar{D}^2}{16\Box(\Box - m^2)} & -\frac{1}{4}\frac{mD^2}{\Box - m^2} \end{pmatrix} \tag{3.51}$$

The vertices are given by the usual functional formula

$$Z(J) = \exp[I_{int}\left(\frac{\delta}{\delta J}\right)]\, Z_0(J). \tag{3.52}$$

In order to evaluate this, we need to know how to take variational derivatives of superfields. This requires the notion of a superspace δ-function. The rules (2.22) and (2.23) for Berezin integration in superspace show that the product $(\theta_1 - \theta_2)^2(\bar\theta_1 - \bar\theta_2)^2$ works like a δ-function for the fermionic coordinates. Multiplying this by $\delta^4(x_1 - x_2)$, we get the full superspace δ-function

$$\delta^8(z_{12}) = \delta^4(x_1 - x_2)(\theta_1 - \theta_2)^2(\bar\theta_1 - \bar\theta_2)^2. \tag{3.53}$$

The variational derivative of one superfield with respect to itself is then

$$\frac{\delta V(z_1)}{\delta V(z_2)} = \delta^8(z_{12}). \tag{3.54}$$

When we take variational derivatives of chiral superfields, it is necessary to account for the fact that an integral over functions of superfields of the same chirality is taken only over the chiral superspace, $\int d^4x\, d^2\theta$. This requires removing two $\bar\theta$'s from the δ-function (3.53). The required expression can be written covariantly as

$$\frac{\delta J(z_1)}{\delta J(z_2)} = -\frac{1}{4}\bar D_1^2\, \delta^8(z_{12}). \tag{3.55}$$

Note that the derivatives can equivalently be considered as acting on z_2.

The vertices can now be written as

$$\begin{aligned} I_{int}(\delta/\delta J)J(z_1)J(z_2)J(z_3) &= -\frac{g}{6}\left[\int d^4x_4 d^2\theta_4 \left(\delta/\delta J(z_4)\right)^3\right] J(z_1)J(z_2)J(z_3) \\ &= -g\int d^8z_4 \delta^8(z_{14})\left[-\frac{1}{4}\bar D_2^2\delta^8(z_{24})\right]\left[-\frac{1}{4}\bar D_3^2\delta^8(z_{34})\right], \end{aligned} \tag{3.56}$$

so that two of the three propagators entering the vertex will be acted upon by factors of $-\frac{1}{4}\bar D^2$.

For general scalar superfields, the Feynman rules are also straightforward to develop: the action must be expanded into a series of interactions, after supplying a gauge-breaking term and adding Faddeev-Popov ghost terms with chiral ghost superfields C, C':

$$I_{GB} = -\frac{1}{16}\text{tr}\int d^4x\, d^4\theta (D^2V)(\bar{D}^2V) \tag{3.57}$$

$$I_{FP} = \text{tr}\int d^4x\, d^4\theta (\bar{C}' - C')L_{gV/2}[(\bar{C}+C) + (\coth L_{gV/2})(C-\bar{C})] \tag{3.58}$$

where $L_X[Y] = [X, Y]$. For more details, see Ref. 12. For our present purposes, it is sufficient to quote the resulting Feynman rules from Ref. 12:

(i) Massless propagators are $\pm(1/p^2)\delta^4(\theta_{12})$, with + for the $\phi\bar{\phi}$ propagators of chiral superfields (both physical and ghost chiral multiplets) and — for real superfields (gauge multiplets). In addition, the massive chiral multiplet has $+(p^2+m^2)^{-1}\delta^4(\theta_{12})$ for the $\phi\bar{\phi}$ propagator and a $\phi\phi$ propagator $m\{p^2(p^2+m^2)\}^{-1}\frac{1}{4}D^2(p, \theta_1)\delta^4(\theta_{12})$.

(ii) Vertices are read directly from I, with an extra $-\frac{1}{4}\bar{D}^2$ (for each antichiral one), but omitting one factor of $-\frac{1}{4}\bar{D}^2$ for converting $d^2\theta$ into $d^4\theta$ (a factor of $-\frac{1}{4}D^2$ for antichiral superfields).

(iii) There are the usual combinatorial factors, and -1 for each ghost loop.

(iv) The amputated one-particle-irreducible graphs should have each amputated external line multiplied by the appropriate superfield. For each external chiral superfield, a factor of $-\frac{1}{4}D^2$ should be omitted from a vertex (a factor $-\frac{1}{4}\bar{D}^2$ for an antichiral external superfield).

(v) The diagram is to be integrated over

$$\int \left[\prod_{loops} \frac{d^4k}{(2\pi)^4}\right]\left[\prod_{ext} \frac{d^4p}{(2\pi)^4}\right](2\pi)^4\delta^4(\sum_{ext} p)\int \prod_{vertices} d^4\theta.$$

The superspace Feynman rules have an immediate important consequence. Since all vertices are to be integrated over $\int d^4\theta$, all contributions to the Green's functions must be written as full superspace integrals. Moreover, one can integrate by parts to move fermionic covariant derivatives on any propagator off from the corresponding $\delta^4(\theta_{mn})$. One can then perform the Berezin integration $\int d^4\theta_n$ on one connecting vertex, thus setting $\theta_n = \theta_m$. This can be performed within any loop until only one fermionic integration is left, with a factor $D \ldots D\delta^4(\theta_{n1})_{\theta_n=\theta_1}$. The only such factor that is nonvanishing is $D^2\bar{D}^2\delta^4(\theta_{n1})_{\theta_n=\theta_1} = 16$. Repeating this procedure for all loops in a diagram, the result is an expression with a single remaining $\int d^4\theta$ integral: all contributions to the effective action have the form

$$\int d^4x_1 \ldots d^4x_n \, F_1(x_1, \theta) \ldots F_n(x_n, \theta) G(x_1 \ldots x_n) \tag{3.59}$$

where the F_n are products of superfields and their derivatives and G is translationally invariant, so that (3.39) is invariant under the supersymmetry transformations

$$\theta^\alpha \to \theta^\alpha + \epsilon^\alpha, \quad \bar{\theta}^{\dot{\alpha}} \to \bar{\theta}^{\dot{\alpha}} + \bar{\epsilon}^{\dot{\alpha}}, \quad x_n^\mu \to x_n^\mu - i(\epsilon^\alpha \bar{\theta}^{\dot{\beta}} + \bar{\epsilon}^{\dot{\beta}} \theta^\alpha) \sigma^\mu_{\alpha\dot{\beta}}. \tag{3.60}$$

In calculating the quantum corrections to Green's functions, it is also necessary to regularize the diagrams. While this is still a subject of some controversy, the technique of regularization by dimensional reduction[14)] works in practice, at least at low orders in the loop expansion. In this technique, all the algebraic manipulations of the covariant derivatives are to be carried out first in four-dimensional space, while the momentum integration is carried out subsequently in d-dimensional space.

Since the overall infinite parts of the contributions to the effective action are local expressions, *i.e.*, involve only one overall $\int d^4x$ integration, their structure is constrained in an important way by the result (3.59). Divergences due to infinite subdiagrams are nonlocal in structure, but are removed by the appropriate renormalizations at the corresponding lower loop level. Counterterms that cannot be rewritten locally as full superspace integrals according to (3.59) cannot arise, even if they are supersymmetric in structure.

Thus, we see that in the Wess-Zumino model, mass and interaction terms like (2.74) and (2.76) cannot occur as counterterms, *i.e.*, these terms in the classical action are not renormalized. For this reason, the result (3.59) is known as the non-renormalization theorem of $N = 1$ supersymmetry. The only independent renormalization in the Wess-Zumino model is the wavefunction renormalization coming from the allowed kinetic counterterm (2.72).

The behavior of a quantum field theory at high momenta is determined by the trajectories of the coupling constants as given by the solution to the renormalization group equations. In supersymmetric theories, there are important relations between the various renormalization constants and the associated β and γ functions. For example, as we have just seen, the interaction terms (2.76) in the Wess-Zumino model cannot appear as a counterterm.

This imposes the constraint

$$Z_\phi^{3/2} Z_g = 1 \tag{3.61}$$

between the wavefunction and coupling constant renormalization constants, where

$$\phi_0 = Z_\phi^{3/2} \phi \tag{3.62}$$

$$g_0 = Z_g g \tag{3.63}$$

give the unrenormalized (ϕ_0, g_0) in terms of the renormalized (ϕ, g) field and coupling constant. In the massless Wess-Zumino model, this has the consequence that if there were a nontrivial fixed point g^∞ with $\beta(g^\infty) = 0$, then the γ function would have to vanish at that value as well, and all the fields would have their canonical dimensions, so conformal invariance would be preserved.[15] In such a model without gauge fields, it has been shown that the preservation of conformal invariance is only possible in a free theory,[16] so $g^\infty = 0$. Whether such conclusions can be drawn in a gauge theory is unknown.

Further restrictions on the quantum corrections to a theory are given by the constraints of supersymmetry, and gauge invariance in a gauge theory, upon the quantum supermultiplet containing the stress tensor. From our analysis of representation in Chapters 1 and 2, it can be seen that there is only one irreducible multiplet of fields containing a single spin 2 object like $T_{\mu\nu}$. This irreducible multiplet contains also a spin 3/2 object $(Q_\mu^\alpha, \bar{Q}_\mu^\alpha)$ and an axial spin 2 object J_μ^5. Irreducibility requires the following conditions on the components of the multiplet:

$$\partial^\mu T_{\mu\nu} = \partial^\mu Q_\mu^\alpha = 0 \tag{3.64}$$

$$T_\mu^\mu = \sigma^\mu_{\alpha\dot{\alpha}} Q_\mu^\alpha = \partial^\mu J_\mu^5 = 0. \tag{3.65}$$

In the case of a conformally invariant theory, supersymmetry thus requires the existence of a conserved vector-spinor current Q_μ^α which is "σ-traceless" and a conserved axial vector current J^5_μ. The vector-spinor current is none other than the supersymmetry current which we have already met in Chapters 1 and 2. The axial current is conserved due to the existence of a symmetry which rotates spinors by γ_5 transformations and rotates scalars into pseudoscalars. The supermultiplet containing these three currents was found by Ferrara and Zumino,[17)] who gave the multiplet the name "supercurrent."

In superfields, we may describe the conformal supercurrent multiplet by an axial vector superfield $V_\mu(x,\theta,\bar\theta) = \sigma^{\alpha\dot\alpha}_\mu V_{\alpha\dot\alpha}$, satisfying the condition

$$D^\alpha V_{\alpha\dot\alpha} = \bar D^{\dot\alpha} V_{\alpha\dot\alpha} = 0. \tag{3.66}$$

It may easily be checked using the rules given in chapter 2 that this superfield contains the components

$$J^5_\mu = \sigma^{\alpha\dot\alpha}_\mu V_{\alpha\dot\alpha}\big|_{\theta=\bar\theta=0} \tag{3.67}$$

$$Q^\alpha_\mu \equiv \sigma^{\beta\dot\beta}_\mu Q^\alpha_{\beta\dot\beta} \equiv \sigma^{\beta\dot\beta}_\mu (D_\beta V^\alpha_{\dot\beta})\Big|_{\theta=\bar\theta=0} \tag{3.68}$$

$$T_{\mu\nu} = \sigma^{\alpha\dot\alpha}_\mu \sigma^{\beta\dot\beta}_\nu T_{\alpha\beta\dot\alpha\dot\beta}$$

$$T_{\alpha\beta\dot\alpha\dot\beta} = \frac{1}{2}([\bar D_{\dot\alpha}, D_\beta]V_{\alpha\dot\beta} + [\bar D_{\dot\beta}, D_\alpha]V_{\beta\dot\alpha})\Big|_{\theta=\bar\theta=0} \tag{3.69}$$

satisfying the conditions (3.64) and (3.65).

In non-conformally-invariant theories, the conditions (3.45) must be relaxed so as to allow a trace for the stress tensor. In superfields, this means relaxing the condition (3.66), but not totally, for (3.66) is also responsible for the conservation conditions (3.64), which we must keep. Since we are trying to add lower spin projections to the components $T_{\mu\nu}$, Q^α_μ, andJ^5_μ, we need to combine the conformal supercurrent with a multiplet containing lower spins. There are two basic ones available: the chiral multiplet (2.9) and the linear multiplet (2.10).

If we use a chiral superfield C to relax (3.66), we write

$$\bar D^{\dot\alpha} V_{\alpha\dot\alpha} = D_\alpha C, \qquad \bar D_{\dot\alpha} C = 0. \tag{3.70}$$

Expanding the superfield in components, we find that $\sigma^\mu_{\alpha\dot\alpha} Q^\alpha_\mu$, T^μ_μ, and $\partial^\mu J^5_\mu$ are no longer zero, but are given by the spinor and by the real and imaginary parts of the highest dimension complex scalar in C, respectively. In addition, there are now the real and imaginary parts of the lowest dimension scalar in C, which are not related to conserved currents. The multiplet described by (2.50) is an example of a reducible supermultiplet that is not given by a single superfield, since the lowest dimension components of C are not found in $V_{\alpha\dot{\alpha}}$ (although the gradients of these components are). This is not an unusual feature in supersymmetry. Irreducible reprsentations can always be described by single superfields subject to constraints, while reducible representations may be described by single superfields with weakened constraints [such as the general scalar (2.42) with no constraints] or by several superfields linked by constraints [like (3.70)]. The sense in which (3.70) describes a reducible representation is the same as that in which the nonconformal $T_{\mu\nu}$ is a reducible Poincaré representation. The trace may obviously be extracted by a local operation, but the traceless part may not, since the conserved traceless part is given by the nonlocal expression

$$\left(\eta_{\mu\alpha} - \frac{\partial_\mu \partial_\alpha}{\Box}\right)\left(\eta_{\nu\beta} - \frac{\partial_\nu \partial_\beta}{\Box}\right)\left(T^{\alpha\beta} - \frac{1}{3}\eta^{\alpha\beta} T^\rho_\rho\right). \tag{3.71}$$

Similarly, (3.70) is not fully locally reducible, but it is non-locally reducible.

If we use the real linear multiplet to relax (3.66), we write

$$\bar{D}^{\dot{\alpha}} V_{\alpha\dot{\alpha}} = \bar{D}^2 D_\alpha L \tag{3.72}$$

where in order for the real superfield L to be linear, $D^2 L = \bar{D}^2 L = 0$. These latter conditions on L do not actually need to be applied in order to have a viable supercurrent, since the operator $\bar{D}^2 D_\alpha$ projects out only the linear multiplet ($J_\Xi = \frac{1}{2}$) part of a real general scalar superfield, so L may be taken to be a real general scalar without affecting the currents in $V_{\alpha\dot{\alpha}}$. The main difference between the form (3.70) and the form (3.72) is that in (3.72) the axial current J^5_μ remains conserved, due to the reality of L and the identity

$$D^\alpha \bar{D}^2 D_\alpha L = \bar{D}_{\dot{\alpha}} D^2 \bar{D}^{\dot{\alpha}} L. \tag{3.73}$$

Thus the tracelessness of $T_{\mu\nu}$ and Q^α_μ is always broken in non-conformal theories, but the axial vector current can, in some cases, be preserved.

In order to make the above discussion more concrete, we give some examples of the two types of supercurrent at the level of classical physics. In the free massless Wess-Zumino model, the "improved" supercurrent is

$$V^{cl}_{\alpha\dot{\alpha}} = D_\alpha \phi \bar{D}_{\dot{\alpha}} \bar{\phi} + 2i(\phi \partial_{\alpha\dot{\alpha}} \bar{\phi} - \partial_{\alpha\dot{\alpha}} \phi \bar{\phi}). \tag{3.74}$$

This form of the supercurrent contains the "improved" form of the massless scalar stress tensor, and satisfies (3.66) using the field equations. "Non-improved" forms of the supercurrent also exist, both of the type (3.70) and of the type (3.72), but for a more definite example of a non-conformal supercurrent, we consider the massive Wess-Zumino model. In that case, the free field equations are

$$\bar{D}^2 \bar{\phi} = m\phi \tag{3.75}$$

and the conditions (3.46) becomes

$$\bar{D}^{\dot{\alpha}} V^{cl}_{\alpha\dot{\alpha}} = \frac{m}{2} D_\alpha(\phi^2), \tag{3.76}$$

which is clearly of the chiral type (3.70).

An example of the second type of non-conformal supercurrent (3.52) is the massless model with the same spin content as the Wess-Zumino model, but with the pseudoscalar represented by a gauge antisymmetric tensor.[18] The gauge invariant fields of this non-conformal model are given by a real linear superfield G, $D^2 G = \bar{D}^2 G = 0$. As we saw in (2.10), this multiplet contains a conserved vector. In the present application, this is the field strength for the antisymmetric tensor,

$$G_\mu = \partial^\nu A_{\mu\nu}. \tag{3.77}$$

The other two fields in the multiplet are the physical scalar and spinor; there are no auxiliary fields in this model. The supercurrent for the model is

$$V^{cl}_{\alpha\dot{\alpha}} = D_\alpha G \bar{D}_{\dot{\alpha}} G. \tag{3.78}$$

Using the field equations

$$\partial^{\alpha\dot{\beta}} \bar{D}_{\dot{\beta}} G = 0 \tag{3.79}$$

we find that instead of (3.46), we have

$$\bar{D}^{\dot{\alpha}} V^{cl}_{\alpha\dot{\alpha}} = -\frac{1}{2}\bar{D}^2 D_\alpha (G^2), \tag{3.80}$$

which is of the type (3.72) with $L = -\frac{1}{2}G^2$, a real general scalar superfield.

The supercurrent for a gauge theory is restricted by the additional requirement of gauge invariance. In the supersymmetric Yang-Mills theory (3.21), the only choice for a classical level supercurrent is

$$V^{cl}_{\alpha\dot{\alpha}} = \mathrm{tr} W_\alpha \bar{W}_{\dot{\alpha}}. \tag{3.81}$$

Using the field equations

$$\nabla^\alpha W_\alpha = \bar{\nabla}_{\dot{\alpha}} \bar{W}^{\dot{\alpha}} = 0, \tag{3.82}$$

we obtain

$$\bar{D}^{\dot{\alpha}} V^{cl}_{\alpha\dot{\alpha}} = -\mathrm{tr} W_\alpha \bar{\nabla}^{\dot{\alpha}} \bar{W}_{\dot{\alpha}} = 0 \tag{3.83}$$

as is required of a conformally invariant theory.

In quantum theories derived from classically conformally invariant actions, the scale dependence of the renormalized coupling constants upon the renormalization point causes a violation of conformal invariance, the well-known conformal anomaly. In supersymmetric theories, the conformal anomaly is given by a relaxation of (3.66) for the quantum supercurrent. The particular form this relaxation takes may be determined by details of the quantization procedure, in particular by the various subtraction schemes employed in renormalization. The particularities of a given subtraction scheme are frequently implied by the regularization method chosen. Thus, in the Wess-Zumino model if one uses Pauli-Villars regularization,[19] conformal invariance is broken by the regularization together with σ-tracelessness of Q^α_μ and conservation of the current J^5_μ for the axial $U(1)$ symmetry of the model. In this case, the anomaly equation takes the form (3.70)

$$\bar{D}^{\dot{\alpha}} V^{qu}_{\alpha\dot{\alpha}} = c\, D_\alpha(\phi \bar{D}^2 \bar{\phi}), \tag{3.84}$$

where c is the anomaly coefficient.

On the other hand, if one uses higher derivative regularization, the kinetic term (2.72) is supplemented by the regulator term

$$I_{reg} = -\frac{1}{m^2}\int d^4x\, d^4\theta \bar{\phi}\Box\phi. \tag{3.85}$$

The regulator term (3.65) is invariant under the axial $U(1)$ transformations just like the kinetic term (2.72), from which it differs only by the extra, axially inert, D'Alembertian. Thus, higher derivative regularization preserves the axial $U(1)$ invariance and the natural subtraction scheme which is suggested by it does too. Thus, with higher derivative regularization, there can be no J^5_μ anomaly, and the anomaly equation takes the form (3.72)

$$\bar{D}^{\dot{\alpha}} V^{qu}_{\alpha\dot{\alpha}} = c\,\bar{D}^2 D_\alpha(\phi\bar{\phi}). \tag{3.86}$$

In both (3.64) and (3.66), the anomaly multiplet is given by the kinetic part of the superspace Lagrangian multiplet, but in forms differing by a total derivative, corresponding to the two ways of writing the kinetic action given in (2.72). Thus, the J^5_μ anomaly that seems to be present in (3.84) is, in fact, illusory, for it can be eliminated by a supersymmetric redefinition of the quantum supercurrent $V^{qu}_{\alpha\dot{\alpha}}$ similar to that involved in going from a "non-improved" to an "improved" form of the supercurrent in the classical theory.

In the supersymmetric Yang-Mills theory, there is no arbitrariness in the non-conformal part of the supercurrent such as there is in the Wess-Zumino model, for we have the additional constraint of gauge invariance. Again, the anomaly multiplet is the multiplet containing the kinetic superspace Lagrangian, which must be written as the chiral superspace integral (3.21) if manifest gauge invariance is to be preserved.

Thus, the only possible form for the quantum supercurrent anomaly is the form (3.70),

$$\bar{D}^{\dot{\alpha}} V^{qu}_{\alpha\dot{\alpha}} = d\,D_\alpha(\mathrm{tr}W^\beta W_\beta). \tag{3.87}$$

In this case, there is no way consistent with gauge invariance to redefine $V^{qu}_{\alpha\dot{\alpha}}$ so as to eliminate the J^5_μ anomaly. The attempt to derive the anomaly by using higher derivative regularization and subtraction does not allow us to circumvent this conclusion, for the requirements of gauge invariance now force us to add higher derivative vertices as well as kinetic-type terms, and these prevent the higher derivative technique from regularizing at the one-loop level of perturbation theory. Only in a gauge theory which turns out to be finite at the one-loop level can higher derivative regularization be used.

Apart from the group-theoretical aspects of unifyiing space-time and internal symmetries, the most striking characteristic of supersymmetric theories is their greatly improved ultraviolet behavior. Many cancellations take place between the infinities in diagrams with loops of bosons and those with loops of fermions. This is seen in the case of $N = 1$ supersymmetry in the absence of interaction or mass counterterms in the Wess-Zumino model, requiring that the corresponding Z_g and Z_m just cancel the wave function renormalizations in these terms.

A basic tool for the analysis of such situations is the background field method.[20] The essential features of the method can be illustrated by ordinary Yang-Mills theory. The first step is to split the gauge field A_μ into a background and a quantum part

$$A_\mu = A_\mu^B + a_\mu, \tag{4.1}$$

where the gauge fields are contracted with the Yang-Mills hermitian generators T_r

$$A_\mu = A_\mu^r T_r. \tag{4.2}$$

The standard gauge transformation

$$A_\mu \to \bar{e}^{igK} A_\mu e^{igK} + \frac{i}{g} \bar{e}^{igK} \partial_\mu e^{igK} \tag{4.3}$$

can be interpreted in two ways. It can be thought of as a background transformation under which the background field transforms as a connection and the quantum field transforms covariantly,

$$\begin{aligned} A_\mu^B &\to \bar{e}^{igK} A_\mu^B e^{igK} + \frac{i}{g} \bar{e}^{igK} \partial_\mu e^{igK} \\ a_\mu &\to \bar{e}^{igK} a_\mu e^{igK} . \end{aligned} \tag{4.4}$$

Alternatively, the gauge transformation can be interpreted as a "quantum" transformation in which the role of the connection is played by the quantum field. As long as the correct form for the *total* transformation is preserved, the assignment of the homogeneous part of the transformation to the background and quantum fields is arbitrary. One possibility is to leave the background unchanged under a quantum gauge transformation:

$$\begin{aligned} A^B_\mu &\rightarrow A^B_\mu \\ a_\mu &\rightarrow \bar{e}^{igK_{(q)}} a_\mu e^{igK_{(q)}} + \frac{i}{g} \bar{e}^{igK_{(q)}} \mathcal{D}^B_\mu e^{igK_{(q)}} \end{aligned} \tag{4.5}$$

where

$$\mathcal{D}^B_\mu e^{igK_{(q)}} = \partial_\mu e^{igK_{(q)}} - ig[A^B_\mu, e^{igK_{(q)}}]. \tag{4.6}$$

The theory is quantized by integrating over the quantum field only, and the background field plays the role of a source. In order to factor out the volume of the gauge group at each space-time point, we have to do gauge fixing for the quantum gauge transformations (4.5) and supply a Faddeev-Popov ghost action. An essential point of the background field method is to pick a quantum gauge fixing term that is *covariant* with respect to the background gauge transformations, *e.g.*,

$$I_{GF} = \int d^4x \left(-\frac{1}{2\alpha} tr\, F^2 \right) \tag{4.7}$$

where

$$F = \mathcal{D}^B_\mu a^\mu = \partial_\mu a^\mu - ig[A^B_\mu, a^\mu]. \tag{4.8}$$

In this case, the full quantum formalism remains manifestly covariant with respect to the background gauge transformations.

The basic object that one wishes to calculate in perturbation theory is the effective action Γ, which is calculated from the one-particle-irreducible (1PI) graphs. From the effective action may be derived the connected Green's functions by the standard procedure, writing a series of tree graphs with quantum-corrected vertices obtained by expanding Γ. In the background field method, the object to be calculated is $\Gamma(A^B_\mu)$, *i.e.*, from 1PI graphs with only external background legs. That this is sufficient for the renormalization program is *not* manifest, but follows from the Slavnov-Taylor identities for the Q-gauge invariances. Thus, in three-loop and higher order diagrams where there are divergent subdiagrams involving quantum fields a_μ, the renormalizations of the coupling constants for vertices involving quantum fields are correctly taken from lower order divergent diagrams with only external background fields. Due to the presence of the background field A^B_μ in the gauge-fixing term (4.7), the Green's functions derived in the background field method are not the same as those of ordinary field theory, but they give rise to the same results for gauge-invariant matrix elements[21] (such as the S-matrix, if one sets aside the usual problems of infrared divergences in Yang-Mills theories).

In order to calculate the effective action, we write down the classical action together with its gauge fixing and ghost action terms, and separate this into terms homogeneous in the quantum fields. From the terms quadratic in the quantum fields are derived the propagators and background-quantum vertices with two quantum legs; terms of higher order in the quantum fields occur as vertices in two- and higher-loop diagrams. Terms involving only the background field or linear in quantum fields are not used in the Feynman rules for calculating the quantum corrections to the effective action. Of course, the *tree-level* contribution to the effective action consists in just the terms in the action without quantum fields. In a renormalizable theory such as Yang-Mills theory, the counterterms must be of the same form as this tree-level contribution to $\Gamma(A^B_\mu)$. The possibility of non-renormalization theorems arises from the fact that the Feynman rules are derived from different terms in the expansion of the action from those which must serve as counterterms.

Since the background field transformations (4.4) must remain a manifest invariance of $\Gamma(A^B_\mu)$, the counterterm in Yang-Mills theory can only be some divergent coefficient times the original action. This can only be achieved by a multiplicative renormalization of fields and coupling constants such that the product gA^B_μ remains unrenormalized, so

$$Z_g = Z_A^{-1/2}. \tag{4.9}$$

Another consequence of the background field method in Yang-Mills theory is the non-renormalization of the Chern-Simons form,

$$K_\mu = \epsilon^{\mu\nu\rho\sigma} tr\ (A_\nu \partial_\rho A_\sigma - \frac{i}{3} g\, A_\nu [A_\rho, A_\sigma]), \tag{4.10}$$

the operator whose divergence is $tr\ F^*_{\mu\nu} F^{\mu\nu}$. Green's functions with insertions of this operator may be calculated from Feynman rules derived from an action including the term

$$I_{CS} = \int d^4x\, \xi^\mu K_\mu. \tag{4.11}$$

While this is clearly not locally gauge invariant, it is nonetheless renormalizable. One might expect it to mix with other non-gauge-invariant operators of dimension three. However, the term (4.11) does become locally gauge invariant when the source ξ^μ is taken to be a longitudinal vector, and this requires the two terms in (4.10) to be renormalized with the same coefficient. Consideration of this special case with longitudinal ξ^μ would not be sufficient if there were a purely transverse operator of dimension three, but there is no such operator that is a rigid gauge scalar.

If we insert the background-quantum splitup into (4.10), the relative factor of -1/3 is just what is needed to make the terms quadratic and cubic in quantum fields background gauge invariant:

$$I_{CS}^{(2)} = \frac{1}{2}\int d^4x \xi^\mu \epsilon_{\mu\nu\rho\sigma}\, tr\{a^\nu(\mathcal{D}_B^\rho a^\sigma - \mathcal{D}_B^\sigma a^\rho)\} \tag{4.12}$$

$$I_{CS}^{(3)} = -\frac{ig}{3}\int d^4x \xi^\mu \epsilon_{\mu\nu\rho\sigma}\, tr\{a^\nu[a^\rho, a^\sigma]\}. \tag{4.13}$$

Thus, the Feynman rules derived in the background field method will be fully background covariant, despite the lack of local invariance of the total term (4.11). The lack of local gauge invariance of (4.11) does show up, however, in the parts of the action of zeroth and first order in the quantum fields. Since (4.11) is a renormalizable term, the counterterm for renormalizing $\Gamma(A_B^\mu, \xi^\nu)$ would be just the part of (4.11) dependent only on background fields. But since this part is not background gauge invariant, it cannot occur as a counterterm, so (4.11) must remain unrenormalized.[22)]

As in the mass and interaction terms in the Wess-Zumino model, the non-renormalization of the term (4.11) happens because the renormalization of the source field ξ^μ just cancels the wave function renormalizations of the background gauge fields. Alternately, since the product gA_μ^B is unrenormalized, the product g^2K_μ is finite. Thus, in a minimal subtraction scheme where renormalization constants are taken only from the poles, *e.g.*, in dimensional regularization, g^2K_μ has no anomalous dimension. Writing the renormalization group equation for this operator, we have

$$\left(\mu\frac{\partial}{\partial\mu} + \beta(g)\frac{\partial}{\partial g}\right)(g^2K_\mu) = 0 \tag{4.14}$$

where μ is the renormalizaton scale. Letting the anomalous dimension of K_μ be γ_K, we have

$$-\gamma_K g^2 K_\mu + 2\beta(g) g K_\mu = 0 \tag{4.15}$$

or

$$\gamma_K = \frac{2\beta(g)}{g}. \tag{4.16}$$

The result (4.16) for the anomalous dimension of the Chern-Simons term is the basis for the well-known Adler-Bardeen theorem[23)] for the anomalous divergence of the $U(1)$ chiral current in a Yang-Mills theory with spinors. If one couples the Yang-Mills field to spinors, then the Chern-Simons form can mix under renormalization with the chiral current

$$J^5_\mu = \bar{\lambda}^b \gamma_5 \gamma_\mu \lambda_b \tag{4.17}$$

where b is a representation index for the Yang-Mills group. Due to the possibility of mixing, we now have a matrix of anomalous dimensions

$$\gamma \begin{pmatrix} J^5_\mu \\ K_\mu \end{pmatrix} = \begin{pmatrix} \gamma_J & 0 \\ \gamma_{KJ} & \gamma_K \end{pmatrix} \begin{pmatrix} J^5_\mu \\ K_\mu \end{pmatrix}, \tag{4.18}$$

where the zero in the upper right-hand corner is due to the lack of local gauge invariance of K_μ. J^5_μ may mix into K_μ, on the other hand, so there is, in general, an off-diagonal γ_{KJ}.

If there is a chiral anomaly, the divergence of J^5_μ is

$$\partial^\mu J^5_\mu = RF^*_{\mu\nu}F^{\mu\nu} = R\partial^\mu K_\mu. \tag{4.19}$$

Although not gauge invariant, the current $J^5_\mu - RK_\mu$ is nonetheless conserved, and thus must have vanishing anomalous dimension if it is to correctly generate transformations on fields. Thus,

$$\left(\mu\frac{\partial}{\partial\mu} + \beta(g)\frac{\partial}{\partial g}\right)(J^5_\mu - RK_\mu) = 0. \tag{4.20}$$

Substituting the renormalization group operators for J^5_μ and K_μ, we have

$$(-\gamma_J + R\gamma_{KJ})J^5_\mu + \left(-\beta(g)\frac{\partial R}{\partial g} + R\gamma_K\right)K_\mu = 0, \tag{4.21}$$

and since J_μ^5 and K_μ are independent, we have

$$\gamma_J = R\gamma_{KJ} \tag{4.22}$$

$$\frac{\beta(g)}{R}\frac{\partial R}{\partial g} = \gamma_K. \tag{4.23}$$

Relation (4.22) provides a convenient way to calculate the anomalous dimension γ_J by calculating γ_{KJ} to one loop order less. In a supersymmetric theory, if J_μ^5 is the chiral current component of the supercurrent, which contains other component currents that remain conserved, then it must have zero anomalous dimension and similarly for γ_{KJ}. Anomalous chiral currents that are not in the supercurrent may have a nonvanishing anomalous dimension.

The result (4.16) for the anomalous dimension γ_K still holds in the presence of coupling to spinors, for the Chern-Simons term must still remain unrenormalized. Inserting (4.16) into (4.23), we obtain the Adler-Bardeen theorem

$$\left(g\frac{\partial R}{\partial g} - 2R\right) = 0 \tag{4.24}$$

i.e.,

$$R = cg^2 \tag{4.25}$$

where

$$c = \frac{C_2(G)}{16\pi^2} \tag{4.26}$$

given by the well-known one-loop result.

A second example of a non-renormalization theorem in the background field method is the familiar non-renormalization of chiral superpotentials in $N = 1$ supersymmetry. Here one splits up the chiral superfield Φ into background and quantum parts

$$\Phi = \Phi_B + \phi. \tag{4.28}$$

All these superfields satisfy the chirality constraint

$$\bar{D}_{\dot{\alpha}}\Phi_B = \bar{D}_{\dot{\alpha}}\phi = 0. \tag{4.29}$$

In the case of chiral superfields, it is simple enough to work directly with the constrained superfield[12] ϕ by going to a chiral basis where it depends only on $z^\mu = x^\mu - i\theta^\alpha \sigma^\mu_{\alpha\dot{\alpha}} \bar{\theta}^{\dot{\alpha}}$ and θ^α.

Working directly with constrained superfields is not something that is generally feasible with the more complicated constraints that occur in extended supersymmetry. Therefore, we proceed to solve the constraint on ϕ[13,24] and work in terms of unconstrained superfields. The chirality constraint on ϕ is solved by

$$\phi = \bar{D}^2 X \tag{4.30}$$

where X is a dimension zero "prepotential" superfield. Since not all of the superspins in X contribute to ϕ, there is a gauge transformation

$$\delta X = \bar{D}^{\dot{\alpha}} \Lambda_{\dot{\alpha}} \tag{4.31}$$

that leaves ϕ unchanged.

To develop the Feynman rules, we insert the background-quantum split into the action and keep the terms quadratic or higher in the quantum fields. Taking the example of a single chiral superfield with mass and interaction terms, we have

$$I^B = \int d^4x d^4\theta \bar{\Phi}_B \Phi_B - \mathrm{Re} \int d^4x d^2\theta (m\Phi_B^2 + \frac{1}{3} h \Phi_B^3) \tag{4.32}$$

$$I^2_{(kinetic)} = \int d^4x d^4\theta (D^2 \bar{X} \bar{D}^2 X) \tag{4.33}$$

$$I^2_{(mass)} + I^3 = -\mathrm{Re} \int d^4x d^2\theta (m\bar{D}^2 X \bar{D}^2 X + \frac{h}{3}(3(\bar{D}^2 X)^2 \Phi_B + (\bar{D}^2 X)^3). \tag{4.34}$$

The kinetic term (4.33) still possesses the gauge invariance (4.31), which needs to be fixed. A convenient gauge fixing function is

$$F_\alpha = D_\alpha X, \tag{4.35}$$

and the most convenient gauge fixing term is[13)]

$$I_{GF} = \int d^4x d^4\theta \bar{F}^{\dot\alpha}(D_\alpha \bar{D}_{\dot\alpha} + \frac{3}{4}\bar{D}_{\dot\alpha} D_\alpha)F^\alpha \tag{4.36}$$

so that the resulting kinetic term is

$$I^2_{(kinetic)} + I_{GF} = \int d^4x d^4\theta \; \bar{X}\Box X. \tag{4.37}$$

There is no ghost action, since the transformation (4.31) is Abelian.

The mass and interaction terms (4.34) may be rewritten as a full superspace integral:

$$I^2_{(mass)} + I^3 = -\mathrm{Re}\int d^4x d^4\theta \left(mX\bar{D}^2X + \frac{h}{3}(3X(\bar{D}^2X)\Phi_B + X(\bar{D}^2X)^2) \right). \tag{4.38}$$

Notice that the background field enters as the constrained field Φ_B; it was never necessary to solve the constraint in the background.

The non-renormalization theorem is now apparent, for the Feynman rules derived from (4.37) and (4.38) give quantum corrections to the effective action that involve only full superspace integrals and couplings to the background in the constrained form Φ_B. Since it is not possible to rewrite the mass and interaction counterterms depending only on Φ_B (cf. 4.32) as full superspace integrals using only local operators, these counterterms are ruled out.

Several points need to be added to the above discussion for completeness, which we shall not dwell upon in detail here. Regularization of the quantum theory is a crucial issue that has received much debate. In practice, regularization by dimensional reduction[14)] has been used, guaranteeing supersymmetry in the counterterms at least at low-loop orders. For the pure Wess-Zumino model, supersymmetric Pauli-Villars or higher derivative regularization can also be used. If dimensional reduction is used, care must be taken in massless theories to separate the ultraviolet from the infrared divergences, requiring the introduction of some form of infrared regularization.

When we come to treat supersymmetric gauge theories, we must make the background-quantum split in such a way as to preserve the maximal manifest background symmetry. Although $N = 1$ super Yang-Mills theory is ultimately expressed in terms of a dimension zero prepotential S, the best way to make the background-quantum split is at the level of the superspace gauge connection A_α [13]):

$$\nabla_C = D_C - igA_C, \quad C = \alpha, \dot\alpha; \mu \tag{4.39}$$

$$A_\alpha = A^B_\alpha + a_\alpha, \ etc. \tag{4.40}$$

In order to allow coupling to constrained matter superfields, we must still impose the representation-preserving constraints

$$F_{\alpha\beta} = F_{\dot\alpha\dot\beta} = 0 \tag{4.41}$$

and we shall continue to express the vector gauge connection A_μ in terms of $A_\alpha, A_{\dot\alpha}$ by solving the conventional constraint

$$F_{\alpha\dot\beta} = 0. \tag{4.42}$$

In order to maintain background gauge covariance, the background field must be constrained by

$$F^B_{\alpha\beta} = F^B_{\dot\alpha\dot\beta} = F^B_{\alpha\dot\beta} = 0 \tag{4.43}$$

where

$$F^B_{\alpha\beta} = \frac{i}{g}\{\mathcal{D}^B_\alpha, \mathcal{D}^B_\beta\} \tag{4.44}$$

$$\mathcal{D}^B_\alpha = D_\alpha - ig\, A^B_\alpha, \tag{4.45}$$

etc. The quantum fields a_α, $a_{\dot\alpha}$ must then be constrained by

$$\mathcal{D}^B_\alpha a_\beta + \mathcal{D}^B_\beta a_\alpha - ig\{a_\alpha, a_\beta\} = 0, \tag{4.46}$$

and similarly for $a_{\dot\alpha}$. As in the case of the chiral superfield, we solve only the constraints on the quantum fields in terms of a dimension zero complex prepotential U:

$$a_\alpha = \frac{i}{g} e^{-gU} \mathcal{D}^B_\alpha e^{gU} \tag{4.47}$$

$$a_{\dot\alpha} = \frac{i}{g} e^{g\bar U} \bar{\mathcal{D}}^B_{\dot\alpha} e^{-g\bar U}. \tag{4.48}$$

When the super Yang-Mills multiplet is coupled to matter, the constraints on the matter fields must be written with covariant derivatives. Since we wish to work with background covariant objects, we use the quantum gauge prepotential to define a background covariantly constrained chiral field:

$$\tilde\Phi = e^{-g\bar U}\Phi \tag{4.49}$$

where

$$\bar{\mathcal{D}}^B_{\dot\alpha}\tilde\Phi = 0. \tag{4.50}$$

The intermediate field $\tilde\Phi$ can now be separated into background and quantum parts, generalizing (4.28):

$$\tilde\Phi = \Phi_B + \bar{\mathcal{D}}^2_B X. \tag{4.51}$$

As before, there is a new gauge invariance for the matter system that arises from the solution (4.51),

$$\delta X = \bar{\mathcal{D}}^{\dot\alpha}_B \Lambda_{\dot\alpha}, \tag{4.52}$$

which leaves $\tilde\Phi$ invariant by virtue of the background constraint (4.43).

In order to derive the Feynman rules used to calculate the effective action $\Gamma(A^B_\alpha, \Phi_B)$, we must insert the background-quantum split into the classical action and keep terms quadratic or higher in quantum fields:

$$I^Q_{Y.M.} = \int d^4x d^4\theta \, \{\bar a^{\dot\alpha}(2\mathcal{D}^2_B \bar a_{\dot\alpha} + \mathcal{D}^B_\alpha \bar{\mathcal{D}}^B_{\dot\alpha} a^\alpha + i\mathcal{D}^B_{\alpha\dot\alpha} a^\alpha$$

$$-ig[a^\alpha, \bar{\mathcal{D}}^B_{\dot\alpha} a_\alpha] + \frac{g^2}{4}[a^\alpha, \{a_\alpha, \bar a_{\dot\alpha}\}]) + c.c.\} \tag{4.53}$$

$$I_M^Q = \int d^4x d^4\theta \{ (\bar{\Phi}^B + \mathcal{D}_B^2 \bar{X}) e^{2gV} (\Phi^B + \bar{\mathcal{D}}_B^2 X) - \bar{\Phi}^B (1 + 2gV) \Phi^B - \mathcal{D}_B^2 \bar{X} \Phi_B - \bar{\Phi}^B \bar{\mathcal{D}}_B^2 X + \left(mX \bar{\mathcal{D}}_B^2 X + \frac{h}{3} (3 (\bar{\mathcal{D}}_B^2 X) \Phi_B + X (\bar{\mathcal{D}}_B^2 X)^2) \right) + c.c. \} \tag{4.54}$$

where

$$e^{2gV} = e^{gU} e^{g\bar{U}} . \tag{4.55}$$

The essential point to note in (4.53),(4.54) is that all integrals are over the full superspace, with the background fields appearing only through the constrained superfields A_α^B, Φ_B. The quantum corrections to the effective action must be written in the same way, and the ultraviolet counterterms must be gauge invariant local expressions written in this fashion.

Before stating the non-renormalization theorem, we must discuss a special feature of the background field method that occurs only in the one-loop graphs. In (4.53),(4.54), we have not included the gauge fixing terms for either the super Yang-Mills action or for the matter action. The Λ-gauge fixing function for the Yang-Mills action may be chosen to be

$$F = \bar{\mathcal{D}}_B^2 V, \tag{4.56}$$

which remains covariant under background K-gauge transformations. Since the parameter Λ of a quantum gauge transformation is now background covariantly chiral, so are the ghost and antighost. In order to quantize in terms of unconstrained superfields, it is then necessary to solve the background covariant constraints on the ghost and antighost in the same way as for a chiral matter field. Note that since ghost fields never occur on the external legs of diagrams, we do not split them into background plus quantum parts. In addition to the Faddeev-Popov ghosts, the functional averaging of the gauge condition required to obtain the standard gauge fixing term introduces further chiral ghosts, the Nielsen-Kallosh ghosts,[25] which may be treated in the same way as the Faddeev-Popov ghosts.

For background covariantly chiral fields, whether matter fields or ghosts, we can fix the gauge as in (4.35),(4.36). However, we now have to include a ghost action, for the background covariant gauge fixing function

$$F_\alpha = \mathcal{D}_\alpha^B X \tag{4.57}$$

and the gauge transformation (4.52) cause the ghost $C^{(1)}_{\dot{\alpha}}$ to couple to the background field. Since this ghost occurs only in the combination $\bar{\nabla}^{\dot{\alpha}}_{B} C^{(1)}_{\dot{\alpha}}$, the ghost action has its own gauge invariance,

$$\delta C^{(1)}_{\dot{\alpha}} = \bar{\nabla}^{\dot{\beta}}_{B} \Lambda_{(\dot{\alpha}\dot{\beta})}. \tag{4.58}$$

Fixing the gauge invariance (4.58) and providing a second level ghost action with $C^{(2)}_{\dot{\alpha}\dot{\beta}}$, we find another gauge invariance with parameter $\Lambda_{(\dot{\alpha}\dot{\beta}\dot{\gamma})}$, and so forth ad infinitum. This problem is known as that of "ghosts for ghosts."[26)]

In order to stop the above proliferation of ghosts, it is necessary to introduce a prepotential for the background gauge connnection. Thus, for this purpose only, we introduce the dimension zero background prepotential u^B:

$$A^B_{\alpha} = \frac{i}{g} e^{-gU^B} D_{\alpha} e^{gU^B} \tag{4.59}$$

$$A^B_{\dot{\alpha}} = \frac{i}{g} e^{g\bar{U}^B} \bar{D}_{\alpha} e^{-g\bar{U}^B}. \tag{4.60}$$

Using $e^{g\bar{U}^B}$, the $C^{(1)}_{\dot{\alpha}}$ transformation can be written

$$\delta C^{(1)}_{\dot{\alpha}} = e^{g\bar{U}^B} \bar{D}^{\dot{\beta}} \tilde{\Lambda}_{(\dot{\alpha}\dot{\beta})} e^{-g\bar{U}^B}. \tag{4.61}$$

The next ghost then appears only in the combination $\bar{D}^{\dot{\beta}} \tilde{C}^{(2)}_{\dot{\alpha}\dot{\beta}}$, and its gauge transformation involves only *flat* derivatives:

$$\delta \tilde{C}^{(2)}_{\dot{\alpha}\dot{\beta}} = \bar{D}^{\dot{\gamma}} \tilde{\Lambda}_{(\dot{\alpha}\dot{\beta}\dot{\gamma})}. \tag{4.62}$$

Then the gauge freedom (4.62) may be fixed using *flat* derivatives and all further ghosts decouple.

The price of decoupling the infinite series of ghosts has been the introduction of a background prepotential for the Yang-Mills multiplet. It was never necessary to introduce a prepotential for the background chiral superfields. Since it was introduced solely for the purpose of converting background covariant derivatives to flat ones in ghost actions, the background Yang-Mills prepotential can occur only in one-loop diagrams. The interactions with U^B occur only in ghost actions like that for $\tilde{C}^{(2)}_{\dot\alpha\dot\beta}$, derived using (4.61). Since the ghost $\tilde{C}^{(2)}_{\dot\alpha\dot\beta}$ couples *only* to background fields, it can occur only in one-loop diagrams.

The non-renormalization theorem for general $N = 1$ gauge and gauge-coupled matter theories can now be stated. The leading infinite parts of the effective action must be local in structure. Nonlocal infinities are removed by renormalizations at lower loop orders. At one loop, the divergent parts of Γ must be written as background gauge invariant integrals over a single $d^4\theta$ and the loop momentum of an expression involving the background super Yang-Mills prepotential U^B and the covariantly constrained matter fields Φ_B. At two loops and higher, the local divergent parts of Γ must be written as a single $d^4\theta$ integral and integrals over the loop momenta of a background gauge invariant expression involving the background connection A^B_α and the constrained matter fields Φ_B. Terms that cannot be written in accordance with these rules must remain unrenormalized, such as mass and interaction terms for chiral matter multiplets.

In extended supersymmetric theories, there are even more powerful non-renormalization theorems that can be derived along the lines detailed above. A basic problem that arises is the fact that, in many cases, the full supersymmetry of the field equations cannot be manifestly realized in the quantum formalism, at least with a finite number of component fields. Thus, one must first determine what is the maximal symmetry that can be manifestly realized. The chief example of a theory with this sort of problem is the maximally supersymmetric Yang-Mills theory,[27], which in four dimensions has $N = 4$ supersymmetry.

The classical dynamics of the $N = 4$ super Yang-Mills theory may be derived from the component action

$$
\begin{aligned}
I_{Y.M.4} = \frac{1}{g^2}\int d^4x \mathrm{tr}\Big\{ & -\frac{1}{4}F_{\mu\nu}F^{\mu\nu} + \frac{1}{2}\nabla^\mu\bar\phi^{ij}\nabla_\mu\phi_{ij} - i\lambda_i\sigma_\mu\nabla^\mu\bar\lambda^i \\
& -\frac{i}{2}\bar\lambda_{\dot\beta}[\bar\lambda^{\dot\beta j},\phi_{ij}] - \frac{i}{2}\lambda^\alpha_i[\lambda_{\alpha j},\bar\phi^{ij}] \\
& + \frac{1}{4}[\phi_{ij},\phi_{k\ell}][\bar\phi^{ij},\bar\phi^{k\ell}]\Big\}
\end{aligned}
\tag{4.63}
$$

where $i, j, k, \ell : 1 \to 4$, and the four Weyl spinors λ^i_α and the six scalars $\phi_{ij} = \frac{1}{2}\epsilon_{ijk\ell}\bar\phi^{k\ell}$ are in the adjoint representation of the gauge group. The action (4.63) has a manifest $SU(4)$ invariance, but supersymmetry is not manifest and only forms a closed algebra on the fields present in (4.63) subject to the equations of motion.

In order to see the problem with deriving a full linearly realized supersymmetric formalism for the $N = 4$ theory,[28,24)] suppose that there were an off-shell version involving a multiplet of fields Φ. The linearized action could then be written schematically in the form

$$I = \int d^4x \; \Phi \mathcal{O} \Phi \tag{4.64}$$

where $\mathcal{O}$ is some differential operator. With linearly realized supersymmetry, we can also form the higher derivative invariant

$$I' = \frac{1}{m^2} \int d^4x \Phi \mathcal{O} \Box \Phi \tag{4.65}$$

and analyze the spectrum of $I + I'$, which must fall into linear representations of $N = 4$ supersymmetry, with either physical or ghost signs for the residues of the propagators. Each massless physical fermion in the theory derived from I gives rise to two massless physical helicity states and four massive ghost states, while each auxiliary fermion pair in I contributes four massive physical states and four massive ghost states. It is assumed here that auxiliary fermions always occur in pairs. Since each massive linear $N = 4$ multiplet contains an integral multiple of 128 fermionic states (cf. the superspin zero multiplet given in Table I), we have for n auxiliary fermion pairs and the four original fermion fields,

$$\begin{aligned} 4n + 4.4 &= 128p \quad \text{massive physical states} \\ 4n &= 128q \quad \text{massive ghost states} \end{aligned} \tag{4.66}$$

where p and q are integers. Subtracting, one gets

$$p - q = \frac{1}{8} \tag{4.67}$$

which is not possible for finite integers p and q. It is possible to evade the above "no-go" theorem if one admits an infinite number of component fields, as in the harmonic superspace formalism.[29] To date, this formalism has not been developed to give a full quantization procedure.

In order to proceed with a finite number of component fields in a theory such as $N = 4$ super Yang-Mills, it is necessary to pick some part of the total symmetry of the theory that can be manifestly realized. One approach[24,30] that has been applied to the analysis of the $N = 4$ theory's infinities uses manifest Lorentz invariance and $N = 2$ supersymmetry. Another approach uses the light-cone gauge,[31] in which manifest Lorentz invariance is given up, but one can keep the $SU(4)$ invariance manifest. In both of these approaches, half of the total number of components of the supersymmetry transformations are linearly realized. Since the light-cone gauge precludes use of the background field method, it has not proven as useful in understanding ultraviolet cancellations where delicate issues involving gauge invariance arise.

When decomposed into representations of $N = 2$ supersymmetry, the propagating fields of $N = 4$ super Yang-Mills theory fall into two $N = 2$ massless supermultiplets, each carrying the adjoint representation of the gauge group. The two multiplets are the $N = 2$ Yang-Mills multiplet[32] ($A_\mu, \lambda_{\alpha i}(i = 1,2), \phi$ (complex)) and the $N = 2$ matter representation, the hypermultiplet[33] ($\psi_{\alpha i}$, $L^{(ij)} = \epsilon^{ik}\epsilon^{jm}L_{km}(i,j,k,m = 1,2)$, S (real)). In this formalism, there is a manifest $SU(2) \times U(1)$ symmetry. It is not necessary to discuss the $N = 4$ theory specifically in this context, for it turns out to be but one of a whole class of ultraviolet finite theories with $N = 2$ supersymmetry.

The superspace for $N = 2$ supersymmetry has fermionic coordinates θ^i_α, where i is an $SU(2)$ doublet index. In order to carry out background field quantization as outlined in the $N = 1$ case above, it is necessary to have a formulation of the theory in unconstrained $N = 2$ superfields. This formulation is considerably more complicated algebraically than the $N = 1$ formalism. For practical calculations, the $N = 1$ superfield formalism appears to be optimal. It has allowed a calculation by hand of the $N = 4$ theory's divergences to the three-loop order,[34] a result which was originally obtained by a component field calculation[35] starting from (4.63), but needing a computer. In contrast, in the $N = 2$ formalism, the non-renormalizaton theorems become transparent, although its complexity makes practical calculation difficult. Here, we shall only set out the selient features of the $N = 2$ quantization program. For more details, we refer the reader to Ref. 24.

The action for the $N = 2$ gauge multiplet can be written as a chiral integral over the square of the field strength superfield W[36]:

$$I_{Y.M.2} = \mathrm{Re} \int d^4x d^4\theta \mathrm{tr} W^2. \tag{4.68}$$

The chiral integral for $N = 2$ supersymmetry is now over the four $d\theta_i^\alpha$; a full superspace integral would be over $d^8\theta$. The field strength superfield W satisfies the $N = 2$ covariant chirality condition

$$\bar{\nabla}_{\dot{\alpha}i} W = 0, \tag{4.69}$$

where

$$\bar{\nabla}_{\dot{\alpha}i} = \bar{D}_{\dot{\alpha}i} - igA_{\dot{\alpha}i} = -\frac{\partial}{\partial\bar{\theta}^{\dot{\alpha}i}} - i\theta_i^\alpha\sigma^\mu_{\alpha\dot{\alpha}}\partial_\mu - igA_{\dot{\alpha}i}. \tag{4.70}$$

The $\theta = \bar{\theta} = 0$ component of W is the complex scalar field of the multiplet. As in the $N = 1$ case, the gauge connection A_α^i must satisfy certain integrability conditions in order to allow gauge coupling to $N = 2$ matter multiplets (hypermultiplets). These representation-preserving constraints are

$$F_{\alpha\beta}^{(ij)} = F_{\dot{\alpha}(i\dot{\beta}j)} = 0 \tag{4.71}$$

$$F^i_{\alpha\dot{\beta}j} - \frac{1}{2}\delta^i_j F^k_{\alpha\dot{\beta}k} = 0. \tag{4.72}$$

In addition, there is a conventional constraint that determines the vectorial part of the connection, A_μ, in terms of $A_{\alpha i}$, $A^i_{\dot{\alpha}}$:

$$F^k_{\alpha\dot{\beta}k} = 0.. \tag{4.73}$$

The unconstrained part of $F^{ij}_{\alpha\beta}$ is the field strength superfield W:

$$F^{ij}_{\alpha\beta} = \epsilon^{\alpha\beta}\epsilon^{ij}W. \tag{4.74}$$

In the case of an Abelian gauge group, the constraints (4.71),(4.72) are solved by[37)]

$$A^i_\alpha = D^i_\alpha P + D^{3i}_\alpha \bar{D}^{jk}V_{jk} - \frac{1}{2}D^i_\alpha\{D^j_k, \bar{D}^{\ell k}\}V_{j\ell} \tag{4.75}$$

where P is real and may be eliminated by a K-gauge transformation. The pre-potential $V_{(jk)}$ is an $SU(2)$ triplet of dimension -2. The quadratic and cubic fermionic derivatives in (4.74) are defined by

$$D^i_\alpha = \epsilon_{\alpha\beta} D^{ij} + \epsilon^{ij} D_{\alpha\beta} \tag{4.76}$$

$$D^i_\alpha D^{jk} = \epsilon^{i(j} D^{3k)}_\alpha . \tag{4.77}$$

Extending the solution (4.75) for the constraints (4.71),(4.72) to the non-Abelian case is algebraically very complicated. However, since all we require in order to develop Feynman rules is an expansion of the solution in powers of the coupling constant g, we may employ the following trick.[38,24] Differentiating the constraints (4.71),(4.72) with respect to the coupling constant g, we obtain

$$\nabla^{(i}_\alpha B^{j)}_\beta + \nabla^{(j}_\beta B^{i)}_\alpha = 0 \tag{4.78}$$

$$\nabla^{(i}_\alpha B^{j)}_{\dot\beta} + \bar\nabla^{(j}_{\dot\beta} B^{i)}_\alpha = 0 \tag{4.79}$$

where

$$B^i_\alpha = \frac{\partial}{\partial g}(g A^i_\alpha) \tag{4.80}$$

$$B^i_{\dot\alpha} = \epsilon^{ij} B_{\dot\alpha j} . \tag{4.81}$$

Since (4.78),(4.79) have a similar form to the constraints in the Abelian gauge theory, they may be solved analogously:

$$\begin{aligned} B^i_\alpha = {} & \nabla^i_\alpha P + \nabla^{3i}_\alpha \bar\nabla^{jk} V_{jk} \\ & - \frac{1}{2} \nabla^i_\alpha \left(\{\nabla^j_k, \bar\nabla^{\ell k}\} V_{j\ell} + \frac{i}{2}[W, \nabla^{jk} V_{jk}] - \frac{i}{2}[\bar W, \bar\nabla^{jk} V_{jk}] \right) . \end{aligned} \tag{4.82}$$

The gauge connection A^i_α may then be reconstructed as a power series in g from the integral equation

$$A^i_\alpha = \frac{1}{g} \int_0^g dk\ B^i_\alpha(k). \tag{4.83}$$

The split between background and quantum fields in the $N = 2$ case is done at the level of the connection A^i_α, in direct analogy with the $N = 1$ case. For everything but one-loop diagrams, only the quantum part of the connection needs to be expressed in terms of a prepotential. For details, and treatment of the quantum gauge invariances, their gauge fixing, and ghosts, see Ref. 24. The non-renormalization theorem derives from the fact that all superspace integrations in the Feynman rules are over the full superspace $d^8\theta$ and the background field enters only as the dimension 1/2 connection A^{Bi}_α, except in some one-loop ghost diagrams. Since the action (4.68) cannot be written this way as a local expression, it is ruled out as a counterterm at two loops and higher. What happens at the one-loop level must be settled by specific calculation in a given theory.

The hypermultiplet matter sector may also be given an off-mass-shell formulation in terms of $N = 2$ superfields[30] (without a central charge). In the free case, the action is

$$I_{H.M.} = \mathrm{Re} \int d^4x d^8\theta (\rho^\alpha_i D_{\alpha j} L^{ij} + X_{ijk\ell} L^{ijk\ell}) \tag{4.84}$$

where ρ^α_i is a dimension -3/2 prepotential and $X_{(ijk\ell)}$ is a dimension -1 prepotential carrying the pseudo-real $\underset{\sim}{5}$ representation of $SU(2)$. The non-gauge physical and auxiliary fields are contained in the pseudo-real or real superfields $L^{(ij)}$, $L^{(ijk\ell)}$, and S, which satisfy the constraints

$$\begin{aligned} D^{(i}_\alpha L^{jk)} &= D_{\alpha\ell} L^{ijk\ell} \\ D^{(t}_\alpha L^{ijk\ell)} &= \bar{D}^{(t}_{\dot\alpha} L^{ijk\ell)} = 0 \\ D_{\alpha\beta} S &= [D^i_\alpha, \bar{D}_{\dot\alpha i}] S = 0. \end{aligned} \tag{4.85}$$

The solution to the constraints (4.85) is given in terms of the prepotentials ρ^α_i and $X_{ijk\ell}$:

$$\begin{aligned} L^{ij} &= \frac{3}{4} \bar{D}^{ij} D^{3k}_\alpha \rho^\alpha_k + \frac{1}{2} \bar{D}^{(i}_k D^{j)k} D^\ell_\alpha \rho^\alpha_\ell + c.c. \\ L^{ijk\ell} &= \frac{2}{5} D^{(ij} \bar{D}^{k\ell)} [D^m_\alpha \rho^\alpha_m + \bar{D}^{\dot\alpha}_m \bar{\rho}^m_{\dot\alpha}] \\ S &= D^{ij} \bar{D}^{k\ell} X_{ijk\ell}. \end{aligned} \tag{4.86}$$

The equations of motion following from (4.84) are

$$\begin{aligned} L_{ijk\ell} &= 0 \\ D_{\alpha j} L^{ij} &= D^i_\alpha S, \end{aligned} \tag{4.87}$$

which correctly describe the free hypermultiplet.

The $N=2$ hypermultiplet has a non-renormalization theorem for its action (4.84). Making the background quantum split in L^{ij}, $L^{ijk\ell}$, and S analogously to the procedure followed in the chiral $N=1$ multiplet, solving the constraints (4.85) only for the quantum fields, one finds that all quantum corrections to the effective action must be written as full superspace integrals over functionals of L^{ij}, $L^{ijk\ell}$, and S. Since the action (4.84) cannot be locally written in this way, it is ruled out as a counterterm and must remain unrenormalized. This is true at the one-loop and at higher loop orders, since, as with the chiral $N=1$ multiplet, there is never any necessity of introducing a prepotential for the background hypermultiplet fields. The non-renormalization of the hypermultiplet action means that there is no wave function renormalization for the hypermultiplet fields. These conclusions are obtained equally in the case of coupling to the $N=2$ gauge multiplet. For details, see Ref. 24.

The non-renormalization theorems for $N=2$ supersymmetry are summarized by the statements that matter wave function renormalizations never occur, while the only possibility for a leading divergence of the form of the Yang-Mills action is at the one-loop order. As usual in the background field method, if such a divergence does occur, it requires coupling constant and wave function renormalizations such that the product (gA^i_α) remains unrenormalized. Once an infinity has occurred at the one-loop order, it also gives rise to non-leading local divergences in higher orders due to counterterm insertions to remove non-local divergences coming from divergent subgraphs. In regularization by dimensional reduction these non-leading divergences are of quadratic or higher order in $(d-4)^{-1}$, with coefficients that can be obtained from the renormalization group pole equations.

The condition for an $N=2$ supersymmetric theory to be ultraviolet finite is that the one-loop gauge coupling constant renormalization vanish[39)]

$$\sum_a m_a T(R_a) - C_2(G) = 0, \tag{4.88}$$

where there are m_a hypermultiplets in representations R_a of the gauge group, and for generators T^r

$$\begin{aligned} \mathrm{tr} T^r T^s &= C_2(G)\delta^{rs} \\ T^r T^r &= T(R)\mathbf{1}, \end{aligned} \tag{4.89}$$

where the conventional normalization has $T(R)=\frac{1}{2}$ for the fundamental representation. Equation (4.88) has many solutions beside the $N=4$ theory, in which case R is the adjoint and $m = 1$. For example, for $SU(N)$ with hypermultiplets in the fundamental representation, $C_2(G) = N$, so we need $m = 2N$ hypermultiplets for a finite theory. In the formulation of the hypermultiplet that we have discussed, there must be an even number of hypermultiplets if they are to be given a complex representation of the gauge group, since the formulation with L^{ij}, $L^{ijk\ell}$, and S is pseudo-real (*i.e.*, satisfies a reality condition with indices raised and lowered with ϵ_{ij}). By explicit two-loop calculation[40], it has been found that finiteness is obtained even without this restriction, provided (4.88) is satisfied. A proof of finiteness in the general case may be afforded by the harmonic superspace formalism.[29]

In the case of the $N = 4$ theory, the cancellations may also be understood by an attractive heuristic argument[41] based upon the multiplet structure of the anomalies. Due to the pseudo-reality condition $\phi_{ij} = \frac{1}{2}\epsilon_{ijk\ell}\bar{\phi}^{k\ell}$ on the $SU(4)$ $\underset{\sim}{6}$ of scalars, there is no way to nontrivially extend the $SU(4)$ to $U(4)$, unlike the case in the non-self-conjugate multiplets for $N = 2$. Thus, if there were a non-vanishing trace anomaly, supersymmetry would link this with a non-vanishing anomaly in some chiral current of the theory, which could only be one of the 9 chiral generators of $SU(4)$. Since the adjoint representation of $SU(4)$ which is carried by the spinors has been found to be anomaly free in all other contexts, this indicates that the trace anomaly should vanish as well. Applying a similar argument to the $N = 2$ supersymmetric finite theories requires use of the Adler-Bardeen theorem.[42] In this case, one also needs to address the question as to which chiral current is the one that satisfies the Adler-Bardeen theorem.

Starting from one of the massless theories that are finite by virtue of $N = 2$ supersymmetry and (4.88), it is possible to add soft supersymmetry-breaking mass and dimension three interaction terms that preserve the finiteness.[43,44] The soft breaking terms have coefficients carrying dimensions of mass and thus cannot upset the cancellations between the logarithmic divergences which are governed by the original supersymmetry. It must then be arranged for the soft breaking terms themselves not to generate divergences of their own forms or of other forms with dimension less than four. The most straightforward technique[45] for achieving these cancellations is to introduce a background "spurion" superfield in interactions that are protected by a non-renormalization theorem (*e.g.*, the $N = 1$ non-renormalization of chiral multiplet mass and interaction terms). Then one lets the spurion take a non-supersymmetric vacuum expectation value, thus introducing the soft symmetry breaking parameters. In general, soft breaking is characterized by a sum rule for the masses given to the various fields of spin J:

$$\sum_J (-1)^{2J+1}(2J+1)m_J^2 = 0. \qquad (4.90)$$

There is an exception[44] to this sum rule in the case of a soft breaking term that gives mass terms of the form (scalar)^2 + (pseudoscalar)^2. In all of the soft breaking terms, there are also trilinear interactions with coefficients determined by the coefficients of the mass terms.

The finiteness of the $N = 2$ supersymmetric models relies upon an improvement in the ultraviolet power counting for extended supersymmetric theories. In order to test the correctness of this power counting, it is necessary to consider theories that are not renormalizable, so that eventually at some loop order the power counting allows infinities to occur. A remarkable set of calculations[46] of the divergences in higher dimensional extended super Yang-Mills theories enables one to test the predictions of power counting using the background field method against the actually occurring infinities.

In six dimensions, the maximally supersymmetric super Yang-Mills theory has $N = 2$ supersymmetry. $N = 1$ supersymmetric theories can be described using left-handed Weyl spinors carrying an additional $SU(2)$ doublet index and obeying an $SU(2)$ pseudo-reality condition

$$\theta_i^\alpha = \bar{\theta}^{\alpha j}\epsilon_{ji} \qquad \alpha = 1..4,\ i = 1,2. \tag{4.91}$$

For $N = 2$ supersymmetry, we need an additional right-handed spinor carrying a doublet index for another $SU(2)$

$$\theta_\alpha^{i'} = \epsilon^{i'j'}\bar{\theta}_{\alpha j'} \qquad \alpha = 1...4,\ i' = 1,2. \tag{4.92}$$

The rigid internal symmetry group for the six-dimensional $N = 2$ theory's field equations is then $SU(2) \times SU(2)$.

The six-dimensional gauge theories satisfy representation-preserving constraints just as in the four-dimensional case. These are[47]

$$\text{for } N = 1: \qquad F_{\alpha\beta}^{ij} = 0 \tag{4.93}$$

$$\begin{aligned} \text{for } N = 2: \qquad & F_{\alpha\beta}^{ij} = F_{i'j'}^{\alpha\beta} = 0 \\ & F_{\alpha j'}^{i\beta} = \delta_\alpha^\beta W_{j'}^i \\ & \bar{W}_i^{i'} = \epsilon^{i'j'} W_{j'}^j \epsilon_{ji}. \end{aligned} \tag{4.94}$$

In the $N = 1$ case, one may check via the Bianchi identities that (4.93) implies

$$F^i_{\alpha\mu} = (\Sigma_\mu)_{\alpha\beta} W^{\beta i}, \tag{4.95}$$

where $(\Sigma_\mu)_{\alpha\beta}$ is the $d = 6$ analog of $(\sigma_\mu)_{\alpha\dot\beta}$ in $d = 4$. The field strength multiplet is thus described in the $N = 1$ case by a spinor superfield, whose lowest component is the physical spinor $\lambda^{\beta i}$. The superfield $W^{\beta i}$ satisfies

$$\begin{aligned} D^i_\alpha W^{\beta j} =& \delta^\beta_\alpha X^{(ij)} - \epsilon^{ij} (\Sigma^{MN})^\beta_\alpha F_{MN} \\ & \alpha, \beta : 1..4 \quad M, N : 1..6 \quad i, j : 1, 2 \end{aligned} \tag{4.96}$$

where X^{ij} is a pseudo-real triplet of auxiliary fields.

As in the case of the $N = 4$ theory in $d = 4$, the $N = 2$, $d = 6$ theory does not have an action formulation displaying its full supersymmetry with a finite number of auxiliary fields. One can describe the $N = 2$ theory as a coupling of the $N = 1$ theory to an $N = 1$ hypermultiplet matter sector in the adjoint representation of the gauge group. The $d = 6$ hypermultiplet is a straightforward generalization of the $d = 4$ hypermultiplet, with pseudo-real or real constrained superfields L^{ij}, $L^{ijk\ell}$, and S satisfying

$$\begin{aligned} \nabla^{(i}_\alpha L^{jk)} &= \nabla_{\alpha\ell} L^{ijk\ell} \\ \nabla^{(i}_\alpha L^{jk\ell m} &= 0 \\ \epsilon_{ij} \nabla^i_{(\alpha} \nabla^j_{\beta)} S &= 0. \end{aligned} \tag{4.97}$$

The naive degree of divergence of ℓ-loop graphs in d-dimensional Yang-Mills plus matter theories is

$$D = (d - 4)\ell + 4. \tag{4.98}$$

In calculating the power counting weight of a counterterm, derivatives, scalar and vector external field lines count as dimension 1; external spinors count as 3/2 and spinorial derivatives count as 1/2. The generic form of an ℓ-loop counterterm in $d = 6$, $N = 1$ super Yang-Mills is thus required by the non-renormalization theorem to be schematically

$$\Delta\Gamma^\ell = g^{2(\ell-1)} \int_{q=2\cdots 2\ell} d^6x d^8\theta (\nabla^{4\ell-2q} (gW^{\alpha i})^q). \tag{4.99}$$

Enumeration of the possibilities shows that there are no allowable essential counterterms at $\ell = 1$ or 2 loops for the pure $N = 1$ theory.[48] All the possible counterterms at 1 or 2 loops vanish subject to the classical field equations and the corresponding infinities may thus be eliminated by field redefinition renormalizations.

In a theory involving hypermultiplets, new counterterms are possible. For example, a generic theory could have counterterms of the form

$$\Delta\Gamma^{\ell} = g^{2(\ell-1)} \int d^6x d^8\theta (\nabla^{4\ell-2q}(gL)^q). \tag{4.100}$$

At one loop, all the candidates vanish subject to the classical field equations. At two loops, however, there could be a counterterm with $q = 4$. Thus, the generic coupling of $d = 6$, $N = 1$ super Yang-Mills to hypermultiplet matter is expected to have essential divergences in the matter sector from two loops onward.

In the case of the $d = 6$, $N = 2$ theory, counterterms must be off-mass-shell $N = 1$ supersymmetric but must acquire the full $N = 2$ supersymmetry when the background is required to satisfy the classical equations of motion. This links the matter terms that can occur to terms involving the gauge multiplet. Analysis of the only two-loop candidate that is $N = 2$ supersymmetric subject to the field equations but does not vanish subject to them shows that there must be a purely gauge multiplet part that, when written in $N = 1$ superfields, fails the test of the non-renormalization theorem. The first loop order at which an allowable essential divergence can occur is at three loops, where in the on-mass-shell full $N = 2$ superfield language it takes the form

$$\Delta\Gamma^{\ell=3}_{N=2} = g^8 \int d^6x d^6\theta^{[\alpha\beta]}_{(ij)} d^6\theta^{(i'j')}_{[\alpha\beta]} \mathrm{tr}[\bar{W}^k_{i'}, \bar{W}_{kj'}][W^i_{k'}, W^{jk'}]. \tag{4.101}$$

Testing for the occurrence of a divergence of the form (4.101) at the three-loop order would be a highly difficult calculation. On the other hand, analysis of the seven-dimensional theory shows that the counterterm (4.101) is just a dimensional reduction of an allowed $d = 7$ two-loop counterterm for the maximal $d = 7$ super Yang-Mills. This $d = 7$ calculation has been performed,[46] with the anticipated divergence being found. At the two-loop level in the six-dimensional theory, however, there are dramatic cancellations leaving no essential divergence, in accord with the non-renormalization theorem. These calculations constitute the most demanding test that has been met by our analysis of the mechanism of ultraviolet cancellations in supersymmetric theories and confirms its essential completeness.

Acknowledgments

This material is based in part upon research supported by the National Science Foundation under Grant No. PHY77-27084, supplemented by funds from the National Aeronautics and Space Administration.

References

1. S. Coleman and J. Mandula, Phys. Rev. **159** (1967) 1251.

2. Y.A. Gol'fand and E.P. Likhtman, JETP Lett. **13** (1971) 323.

3. D.V. Volkov and V.P. Akulov, Phys. Lett. **46**B (1973) 109.

4. J. Wess and B. Zumino, Nucl. Phys. B**70** (1974) 39; Nucl. Phys. B**78** (1974) 1.

5. R. Haag, J.T. Lopuszanski, and M.F. Sohnius, Nucl. Phys. B**88** (1975) 257.

6. A. Salam and J. Strathdee, Nucl. Phys. B**76** (1974) 4778; Nucl. Phys. B**84** (1975) 127.

7. S. Ferrara, in *Proceedings of the 9th International Conference on General Relativity and Gravitation*, Jena, G.D.R. (1980); CERN preprint TH-2957 (1980).

8. S. Ferrara, C.A. Savoy, and B. Zumino, Phys. Lett. **100**B (1981) 393.

9. B.N. Valuyev, JINR Report P2-11638 (1978) (in Russian).

10. F.A. Berezin *The Method of Second Quantization* (Academic Press, New York, 1966).

11. S. Ferrara and B. Zumino, Nucl. Phys. B**79** (1974) 413; A. Salam and J. Strathdee, Phys. Lett. **51**B (1974) 353; B. de Wit and D. Freedman, Phys. Rev. D**12** (1975) 2286.

12. M.T. Grisaru, W. Siegel, and M. Roček, Nucl. Phys. B**159** (1979) 429; M.T. Grisaru and W. Siegel, Nucl. Phys. B**187** (1981) 149.

13. M.T. Grisaru and W. Siegel, Nucl. Phys. B**201** (1982) 292.

14. W. Siegel, Phys. Lett. **84**B (1979) 193; W. Siegel, P.K. Townsend, and P. van Nieuwenhuizen, in *Superspace and Supergravity*, eds. S.W. Hawking and M. Roček (Cambridge University Press, 1981), p. 165.

15. S. Ferrara, J. Iliopoulos, and B. Zumino, Nucl. Phys. B**77** (1974) 413.

16. K. Pohlmeyer, Comm. Math. Phys. **27** (1972) 247.

17. S. Ferrara and B. Zumino, Nucl. Phys. B**87** (1975) 207.

18. V. Ogievetsky and V. Polubarinov, Sov. J. Nucl. Phys. **4** (1966) 216; M. Kalb and P. Ramond, Phys. Rev. D**9** (1974) 2273.

19. M.T. Grisaru, in *Recent Developments in Gravitation*, eds. M. Levy and S. Deser (Plenum, New York, 1979), p. 577.

20. G. 't Hooft, in *Proc. 12th Winter School of Theoretical Physics in Karpacz*, Acta Univ. Wratisl. No. 38 (1975); B.S. De Witt, in *Quantum Gravity 2*, eds. C.J. Isham, R. Penrose, and D.W. Sciama (Oxford University Press, 1981); D.G. Boulware, Phys. Rev. D**23** (1981) 389; L.F. Abbot, Nucl. Phys. B**185** (1981) 189.

21. L.F. Abbot, M.T. Grisaru, and R.K. Schaefer, Brandeis preprint BRX-TH35 (1983).

22. P. Breitenlohner, D. Maison, and K.S. Stelle, Phys. Lett. **134**B (1984) 63.

23. S. Adler, in Brandeis 1970 Lectures, eds. S. Deser, M.T. Grisaru, and H. Pendleton (MIT Press, 1970).

24. P.S. Howe, K.S. Stelle, and P.K. Townsend, Nucl. Phys. B**236** (1984) 125.

25. N.K. Nielsen, Nucl. Phys. B**140** (1978) 499; R.E. Kallosh, Nucl. Phys. B**141**

(1978) 141.

26. W. Siegel and S.J. Gates, Nucl. Phys. B**201** (1982) 292.

27. F. Gliozzi, J. Scherk, and D. Olive, Nucl. Phys. B**133** (1978) 253; L. Brink, J. Schwarz, and J. Scherk, Nucl. Phys. B**121** (1977) 77.

28. M. Roček and W. Siegel, Phys. Lett. **105**B (1981) 275; V. Rivelles and J.G. Taylor, J. Phys. A: Math. Gen. **15** (1982) 163.

29. A. Galperin, E. Ivanov, S. Kalitzin, V. Ogievetsky, and E. Sokatchev, ICTP preprint IC/84/43; Dubna preprint E2-84-441 (1984).

30. P.S. Howe, K.S. Stelle, and P.K. Townsend, Nucl. Phys. B**214** (1983) 519; K.S. Stelle, in *Proc. 21st Intl. Conf. on High Energy Physics*, eds. P. Petiau and M. Porneuf, J. Phys. **12** (1982) 326.

31. S. Mandelstam, in *Proc. 21st Intl. Conf. on High Energy Physics*, eds. P. Petiau and M. Porneuf, J. Phys. **12** (1982) 331; L. Brink, O. Lindgren, and B.E.W. Nielsen, Nucl. Phys. B**212** (1983) 401; Phys. Lett. **123**B (1983) 328.

32. See the first paper of Ref. 11.

33. P. Fayet, Nucl. Phys. B**113** (1976) 135; M.F. Sohnius, Nucl. Phys. B**138** (1978) 109; A. Galperin, E.A. Ivanov, and V.I. Ogievetsky, JETP Lett. **33** (1981) 176.

34. M.T. Grisaru, M. Rocek, and W. Siegel, Phys. Rev. Lett. **45** (1980) 1063; W. Caswell and D. Zanon, Phys. Lett. **100**B (1980) 152.

35. L.V. Avdeev and O.V. Tarasov, Phys. Lett. **112**B (1982) 356.

36. R. Grimm, M. Sohnius, and J. Wess, Nucl. Phys. B**133** (1978) 275.

37. L. Mezincescu, JINR Report P2-12572 (1979).

38. P.S. Howe, K.S. Stelle, and P.K. Townsend, reported by K.S. Stelle in *Proc. Intl. Seminar on Group Theoretical Methods in Physics in Zvenigorod*, Nov. 1982; Imperial College preprint ICTP 82-83/13; J. Koller, Caltech preprint (1983).

39. P.S. Howe, K.S. Stelle, and P.C. West, Phys. Lett. **124**B (1983) 55.

40. P.S. Howe and P.C. West, King's College preprint (1984).

41. M.F. Sohnius and P.C. West, Phys. Lett. **100**B (1981) 245; S. Ferrara and B. Zumino, unpublished.

42. M.T. Grisaru and P.C. West, Brandeis preprint BRX-TH-141 (1983).

43. P.C. West, Lectuie given at Shelter Island II Conf. June 1983 (King's College preprint, 1983); J.G. Taylor, Phys. Lett. **121**B (1983) 386; S. Rajpoot, J.G. Taylor, and M. Zaimi, Phys. Lett. **127**B (1983); King's College preprint (1983).

44. A. Parkes and P.C. West, King's College preprint (1983).

45. L. Girardello and M.T. Grisaru, Nucl. Phys. B**194** (1982) 65.

46. N. Marcus and A. Sagnotti, Phys. Lett. **135**B (1984) 85; Caltech preprint CALT-68-1129 (1984).

47. P.S. Howe, G. Sierra, and P.K. Townsend, Nucl. Phys. B**221** (1983) 331; J. Koller, Phys. Lett. **124**B (1983) 324; Nucl. Phys. B**222** (1983) 319.

48. P.S. Howe and K.S. Stelle, Phys. Lett. **137**B (1984) 175; A.E.M. van de Ven, Stony Brook preprint ITP-SB-84-26.

THE GROUP MANIFOLD APPROACH TO UNIFIED GRAVITY

T. Regge
Università di Torino, Italy

1. Introduction and historical remarks.

General relativity is no longer a unique theory in its historical role of linking physical and geometrical concepts. In the last decade it has become increasingly clear that many unified theories share this and other interesting features with gravity and, just as gravity, they promise to have a key role in the developement of contemporary physics. None of the present unified theories is final and can be considered as a serious candidate for the phenomenology of elementary particles. All of them, however, have so many interesting features and elegance of construction that it is natural to regard them as preliminary stages for the construction of a realistic theory.

Lie groups have a key role in the developement of unified theories. In these notes I shall use systematically the language of forms instead of the standard tensor calculus; there are many advantages in doing this. Forms yield a more compact notation; they are broadly used in the mathematical literature and afford a more direct link to the standard theory of Lie groups. They are also the natural language in which to express the ideas of topology. Here I shall deal mainly with superforms, that is graded forms on a supermanifold. I shall keep some degree of rigour without succeeding to do so uniformly. Readers who are unsatisfied with my use of supervariables are referred to the standard references where this matter is discussed more at length. In these notes supervariables have purely a formal role: they serve a purpose similar to that of a formal variable in a generating function. Not all purely Bose structures admit a superextension. Therefore supervariables are useful not only as entities in their own right, but also as tools in selecting particular Bose sectors.

General relativity is traditionally introduced through the standard tensor calculus. The world is a 4-dimensional differentiable pseudoriemannian manifold with a metric tensor:

$$(d\tau)^2 = g_{\mu\nu}(x)\, dx^{\mu}\, dx^{\nu} \tag{1.1}$$

and a Christoffel connection:

$$\Gamma^{\alpha}_{\beta\gamma} = \Gamma^{\alpha}_{\gamma\beta} = \frac{1}{2} g^{\alpha\eta}\left(\frac{\partial g_{\eta\beta}}{\partial x^{\gamma}} + \frac{\partial g_{\gamma\eta}}{\partial x^{\beta}} - \frac{\partial g_{\beta\gamma}}{\partial x^{\eta}}\right) \tag{1.2}$$

satisfying the relation

$$\frac{\partial g_{\mu\nu}}{\partial x^\lambda} - \Gamma_{\mu\lambda}^{\ \ \beta} g_{\beta\nu} - \Gamma_{\lambda\nu}^{\ \ \beta} g_{\mu\beta} = 0. \tag{1.3}$$

Covariant derivatives of vectors are given by:

$$V_{\alpha;\mu} = \frac{\partial V_\alpha}{\partial x^\mu} - \Gamma_{\alpha\mu}^{\ \ \beta} V_\beta \ , \ V^\alpha_{\ ;\mu} = \frac{\partial V^\alpha}{\partial x^\mu} + \Gamma_{\beta\mu}^{\ \ \alpha} V^\beta . \tag{1.4}$$

By iteration of (1.4), we find that these derivatives do not commute and that:

$$V_{\alpha;\mu;\sigma} - V_{\alpha;\sigma;\mu} = R_{\alpha\ \ \mu\sigma}^{\ \ \beta} V_\beta \ , \tag{1.5}$$

where:

$$R_{\alpha\ \ \mu\sigma}^{\ \ \beta} = -\frac{\partial \Gamma_{\alpha\mu}^{\ \ \beta}}{\partial x^\sigma} + \frac{\partial \Gamma_{\alpha\sigma}^{\ \ \beta}}{\partial x^\mu} + \Gamma_{\alpha\mu}^{\ \ \tau} \Gamma_{\tau\sigma}^{\ \ \beta} - \Gamma_{\alpha\sigma}^{\ \ \tau} \Gamma_{\tau\mu}^{\ \ \beta} , \tag{1.6}$$

$$R_{\alpha\ \ \mu\sigma}^{\ \ \beta} = -R_{\alpha\ \ \sigma\mu}^{\ \ \beta} .$$

From (1.3) we see that:

$$g_{\mu\beta} R_{\nu\ \ \lambda\sigma}^{\ \ \beta} + g_{\beta\nu} R_{\mu\ \ \lambda\sigma}^{\ \ \beta} = 0. \tag{1.7}$$

The Riemann tensor has four indices whose geometrical meaning is quite different. This can be most clearly seen by noticing that the commutation relation (1.5) is but an infinitesimal form for the change of a vector under Levi-Civita transportation around a closed loop. If the vector V^β is moved from x^α to $x^\alpha + dx^\alpha$, then its components change by:

$$\delta V^\beta = -\Gamma_{\alpha\lambda}^{\ \ \beta} V^\lambda dx^\alpha . \tag{1.8}$$

If we consider a rectangular loop of sides dx^α, δx^σ, we find the final change:

$$\delta V^\beta = R_{\tau\ \ \alpha\sigma}^{\ \ \beta} dx^\alpha \delta x^\sigma V^\tau . \tag{1.9}$$

The forms (1.5) represent the infinitesimal area element. Therefore the antisymmetry in (1.6) simply reflects the anticommutativity in the Grassman algebra of the differentials. The relation (1.7) instead implies conservation of the lenght of the vector V during transportation. Its antisymmetry is related to the structure of the Lie algebra of the group SO(3,1). The pair of indices μ,σ lives in space-time, the other pair in an associated SO(3,1) bundle. In spite of this one finds in the riemannian geometry additional symmetries:

$$R_{\mu\nu\lambda\sigma}=R_{\lambda\sigma\mu\nu}, \quad R_{\mu\nu\lambda\sigma}+R_{\mu\lambda\sigma\nu}+R_{\mu\sigma\nu\lambda}=0, \tag{1.10}$$

which follow from (1.7) and $\Gamma^{\mu}_{\alpha\beta}=\Gamma^{\mu}_{\beta\alpha}$. Let us introduce the forms:

$$\Gamma^{\alpha}{}_{\beta}=\Gamma^{\alpha}_{\beta\delta}\, dx^{\delta}, \tag{1.11}$$

which can be considered as vector forms with values in the Lie algebra of GL(4,R). In this spirit the Riemann tensor can be rewritten as:

$$R^{\alpha}{}_{\beta}=d\Gamma^{\alpha}{}_{\beta}-\Gamma^{\alpha}{}_{\lambda}\wedge\Gamma^{\lambda}{}_{\beta}=\frac{1}{2}R^{\alpha}{}_{\beta\mu\sigma}\, dx^{\mu}\wedge dx^{\sigma}. \tag{1.12}$$

It should be emphasized that (1.12) is a much more general formula than it would appear from the context of riemannian geometry. If we introduce capital letters A, B, etc., instead of the pairs $\binom{\alpha}{\beta}, \binom{\alpha}{\lambda}$, etc., then we can rewrite (1.12) as:

$$R^{A}=d\Gamma^{A}-\Gamma^{B}\wedge\Gamma^{F}C_{FB}{}^{A}, \tag{1.13}$$

where the $C_{FB}{}^{A}$ are the structure constants of GL(4,R). The role of the indices in (1.12) is therefore quite different. δ is a space-time index, while α, β are internal indices labelling a base in the adjoint representation of GL(4,R). A GL(4,R) gauge transformation will act on α, β but not on δ. Viceversa, a generic change of coordinates will act on δ but not on α, β. Of course the jacobian matrix of the derivatives is itself an element of GL(4,R) and this coincidence allowes us to identify, in a riemannian geometry, the space-time differentials with the gauge group. These two groups are correlated by the symmetry $\Gamma^{\mu}_{\alpha\beta}=\Gamma^{\mu}_{\beta\alpha}$, i.e. by the vanishing of torsion.

The "metricity" condition (1.7) treats the indices of the metric tensor as internal indices. In order to have general relativity we must somehow identify these internal indices with the space-time indices. This is achieved through the action principle which contains the scalar curvature where indeed indices of different nature are contracted together. Should we use a different action principle and introduce non-metricity or torsion, then we would be loosening the link between space-time properties and properties of the gauge group. The use of the index $A = \binom{\alpha}{\beta}$ is strictly related to the present formulation in terms of the group GL(4,R).

Of the highest interest is the so-called vierbein approach. Here one introduces a set of n covariant vectors $V^a{}_\mu$ in each point of space-time forming a complete basis:

$$V^a{}_\mu V^b{}_\nu g^{\mu\nu} = \eta^{ab} . \tag{1.14}$$

These vectors correspond to a vector valued one-form:

$$V^a = V^a{}_\mu \, dx^\mu . \tag{1.15}$$

Their orthonormality can be rewritten as:

$$V^a{}_\mu V^b{}_\nu \eta_{ab} = g_{\mu\nu} . \tag{1.16}$$

Let us use (1.3), (1.16) and obtain:

$$0 = g_{\mu\nu;\alpha} = \eta_{ab} V^a{}_{\mu;\alpha} V^b{}_\nu + \eta_{ab} V^a{}_\mu V^b{}_{\nu;\alpha} . \tag{1.17}$$

From this condition it follows that there are coefficients $\omega^{ab}{}_\alpha$ such that:

$$V^a{}_{\mu;\alpha} = \omega^{ab}{}_\alpha V_{b,\mu} \, , \quad \omega^{ab}{}_\alpha = -\omega^{ba}{}_\alpha \, ,$$
$$\frac{\partial V^a{}_\mu}{\partial x^\alpha} - \Gamma^\beta_{\mu\alpha} V^a{}_\beta - \omega^{ab}{}_\alpha V_{b\mu} = 0 . \tag{1.18}$$

If there is no torsion, that is $\Gamma^\alpha_{\beta\gamma} = \Gamma^\alpha_{\gamma\beta}$, then we see that

$$\frac{\partial V^a{}_\mu}{\partial x^\alpha} - \frac{\partial V^a{}_\alpha}{\partial x^\mu} - \omega^{ab}{}_\alpha V_{b\mu} + \omega^{ab}{}_\mu V_{b\alpha} = 0 . \tag{1.19}$$

By introducing the forms ω^{ab}, (1.19) can be rewritten as:

$$DV^a = dV^a - \omega^{ab} \wedge V^b = 0 , \quad \omega^{ab} = \omega^{ab}{}_\mu dx^\mu , \tag{1.20}$$

also called Cartan's first structural equation. By commuting derivatives on V^a_μ according to (1.5) and using (1.18), we obtain:

$$d\omega^{ab} - \omega^{at} \wedge \omega_t{}^b = \frac{1}{2} R^{ab}{}_{\alpha\beta}\, dx^\alpha \wedge dx^\beta = R^{ab}, \tag{1.21}$$

that is, Cartan's second structural equation. The idea now is to replace everywhere g and Γ with V and ω respectively. From (1.21) we can retrieve the Riemann tensor without having to reconstruct the Γ from V and ω. Also it is possible to project systematically all tensors on the vierbein and use only scalar components. The definition of these scalars suffers from an inherent ambiguity in the definition of the basis in terms of the metric tensor. We can always change the basis as:

$$V'^a{}_\mu = \Lambda^a{}_b(x)\, V^b{}_\mu , \tag{1.22}$$

sothat for a controvariant vector we have:

$$A'^a = \Lambda^a{}_b(x)\, A^b . \tag{1.23}$$

If we take the derivative, we find:

$$\frac{\partial A'^a}{\partial x^\lambda} = \Lambda^a{}_b \frac{\partial A^b}{\partial x^\lambda} + \frac{\partial \Lambda^a{}_b}{\partial x^\lambda} A^b , \tag{1.24}$$

and we see that these derivatives do not transform as tensors under changes of the frame. The covariant derivative:

$$D_\lambda A^a = \frac{\partial A^a}{\partial x^\lambda} - \omega^{ab}{}_\lambda A_b , \tag{1.25}$$

is easily seen to transform correctly. Therefore we have traded the problems of general coordinate covariance with those of covariance under change of frame; these are related to the structure of the theory as an SO(3,1) gauge theory, whose Yang-Mills potentials are indeed the ω.

Besides the V and the ω, the theory will contain space-time scalars which however are now tensors with respect to SO(3,1). This formal point can be immediately extended to any pseudoriemannian manifold of any space-time dimension. However it is better to use covariant forms of any degree which correspond somehow to the completely skewsymmetric covariant tensors of rank p. It is convenient to consider for instance such a second rank tensor as a scalar 2 - form instead of a set of space-time scalars forming a SO(3,1) skew tensor. This follows from the natural covariance of the exterior derivative of a form, because of the symmetry (1.2). The curl of a vector T_α is given by:

$$\frac{\partial T_\alpha}{\partial x^\lambda} - \frac{\partial T_\lambda}{\partial x^\alpha} \equiv T_{\alpha;\lambda} - T_{\lambda;\alpha} \, , \tag{1.26}$$

and it follows that d is de facto a space-time covariant operation. Similar comments apply to forms of any degree.

Let us rewrite Cartan's structural equations:

$$d\omega^{ab} - \omega^{at} \wedge \omega_t^{\ b} = R^{ab}$$

$$dV^a - \omega^{at} \wedge V_t = R^a = 0 \; . \tag{1.27}$$

Also these equations admit a form with strong similarity to (1.13). We consider V and ω as Yang-Mills potentials for the Poincaré gauge group and can rewrite (1.27) as:

$$d\omega^A - \frac{1}{2}\, \omega^B \wedge \omega^F C_{FB}^{\ \ A} = R^A \, , \tag{1.28}$$

where $C_{FB}^{\ \ A}$ are now structure constants of the Poincaré group. This formula represents the quantities R^a, the torsion, as the traslational components of the Riemannian curvature. But of course R^a vanishes and this raises a number of interesting formal questions as to how this happens. In many traditional settings the quantity R^a is related to the antisymmetric part of $\Gamma^\alpha_{\beta\gamma}$. In a little known theory of Cartan, the torsion is related to the density of matter spin and is expected to be practically not observable with possibly the exception of the early universe. I like torsion because of its logical desirability and its relationship to a generalized action-reaction principle represented here through a 6-box diagram (for gravity) and a 3-box diagram (for electromagnetism). Without torsion the gravity diagram does not close and achieves full symmetry.

Torsion arises naturally through the action principle in the first order vierbein formalism. It is easily shown that:

$$R\sqrt{-g}\, d^4x = \frac{1}{2} R^{ab} \wedge V^c \wedge V^d \, \varepsilon_{abcd} \, , \tag{1.29}$$

Cartan–Einstein's gravity hexagon

$T^{\mu\nu}$
energy-momentum density
translations in P

whose density
of generators is

is source of

Non integrable
translations
in P

Curvature R^{ab}
through
Einstein's equations

leads to

leads to

Torsion R^a or
translational
curvature
Cartan's equations

Non-integrable
SO(3,1)
parallelism

is source of

whose density
of generators is

Spin density
$T^{\alpha\beta,\mu}$
SO(3,1) in P

Maxwell's triangle

J^μ charge
current density

whose density
of generators is

is source of

Non-integrable
U(1) parallelism

$F^{\mu\nu}$
U(1) curvature

leads to

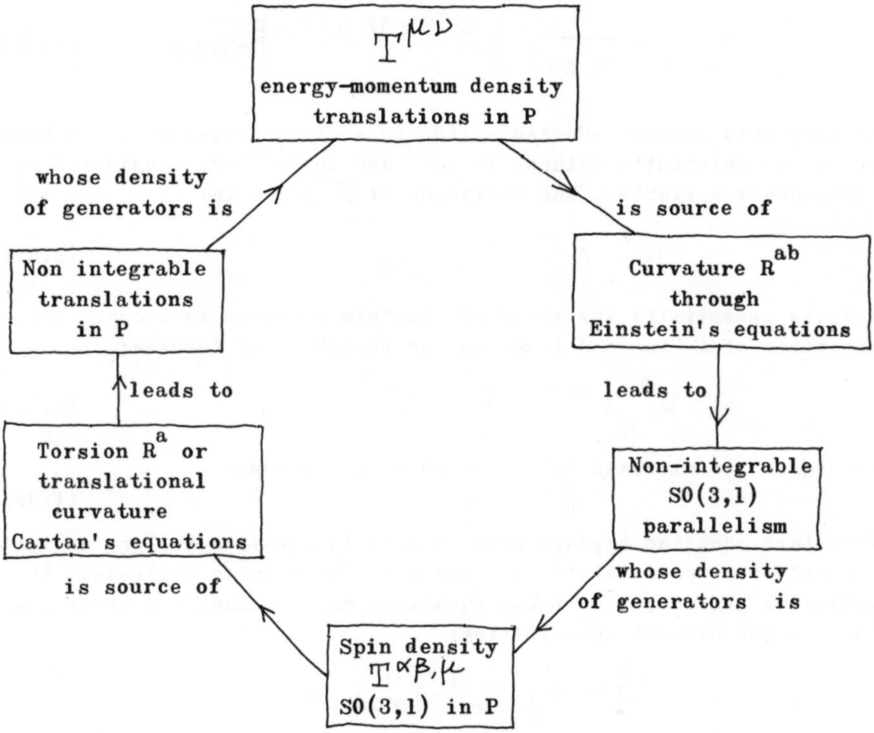

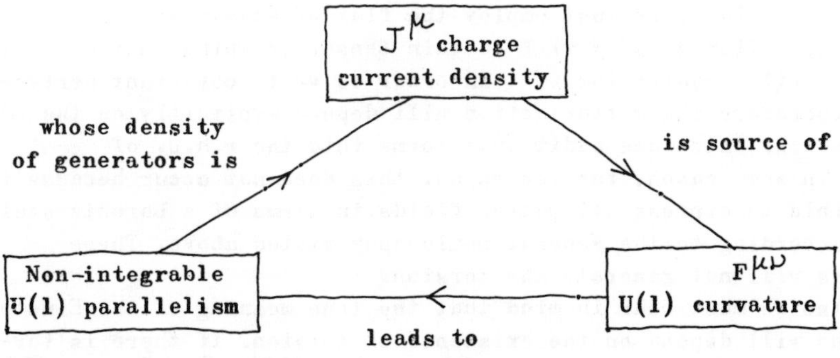

and this suggests the action (c = 1):

$$S_{grav} = \frac{1}{16\pi G}\int_{M^4} R^{ab}\wedge V^c\wedge V^d\,\varepsilon_{abcd}\,. \tag{1.30}$$

We vary this Einstein-Cartan action in a manner similar to the better known Palatini's method. The ω^{ab} and the V^a are considered as independent variables. The variation of V^a leads to:

$$R^{ab}\wedge V^c\,\varepsilon_{abcd} = 0. \tag{1.31}$$

This is essentially the standard Einstein equation in vacuo. The variation of ω instead leads to the vanishing of torsion:

$$R^a\wedge V^b - R^b\wedge V^a = 0; \tag{1.32}$$

indeed if the vierbein is not singular it follows:

$$R^a = 0. \tag{1.33}$$

This last equation appears here as a field equation telling us that our differential manifold is after all a Riemannian manifold. If matter is present, then these equations must be modified accordingly through the minimal substitution:

$$S = S_{grav} + S_{mat}\,, \tag{1.34}$$

where the matter action is derived from its Minkowsky form by replacing everywhere ordinary derivatives with covariant derivatives. Let us notice however that, when using intrinsic components, namely space-time scalars, we must employ the flat Minkowsky metric. The lagrangian matter density will contain tensor or spinor matter fields and these will require the ω in order to write covariant derivatives. Therefore the matter action will depend explicitly on the ω and this will introduce additional terms into the r.h.s. of (1.32). In some cases, for low spins, this does not occur because it is possible to express all matter fields in terms of a Lorentz scalar p-form according to the general philosophy stated above. These couplings will not generate any torsion.

Also we must bear in mind that the true meaning of the Einstein equations will depend on the existence of torsion. If there is torsion then we can always rewrite ω^{ab} as $\omega^{ab} = \tilde{\omega}^{ab} + \delta\omega^{ab}$, where $\tilde{\omega}^{ab}$ is torsionless and deduced from the vierbein univocally through Cartan's first structural equation. Therefore the curvature R^{ab} can be written as $\tilde{R}^{ab} + \delta R^{ab}$ where $\tilde{R}^{ab}$ is the conventional Riemann curvature. If we bring the corrective term in the r.h.s., it will appear as a source term in the Einstein equations. Therefore the energy-matter tensor of a matter field may appear in this formalism as mixed with Einstein's tensor in the l.h.s. This mechanism is present in supergravity; the field acting as matter field is of course that of the gravitino. But clearly this is only very natural if one takes the view that gravitinos are just a different component of the gravitational field.

I introduce now the concept of factorization. On the principal bundle having space-time as basis and SO(3,1) as fibre, it is possible to define an "extended" field configuration as:

$$\omega^{ab}_{ext} = -(\Lambda^{-1} d\Lambda)^{ab} + (\Lambda^{-1})^a{}_s\, \omega^s{}_t\, \Lambda^{tb} \tag{1.35}$$
$$V^a_{ext} = (\Lambda^{-1} V)^a .$$

Notice that here Λ is an independent coordinate, similar to the x of space-time. The definition (1.35) represents in fact a generic gauge transformation, two sections of the bundle being related by one such transformation. Accordingly we have:

$$R^{ab}_{ext} = (d\omega^{ab} - \omega^{at} \wedge \omega_t{}^b)_{ext} = (\Lambda^{-1})^a{}_s\, R^s{}_t\, \Lambda^{tb}, \tag{1.36}$$
$$R^a_{ext} = (\Lambda^{-1} R)^a = (\Lambda^{-1})^a{}_s\, R^s ,$$

sothat the action is written in the same way for the extended fields:

$$S_{grav} = -\frac{1}{16\pi G} \int_{M^4} R^{ab}_{ext} \wedge V^c_{ext} \wedge V^d_{ext}\, \varepsilon_{abcd}. \tag{1.37}$$

Notice however that now the V_{ext} and ω_{ext} form together a complete and independent basis on a 10-dimensional manifold. One could conceive of carrying out the variation in the fields V_{ext}, ω_{ext} considered as independent. The new variational equations are again the same as before, but in the fields V_{ext} and ω_{ext} instead of V, ω. Curiously enough, the final content of both variational principles is the same and leads to the same manifold of solutions. In particular the factorization property for the ω and V follows from the variational principle. The physical content of both restricted and extended action principles is exactly the same. This fact can be traced back to the gauge invariance of the theory under the Lorentz group. A similar mechanism is at work in more general theories, in which one starts from a gauge group G and a theory which is gauge invariant only under a subgroup $H \subset G$. In general the final fields will live only on the factor G/H. And this brings us to a final point. The original theory started with a full set of Poincaré valued gauge fields but the action, purposedly not invariant under all gauge transformations, is invariant only under the Lorentz gauge. This fact was noted first by Mac Dowell and Mansouri, who remarked also that an action built as a polynomial in the exterior algebra and fully invariant would be in general divergent and unsuited for an action principle.

The problem now is that of characterizing all interesting gravity theories through group theoretical concepts but without actually introducing full group invariance or without using this concept at all. Many interesting unified gravity theories satisfy the following conditions:

A) The starting point is a gauge group G, for instance the Poincaré group, the graded Poincaré group, or a generic solvable group containing these groups. One can replace the Poincaré group with a De Sitter group.

B) We consider the Lie algebra $\mathcal{G}$ of the group G and a 1-form with values in $\mathcal{G}$. In components we may write these forms as ω^A where A is a (possibly) superindex, labeling both Bose and Fermi components and both V and ω components at the same time. From the ω^A we calculate the curvatures:

$$R^A = d\omega^A - \frac{1}{2}\, \omega^D \wedge \omega^E\, C_{ED}{}^{A} . \tag{1.38}$$

The left-invariant Cartan's forms on G will yield vanishing curvatures and correspond to gauge fields. The curvatures are the geometrical object which must be identified with the field strengths of physics.

C) The lagrangian density is a polynomial in the forms ω^A and in the curvatures R^A built by using only the operations d and $\wedge$ of the exterior calculus of forms. The duality operator $*$ is explicitly excluded.

D) The field equations must admit the gauge null solutions as a particular solution (called flat space). Flatness will depend on the choice of the group G. Minkowsky space is of course the flat space of the Poincaré group but De Sitter space is flat under the De Sitter group. The field equations must be at least linear in the curvatures. From our discussion we see that general relativity satisfied all our conditions. Also other less interesting theories satisfy them; they must be excluded by introducing additional conditions.

2. The supermanifolds.

At the moment the analogies linking supermanifolds to ordinary manifolds are essentially of a formal nature in the structure of the equations. Questions like convergence and continuity do not seem to extend naturally, nor to have any essential role in the theory. I shall give here a set of formal rules intended as manipulations of objects called "superforms", or briefly just forms, "superfunctions" or " supervariables". The deeper questions, as to what their "true meaning" is, are not analyzed here. The status of supermanifolds is as yet unclear and they may acquire in physics an importance equal to that of ordinary differentiable manifolds.

In what follows, $\mathcal{A}$ is a generic Grassmann algebra. Elements of this algebra have a Z_2 grading (K = 0, 1). Even elements (K = 0) are denoted with latin letters, odd ones (K = 1) with greek letters. By A we intend a label in the set x^A; A will have a natural grading, written as $|A|$. By abuse of language we write $(-1)^{|A|} = (-1)^A$. If the grading of x^A is $|A|$ then the set is called a Bose set, while if the grading is $1 + |A|$ mod. 2, then we have a Fermi set. For a Bose set x^A we have then:

$$x^A x^B = (-1)^{|A| \times |B|} x^B x^A . \tag{2.1}$$

The set of all even elements in $\mathcal{A}$ will be called $R^{1,0}$, the set of odd elements $R^{0,1}$. The direct sum of b copies of $R^{1,0}$ and of f copies of $R^{0,1}$ will be denoted by $R^{b,f}$, and called superlinear space of Bose dimension b and Fermi dimension f. These spaces are truly linear spaces in the ordinary sense. A Bose variable is for instance a shorthand for an element in a linear space of possibly very high dimensionality.

Functions of supervariables have some similarity to functions of a complex variable. Let:

$$\begin{aligned} x &= \Sigma_\alpha x^\alpha P^\alpha(\zeta) , \\ \theta &= \Sigma_\beta \theta^\beta P^\beta(\zeta) , \end{aligned} \tag{2.2}$$

be generic even and odd variables. A differentiable function $f(x, \theta)$ must be separately differentiable in the real variables x^α, θ^β . But we also request that there are elements $(\partial f/\partial x)$, $(\partial f/\partial \theta)$ in $\mathcal{A}$ such that:

$$\frac{\partial f}{\partial x^\alpha} = P^\alpha(\zeta) \frac{\partial f}{\partial x} , \quad \frac{\partial f}{\partial \theta^\beta} = P^\beta(\zeta) \frac{\partial f}{\partial \theta} \tag{2.3}$$

These conditions are obviously very restrictive and similar in scope to the better known Riemann-Cauchy conditions. Sums, products and derivatives of differentiable functions are again differentiable. We have also:

$$\frac{\partial^2 f}{\partial x \partial \theta} = \frac{\partial^2 f}{\partial \theta \partial x}, \quad \frac{\partial^2 f}{\partial \theta^2} = 0. \tag{2.4}$$

In the case of several supervariables $x^1, \ldots, x^b; \theta_1, \ldots, \theta^f$, we have:

$$\frac{\partial^2 f}{\partial x^A \partial x^B} = (-1)^{AB} \frac{\partial^2 f}{\partial x^B \partial x^A}. \tag{2.5}$$

Functions have a natural grading. Products of functions have a grading which is the sum of the grading of the factors mod 2. I write f^K in the case I need to state explicitly the grading of f. The grading of $\partial f^K / \partial x^A$ is $K + |A|$, mod 2. Besides functions, we have also forms. The differentials can be given a bigraded structure through the product $\wedge$ satisfying the commutation relation:

$$dx^A \wedge dx^B = -(-1)^{AB} dx^B \wedge dx^A. \tag{2.6}$$

A superform is a polynomial in the dx^A with coefficients in the ring of differentiable functions:

$$f = \frac{1}{q!} \sum_{(A)} dx^{A_q} \wedge dx^{A_{q-1}} \wedge \ldots \wedge dx^{A_1} f_{A_1, \ldots, A_q}(x). \tag{2.7}$$

Forms have two distinct gradings. One is the degree q in the dx^A, the other is the global grading K. The grading of the component:

$$f_{(A)}(x) = f_{A_1, \ldots, A_q}(x)$$

is then $K + |A_1| + \ldots + |A_q|$, mod 2. If both gradings must be explicit, we write $f^{p,K}$, and ordinary functions are then $f^{0,K} = f^K$. The coefficient $f_{(A)}$ satisfies the generalized antisymmetry:

$$f_{A_1 \ldots A_t A_{t+1} \ldots A_q} = -(-1)^{A_t A_{t+1}} f_{A_1 \ldots A_{t+1} A_t \ldots A_q}. \tag{2.8}$$

The product of forms is then implemented by distributivity and linearity. We have then:

$$f^{q,K} \wedge g^{p,L} = (-1)^{pq+KL} g^{p,L} \wedge f^{q,K}. \tag{2.9}$$

In writing (2.7) we adhered to the convention of always contracting when possible an upper index A with a lower one immediately following it. Of course this is not always possible, as we see in (2.7). A pair satisfying this convention will be called a neutral pair. In (2.7) we see that if we remove the neutral pair $\left({}^{A_1}{}_{A_1}\right)$, then the pair $\left({}^{A_2}{}_{A_2}\right)$ is again neutral. In this case the formula (2.7) does not need compensating phases generated by improper ordering of the indices. This phase is $+1$ if all contracted pairs can be removed recursively and if the ordering of the remaining indices is the same in both sides of the equation. In this case we have a well formed formula (WFF). The operator d increases the degree p by 1 and is such that $d^2 = 0$. It is defined as:

$$df^{q,K} = \frac{1}{q!}\sum_{(A)} dx^{A_q}\wedge\ldots\wedge dx^{A_0}\,\frac{\partial f_{A_1\ldots A_q}}{\partial x^{A_0}}\,(-1)^q \,. \tag{2.10}$$

The operator d obeys the Leibnitz rule:

$$d\left(f^{q,K}\wedge g^{p,L}\right) = df^{q,K}\wedge g^{p,L} + (-1)^q f^{q,K}\wedge dg^{p,L} \,. \tag{2.11}$$

If $df = 0$, then f is closed, and if exists g such that $f = dg$, then f is exact and of course also closed. The form $dx\, f(x)$ is closed, but $d\theta\, f(\theta)$ is not. We have:

$$d\left(d\theta\, f(\theta)\right) = -\, d\theta\wedge d\theta\,\frac{\partial f}{\partial\theta} \neq 0 \,. \tag{2.12}$$

All these remarks and definitions really apply to the superspace $R^{b,f}$. The set of all nihlpotent elements in $R^{b,f}$ is denoted by $N^{b,f}$. The factor $R^{b,f}/N^{b,f}$ is isomorphic to R^b and $R^{b,f}$ can be considered as a fibre bundle with basis R^b and fibre $N^{b,f}$. This suggests that we may consider a generic differentiable manifold M^b as basis; M^b has covering U^α, maps $\varphi^\alpha : U^\alpha \to R^b$. The transition functions φ^α are differentiable. We assume that the bundle $M^{b,f}$ has projection map π, fibre $N^{b,f}$ and basis M^b. We consider the covering $W = U^\alpha \times N^{b,f}$, with maps $\phi^\alpha : U^\alpha \times N^{b,f} = W^\alpha \to R^{b,f}$. We assume that the maps ϕ^α are differentiable (as superfunctions). Finally the diagram:

$$\begin{array}{ccc} W^\alpha & \longrightarrow & R^{b,f} \\ \pi\downarrow & & \downarrow\pi^\circ \\ U^\alpha & \longrightarrow & R^b \end{array} \tag{2.13}$$

is commutative. This last property guarantees that any generic coordinate transformation on the supermanifold reduces to an ordinary coordinate change on M . This definition is of course purely formal, and leaves aside a number of important questions. It serves us well for a provisional treatment of forms. If we have a superdifferentiable map between supermanifolds

$$M^{b,f} \xrightarrow{\lambda} N^{b',f'}, \tag{2.14}$$

then it must be representable in a local chart by differentiable functions:

$$\begin{aligned} y^B &= \lambda^B(x^1 \ldots x^d), \quad d = b+f \\ B &= 1 \ldots d', \quad d' = b'+f' . \end{aligned} \tag{2.15}$$

If we have a form f on $V^{\beta'}$ given by:

$$f = \frac{1}{q!} \sum_{(B)} dy^{B_q} \wedge \ldots \wedge dy^{B_1} f_{B_1 \ldots B_q}(y), \tag{2.16}$$

the "pull-back" is then defined by:

$$\lambda^* f = \frac{1}{q!} \sum_{(B)(A)} dx^{A_q} \frac{\partial y^{B_q}}{\partial x^{A_q}} \wedge \ldots \wedge dx^{A_1} \frac{\partial y^{B_1}}{\partial x^{A_1}} f_{B_1 \ldots B_q}(\lambda(x)), \tag{2.17}$$

i.e. we replace everywhere the variables y^B as functions of the variables x^A. Pull-backs move in the opposite direction as corresponding maps. If $M^{b,f}$ is a superdifferentiable manifold and we have on each set $\Phi^\alpha(W^\alpha) \subset R^{b,f}$ of the covering a form f^α such that:

$$\left(\Phi^\beta (\Phi^\alpha)^{-1}\right)^* f^\alpha = f^\beta \quad \text{on} \quad W^\alpha \cap W^\beta \neq \emptyset, \tag{2.18}$$

then we say that we have assigned a global form f on $M^{b,f}$.

Pull-backs can be naturally extended to global forms according to standard practice. Closed (exact) forms are mapped into closed (exact) forms under pull-back. If we have the chain of maps:

$$M \xrightarrow{\lambda} M' \xrightarrow{\mu} M'', \tag{2.19}$$

then we can compose pull-backs and maps, and we have: $(\mu\lambda)^* = \lambda^* \mu^*$.

I discuss here briefly tangent vectors, almost synonimous of controvariant vectors of physicists, just as 1-forms are synonimous of covariant vectors. Like forms, also tangent vectors have a grading. In a generic chart W^α we can write :

$$T = T_\alpha{}^A \frac{\partial}{\partial x_\alpha^A} . \tag{2.20}$$

If T has the global grading K, the component T_α has grading $K + |A|$, mod 2. The change to another chart is given by:

$$T_\beta{}^{B} \frac{\partial x_\alpha{}^{A}}{\partial x_\beta{}^{B}} = T_\alpha{}^{A} . \tag{2.21}$$

The commutator of two tangent vectors is defined as:

$$\begin{aligned}\{T^K, S^H) &= -(-1)^{KH}\{S^H, T^K) = T^K S^H - S^H T^K (-1)^{KH} = \\ &= \left(T^{K,A}\frac{\partial S^{H,B}}{\partial x^A} - (-1)^{HK} S^{H,A}\frac{\partial T^{K,B}}{\partial x^A}\right)\frac{\partial}{\partial x^B} ,\end{aligned} \tag{2.22}$$

and satisfies the Jacobi identity:

$$\begin{aligned}&\{\{T^K, S^H), U^L)(-1)^{KL} + \{\{S^H, U^L), T^K)(-1)^{HK} + \\ &+\{\{U^L, T^K), S^H)(-1)^{HL} = 0 ,\end{aligned} \tag{2.23}$$

as the differentiation of a product is given by:

$$T^K f^L g^H = (T^K f^L) g^H + (-1)^{KL} f^L (T^K g^H) . \tag{2.24}$$

Notice that for instance formula (2.22) is WFF in the contracted index A. Whenever the ordering of the remaining indices K, H is changed, then we have a compensating phase $(-1)^{HK}$. These definitions insure that all relevant formulas do not depend on the chart. In the sequel we shall drop the explicit mention of chart if not necessary.

3. The geometry of super-Lie groups.

A supermanifold endowed with a product structure of a group is a supergroup. An interesting example comes from the grading of the Poincaré group P. P can be defined as the set of pairs (Λ, x), where $\Lambda \in SO(3,1)$ and $x \in R^4$. The product is then defined as:

$$(\Lambda, x)\ (M, y) = (\Lambda M, x + \Lambda y) . \tag{3.1}$$

This product can be written in matrix form as:

$$(\Lambda, x) \longleftrightarrow \begin{pmatrix} \Lambda & x \\ 0 & 1 \end{pmatrix} . \tag{3.2}$$

The identity of the group is then obviously $\begin{pmatrix} 1 & 0 \\ 0 & 1 \end{pmatrix}$, the inverse of (Λ, x) being $(\Lambda^{-1}, -\Lambda^{-1}x)$. The spinor representation of SO(3,1) is obtained through the relation:

$$\Lambda_a{}^b \, \gamma_b = S^{-1}(\Lambda) \, \gamma_\alpha \, S(\Lambda) . \tag{3.3}$$

The spinor representation $S^4(\Lambda)$ is in fact an element of SL(2,C), and both $\pm S(\Lambda)$ correspond to the same Λ, sothat $S(\Lambda)$ is really a misnomer and we should write instead: $\Lambda(S) = \Lambda(-S)$. We consider the triplets (Λ, x, ξ) where ξ is an anticommuting Majorana spinor. The product of triplets is then defined as:

$$(\Lambda, x, \xi)(M, y, \eta) = \\ = \left(\Lambda M, x + \Lambda y - \tfrac{i}{2}(\bar{\xi} \gamma S(\Lambda) \eta), \xi + S(\Lambda)\eta \right), \tag{3.4}$$

and satisfies the group axioms. The inverse is then:

$$(\Lambda, x, \xi)^{-1} = \left(\Lambda^{-1}, -\Lambda^{-1}x - \tfrac{i}{2}\Lambda^{-1}(\xi\gamma\xi), -S^{-1}(\Lambda)\xi\right). \tag{3.5}$$

Finally we have the matrix triangular form:

$$\begin{pmatrix} \Lambda & -\frac{i}{2}\bar{\xi}\gamma S(\Lambda) & x \\ 0 & S(\Lambda) & \xi \\ 0 & 0 & 1 \end{pmatrix} . \tag{3.6}$$

We use the notation GP for this set of triplets, the graded Poincaré group or super-Poincaré group.

In general we may introduce in a generic supergroup an atlas of coordinates. In what follows I shall dump the explicit mention of charts. Computations will be carried in a chart containing the group identity. Elements in the group will be denoted by latin letters x, y, z, etc., their coordinates as x^A, $A = 1,\ldots,d$, y^A, z^A, etc. The identity will have vanishing coordinates $e^A = 0$. The multiplication will assume the form:

$$z^A = f^A(x^1 \ldots x^d, y^1 \ldots y^d), \; d = b + f, \; A = 1, \ldots, d, \tag{3.7}$$

which can be differentiated:

$$f^* dz^A = dx^B \, T_B{}^A(x,y) + dy^B \, U_B{}^A(x,y) ; \tag{3.8}$$

here f is the product map: $G \times G \to G$, and f^* is the pull-back under this map. $T_B{}^A$, $U_B{}^A$ are differential coefficients satisfying the conditions (from associativity):

$$T_A{}^B(x,yz) = T_B{}^C(x,y)\; T_C{}^A(xy,z)$$

$$T_B{}^C(y,z)\; U_C{}^A(x,yz) = U_B{}^C(x,y)\; T_C{}^A(xy,z) \qquad (3.9)$$

$$U_B{}^A(xy,z) = U_B{}^C(y,z)\; U_C{}^A(x,yz) \quad .$$

In particular:

$$T_C{}^A(x,e) = \delta_C{}^A \quad , \; U_C{}^A(e,y) = \delta_C{}^A$$

$$T_B{}^C(x,x^{-1})\; T_C{}^A(e,x) = \delta_B{}^A \; , \; U_B{}^C(x^{-1},x)\; U_C{}^A(x,e) = \delta_B{}^A \, . \qquad (3.10)$$

The set of forms:

$$\omega^A = dx^B\; U_B{}^A(x^{-1},x) \qquad (3.11)$$

is left invariant; infact, if the map $L_a : x \longrightarrow ax$ denotes left translation in the group, we have:

$$L_a^{*}\omega^A = d(ax)^B\; U_B{}^A(x^{-1}a^{-1},xa) = dx^C\; U_C{}^B(a,x)\; U_B{}^A(x^{-1}a^{-1},xa) =$$
$$= dx^C\; U_C{}^A(x^{-1},x) = \omega^A \, . \qquad (3.12)$$

Also $d\,\omega^A$ is left invariant and we can expand it into the ω^A themselves:

$$d\omega^A = \frac{1}{2}\; \omega^B \wedge \omega^F\; C_{FB}{}^A \quad ; \qquad (3.13)$$

the C's are now called the structure constants of the Lie group. (3.13) is the so called Maurer-Cartan equation. The relation $dd\,\omega^A = 0$ implies the Jacobi identity:

$$(-1)^{PB}\, C_{PQ}{}^F\, C_{FB}{}^A + (-1)^{QB}\, C_{BP}{}^F\, C_{FQ}{}^A + (-1)^{PQ}\, C_{QB}{}^F\, C_{FP}{}^A = 0. \qquad (3.14)$$

These identities can be proved in an equivalent but more transparent notation. We introduce the dual basis of the ω^A, the tangent vectors:

$$u_A = U_A{}^B(x,e)\, \frac{\partial}{\partial x^B} \quad , \qquad (3.15)$$

sothat:

$$\omega^A u_A = dx^A \frac{\partial}{\partial x^A}, \tag{3.16}$$

and

$$\omega^A u_A f = df. \tag{3.17}$$

By applying again d we find:

$$d\omega^A u_A f = \omega^A \wedge d(u_A f);$$

by using (3.17), and expanding the dx^A in the ^{A}, we obtain:

$$\{u_B, u_F) = C_{BF}{}^A u_A, \tag{3.18}$$

sothat the C's define the commutator of the u's, and the Jacobi identities are the familiar ones for the commutator. Then the geometrical content of the relation $d^2 = 0$ is in cotangent space the same as that of the Jacobi identity in tangent space. One can introduce correspondingly right invariant forms and the dual basis:

$$\pi^A = dx^B T_B{}^A(x, x^{-1}), \tag{3.19}$$

$$t_A = T_A{}^B(e, x) \frac{\partial}{\partial x^B}, \tag{3.20}$$

satisfying the same relations, with opposite C's:

$$\begin{aligned} d\pi^A &= -\tfrac{1}{2}\pi^B \wedge \pi^F C_{FB}{}^A, \\ \{t_A, t_B) &= -C_{AB}{}^F t_F \end{aligned} \tag{3.21}$$

as well as:

$$\{u_A, t_B) = 0. \tag{3.22}$$

The tangent vectors u_A generate the infinitesimal right translations on G, the vectors t_B generate the left translations on G. In all cases of physical interest one has so far:

$$C_{BC}{}^A = 0, \text{ unless } |A| + |B| + |C| = 0 \bmod 2.$$

A generalization allowed under the present formal scheme of the structure constants to Fermi grading is not unconceivable in physical applications.

Left invariant forms can be more easily calculated if one has a faithful matrix representation of the group, like the ones given so far for P and GP. If $\mathfrak{D}(x)$ is the matrix representing x on some linear space, we have then the matrix $\mathfrak{D}(x^{-1})\, d\mathfrak{D}(x)$ whose elements are themselves 1- forms and left invariant on G. They must be linear combinations of the ω^A. In particular if we apply this technique to GP, we find the left invariant forms, in a more traditional notation:

$$\begin{aligned} V^a &= (\Lambda^{-1})^a{}_b\, dx^b - \tfrac{i}{2} (\Lambda^{-1})^a{}_b\, \bar{\xi} \gamma^b d\xi , \\ \Omega^a{}_b &= -(\Lambda^{-1} d\Lambda)^a{}_b , \\ \psi &= S^{-1}(\Lambda)\, d\xi . \end{aligned} \tag{3.23}$$

We have naturally:

$$\Lambda^a{}_t \Lambda^b{}_u \eta^{tu} = \eta^{ab} , \quad \eta^{ab} = (1,-1,-1,-1) , \tag{3.24}$$

and by differentiating:

$$\Omega_{ab} = \eta_{at} \Omega^t{}_b = -\Omega_{ba} . \tag{3.25}$$

The Maurer-Cartan equations are then:

$$\begin{aligned} & d\Omega^a{}_b - \Omega^a{}_c \wedge \Omega^c{}_t = 0 , \\ & dV^a - \Omega^a{}_t \wedge V^t - \tfrac{i}{2} \bar{\psi} \gamma^a \psi = 0 , \\ & d\psi - \tfrac{1}{4} \Omega_{ab} \gamma^a \gamma^b \psi = 0 , \end{aligned} \tag{3.26}$$

having used the identity: $\omega = \frac{1}{4} \Omega_{ab} \gamma^a \gamma^b = - S^{-1} dS .$

If we set $\psi = \xi = 0$, then we obtain the corresponding definitions and identities for P.

The forms ω^A are not right invariant as we have:

$$R^*_a \omega^A = \omega^B \left(U(a^{-1},e)\, T(a^{-1},a)\right)_B{}^A = \omega^B Ad(a)_B{}^A ; \tag{3.27}$$

this defines the adjoint representation of G:

$$Ad(a)\, Ad(b) = Ad(ab), \quad \omega^A = \pi^B Ad(x)_B{}^A . \tag{3.28}$$

If we go back to the product map f, then we find the useful relation

$$f^* \omega^A = \omega_y^A + \omega_x{}^B Ad(y)_B{}^A . \tag{3.29}$$

Most of the present treatment of gauge theories is based upon the replacement of the left invariant forms with generic forms. In this case they do not satisfy the Maurer-Cartan conditions. One defines instead the curvatures:

$$R^A = d\omega^A - \frac{1}{2}\omega^B \wedge \omega^F C_{FB}{}^A . \tag{3.30}$$

In physics the ω's are the Yang-Mills potentials; the curvatures are called field strengths. We call flat a form satisfying the Maurer-Cartan conditions; it is also called gauge null field. If we have a field ϕ^A transforming according to the adjoint transformation, then the covariant derivative is defined as:

$$\mathcal{D}\phi^A = d\phi^A - \omega^B \wedge \phi^F C_{FB}{}^A . \tag{3.31}$$

Notice that the ω^A themselves do not transform like the ϕ^A ; their covariant derivative is of course the curvature. We have the Bianchi identity:

$$\mathcal{D} R^A = 0. \tag{3.32}$$

In general, under a map ψ : G $\rightarrow$ G, we have a gauge transformation of the field ω^A defined by:

$$\omega'^A = \left(T(\psi)\,\omega\right)^A = \psi^* \omega_0^A + \omega^B A\, d_B(\psi)^A , \tag{3.33}$$

and satisfying:

$$T(\Theta)\, T(\psi) = T(\psi\Theta) \;; \tag{3.34}$$

we have also:

$$\left(T(\psi) R\right)^A = R^B A\, d_B(\psi)^A . \tag{3.35}$$

Finally, an infinitesimal gauge transformation is of the form:

$$\begin{aligned} \left(T(\psi)\,\omega\right)^A &= d\psi^A - \omega^B \wedge \psi^Q C_{QB}{}^A + \omega^A = \\ &= \omega^A + (\mathcal{D}\psi)^A . \end{aligned} \tag{3.36}$$

4. Superforms.

Of particular interest in the sequel are superforms which can be written as:

$$M = \frac{1}{p!} \sum_{A_1 \ldots A_p} \omega^{A_p} \wedge \ldots \wedge \omega^{A_1} M_{A_1 \ldots A_p} , \tag{4.1}$$

where the coefficients satisfy the symmetry relations:

$$M_{A_1 \dots A_t A_{t+1} \dots A_p} = -(-1)^{A_t A_{t+1}} M_{A_1 \dots A_{t+1} A_t \dots A_p} . \tag{4.2}$$

The contraction ι_A is then defined as:

$$\iota_A M = \frac{1}{(p-1)!} \sum_{A_1 \dots A_{p-1}} \omega^{A_{p-1}} \wedge \dots \wedge \omega^{A_1} M_{A_1 \dots A_{p-1}, A} , \tag{4.3}$$

and satisfies the properties:

$$\begin{aligned} &\iota_A \iota_B M = -(-1)^{AB} \iota_B \iota_A M , \\ &\iota_A \left(M^{m,K} \wedge N^{n,L} \right) = \iota_A M^{m,K} \wedge N^{n,L} + (-1)^{m+AK} M^{m,K} \wedge \iota_A N^{n,L} . \end{aligned} \tag{4.4}$$

A contraction along the tangent vector is defined as:

$$\iota(W) M = W^A \iota_A M , \tag{4.5}$$

and the Lie derivative as:

$$\ell(W) M = \left(\iota(W) d + d\, \iota(W) \right) M . \tag{4.6}$$

We consider forms with p indices, satisfying the symmetry:

$$M_{A_1 \dots A_p} = (-1)^{A_t A_{t+1}} M_{A_1 \dots A_{t-1} A_{t+1} A_t \dots A_p} , \tag{4.7}$$

and the convenient operator ϕ:

$$\begin{aligned} (\phi M)_{A_1 \dots A_{p+1}} = {} & \iota_{A_1} M_{A_2 \dots A_{p+1}} + \dots + \iota_{A_t} M_{A_{t+1} \dots A_{t-1}} (-1)^{\sum_{1}^{t-1}{}_K A_K \cdot \sum_{t}^{p+1}{}_j A_j} + \\ & + \dots + \iota_{A_{p+1}} M_{A_1 \dots A_p} (-1)^{A_{p+1} \sum_{1}^{p}{}_t A_t} , \end{aligned} \tag{4.8}$$

which satisfies the identity: $\phi^2 = 0$. The derivative of an adjoint form M^A:

$$\mathfrak{D} M^A = d M^A - \omega^B \wedge M^F C_{FB}{}^{A} \tag{4.9}$$

can be used to define that of a co-adjoint one as:

$$\mathfrak{D} M_A = dM_A - \omega^B \wedge C_{BA}{}^F M_F \, , \tag{4.10}$$

where now:

$$d(N^A M_A) = dN^A \wedge M_A + (-1)^m N^A \wedge M_A = \mathfrak{D} N^A \wedge M_A + (-1)^m N^A \wedge \mathfrak{D} M_A \, .$$

All formulas which we have used so far are WFF and this yields a distinct advantage in manipulating indices. In dealing with Fermi variables we must exercise some care. The operator $\dagger$ satisfies the relations:

$$(\theta^1 \theta^2)^\dagger = \theta^{2\dagger} \theta^{1\dagger} \, , (\theta^\dagger)^\dagger = \theta \, , (e^{i\alpha} \theta)^\dagger = e^{-i\alpha} \theta^\dagger \, , \tag{4.11}$$

and is currently used in the literature. The advantage is that the product:

$$(\theta^1 + i\theta^2)(\theta^1 - i\theta^2) = 2i\theta^2\theta^1 = (2i\theta^2\theta^1)^\dagger = -2i\theta^1\theta^2$$

is authomatically real. However the use of $\dagger$ raises some problems with the ordering of the indices in a WFF. If we use instead the operator:

$$\theta^* = i\theta^\dagger \, , \tag{4.12}$$

the $*$ does not alter the ordering of the factors. We have in fact:

$$T^A{}_{BC}{}^* = i^{(|A|+|B|+|C|)} (-1)^{\phi(A,B,C)} (T^A{}_{BC})^\dagger \, . \tag{4.13}$$

In calculating derivatives of forms, one can use repeatedly the definition of curvature:

$$R^A = d\omega^A - \frac{1}{2} \omega^B \wedge \omega^F C_{FB}{}^A \, , \tag{4.14}$$

so that the derivative of forms of the kind (4.7) is then written also as:

$$\begin{aligned} dM &= d_o M + R^A \upsilon_A M = \\ &= \frac{1}{2} \omega^B \wedge \omega^F C_{FB}{}^A \upsilon_A M + R^A \upsilon_A M \, . \end{aligned} \tag{4.15}$$

The covariant derivative of a multi-indexed form is given by:

$$\mathfrak{D} N_{A_1 \dots A_p} = dN_{A_1 \dots A_p} - \sum_{t=1}^{p} \omega^B \wedge C_{BA_t}{}^F N_{A_1 \dots A_{t-1} F \dots A_p} (-1)^{\zeta}$$

where $\zeta = (A_t + F)(A_1 + \dots + A_{t-1})$, 94 (4.16)

and satisfies the identity:

$$\mathcal{D}_0 \Phi + \Phi \mathcal{D}_0 = 0. \tag{4.17}$$

Using these conventions it is somewhat simpler to deal with the so called free differential algebras (FDA). The forms are here replaced by a collection of forms ω^A of generic degree p(A). The Maurer-Cartan equations appear now as equations of the kind:

$$d\omega^A - P^A(\omega) = 0, \tag{4.18}$$

where the polynomials P^A are now homogeneous of degree p(A)+1 . They are chosen as to preserve the identity $d^2 = 0$ and satisfy conditions which are the natural extension of the Jacobi identities. In particular, if we write:

$$dd\,\omega^A = dP^A(\omega) = 0 \tag{4.19}$$

and replace everywhere the derivatives of the ω^A with the r.h.s. in (4.18), then (4.19) should be identically satisfied. In general, the equation:

$$R^A = d\omega^A - P^A(\omega) \tag{4.20}$$

will define the curvatures. An interesting example of FDA is the following:

$$\begin{aligned} R^1 &= d\omega^1 - \omega^2 \wedge \omega^3 \\ R^2 &= d\omega^2 - \omega^3 \wedge \omega^1 \\ R^3 &= d\omega^3 - \omega^1 \wedge \omega^2 \\ R^{\otimes} &= d\omega^{\otimes} - \omega^1 \wedge \omega^2 \wedge \omega^3 . \end{aligned} \tag{4.21}$$

The first three equations are the well-known ones for the group SO(3). The remaining equation satisfies the obvious conditions for the FDA. This example, although simple, has not found any physical application. More interesting is the following one:

ω^{ab} spin connection A a 3-form ,

V^a 11-bein a, b = 0, 1 ,..., 10 , (4.22)

ψ Majorana spinor with 32 components ,

with curvatures:

$$R^{ab} = d\omega^{ab} - \omega^{at} \wedge \omega_t{}^b ,$$

$$R^a = dV^a - \omega^{at} \wedge V_t - \frac{i}{2}\bar{\psi}\Gamma^a \wedge \psi, \tag{4.23}$$
$$\rho = d\psi + \frac{1}{4}\Gamma_{ab}\,\omega^{ab} \wedge \psi,$$
$$R^{\otimes} = dA - \frac{1}{2}\bar{\psi}\,\Gamma^{ab} \wedge \psi \wedge V_a \wedge V_b\,.$$

In particular, the derivative:

$$dR^{\otimes} = -\bar{\rho}\,\Gamma^{ab} \wedge \psi \wedge V_a \wedge V_b - \bar{\psi}\,\Gamma^{ab} \wedge \psi \wedge R_a \wedge V_b - \frac{i}{2}\bar{\psi}\,\Gamma^{ab} \wedge \psi \wedge \bar{\psi}\,\Gamma_a \wedge \psi \wedge V_b \tag{4.24}$$

vanishes if we set $R^A = 0$, because the last term vanishes identically as a consequence of a Fiertz identity. Therefore the integrability condition implicit in the FDA is not always easy to check in a theory operating in a higher dimension of space-time.

5. The general theory.

So far I have given some idea of the tools and formulas needed to develop a unified gravity theory. The general principles can be restated as follows.

A) The action density is a superform given by a polynomial in the ω^A and the curvature R^A.

B) The field equations are obtained by imposing that the density, integrated over a generic submanifold $M^P \subset G$, is stationary under all variations of ω^A and of M^P. It is then requested that there exists a gauge null solution, i.e. $R^A = 0$.

C) Other solutions, besides the "flat" space $R^A = 0$, do exist.

These principles imply the use of an action of the form:

$$\mathcal{L} = \sum_s \frac{1}{s!} R^{A_1} \wedge \ldots \wedge R^{A_s}\, \nu^s_{A_s \ldots A_1}\,; \tag{5.1}$$

here we have an arbitrary polynomial in the curvature, but actual applications will be limited to quadratic polynomials; $\nu^s_{(A)}$ is a (p-2s) superform. If we carry out the variation of this action, we find the variational equations:

$$\sum_s \frac{1}{(s-1)!} R^{A_2} \wedge \ldots \wedge R^{A_s} \left(\mathcal{D}_0 \nu^s_{A_s \ldots A_1} + (\phi\nu)^{s-1}_{A_s \ldots A_1}\right) = 0, \tag{5.2}$$

rom B) we get the equation:

$$\mathcal{D}_0 \nu^1{}_A + (\phi \nu^0)_A = 0. \tag{5.3}$$

it can be checked that the same field equations follow from a density obtained by adding an exact form dU to the expansion (5.1), where

$$U = \sum_s \frac{1}{s!} R^{A_1} \wedge .. \wedge R^{A_s} \lambda^s_{A_s .. A_1}, \tag{5.4}$$

λ^s being a (p-2s-1) form. This means that the same theory may assume several forms and that the density (5.1) is defined up to equivalence classes modulo exact differentials. A few concrete examples will be now discussed.

General relativity can be cast into the form (5.1) by setting:

$$\begin{aligned} &\nu^s = 0 \qquad s \neq 1, \\ &\nu^1{}_A = \varepsilon_{abcd} V^c \wedge V^d \quad \text{if } A = (a,b), \\ &\nu^1{}_A = 0 \quad \text{if } A \neq (a,b), \end{aligned} \tag{5.5}$$

and is seen to obey (5.3); also pure gravity in higher dimensions obeys these equations. It stands to reason to say that also C) is satisfied, with the exception of gravity in three-dimension, where in fact the field equations imply that space is flat. I refer to this condition as "rigidity" as opposed to "softness".

A second example comes from the celebrated supergravity N = 1, d = 4. This theory is based on GP gauge fields, already discussed in Section 3. The curvatures are:

$$\begin{aligned} R^{ab} &= d\omega^{ab} - \omega^{at} \wedge \omega_t{}^b, \\ R^a &= dV^a - \omega^a{}_t \wedge V^t, \\ \rho &= d\psi - \frac{i}{2} \Sigma_{ab} \omega^{ab} \wedge \psi, \end{aligned} \tag{5.6}$$

where ψ is a Majorana spinor one-form. Supergravity must contain ordinary gravity when $\psi = 0$, and it must be invariant under SO(1,3) gauge transformations. This suggests an action of the kind:

$$\begin{aligned} S = \int \big(&R^{ab} \wedge V^c \wedge V^d \varepsilon_{abcd} + A \bar{\psi} \gamma_5 \gamma_a \rho V^a + \\ &+ B R^{ab} \wedge \bar{\psi} \Sigma^{cd} \psi \varepsilon_{abcd} \big) \end{aligned} \tag{5.7}$$

that is, again $\nu^s = 0$ unless s = 1. It is better to see ν_A as a co-adjoint multiplet:

$$\mathcal{L} = \frac{1}{2} R^{ab} \wedge \nu_{ab} + R^a \wedge \nu_a + \bar{\rho} \wedge n + \bar{n} \wedge \rho, \tag{5.8}$$

with the SO(1,3) covariant derivative ∇:

$$(D\nu)_a = \nabla \nu_a, \quad Dn = \nabla n - \frac{i}{2} \gamma^a \psi \nu_a,$$
$$(D\nu)_{ab} = \nabla \nu_{ab} + \frac{1}{2}(V_a \wedge \nu_b - V_b \wedge \nu_a) - \frac{i}{2}(\bar{\psi} \Sigma_{ab} n - \bar{n} \Sigma_{ab} \psi), \tag{5.9}$$

in addition to: $(D_o \nu)_A = 0$, which follows from B). By computation, we see that this derivative vanishes only if A = 4, but says nothing on B. The variational equations become now:

$$\varepsilon_{abcd} R^c \wedge V^d - B \bar{\psi} \Sigma^{cd} \rho \varepsilon_{abcd} = 0,$$
$$\varepsilon_{abcd} R^{bc} \wedge V^d + 4 \bar{\psi} \gamma_5 \gamma_a \rho = 0, \tag{5.10}$$
$$2 \gamma_5 \gamma_a \rho V^a - \gamma_5 \gamma_a \psi R^a + \frac{B}{4} \varepsilon_{abcd} \Sigma^{cd} R^{ab} = 0.$$

It is remarkable that if one expands independently in V^a, ω^{ab}, ψ, one obtains $R^A = 0$, unless B = 0. If $B \neq 0$, the theory is rigid, while, if B = 0, the theory factorizes the Lorentz group, reducing to ordinary supergravity. The rigidity of the $B \neq 0$ theory can be check ed by noticing that (5.6) scale under the replacement:

$$d \to \tau d, \quad \omega^{ab} \to \tau \omega^{ab}, \quad V^a \to k V^a,$$
$$\psi \to \sqrt{k\tau} \psi, \quad R^{ab} \to \tau^2 R^{ab}, \quad \rho \to \tau^{3/2} k^{1/2} \rho, \tag{5.11}$$

and from (5.7) B scales as:

$$B \to \frac{\tau}{k} B. \tag{5.12}$$

Therefore, a variation of τ and k such that $\tau k = 1$, yields:

$$R^{ab} \wedge \bar{\psi} \Sigma^{cd} \psi \varepsilon_{abcd} = 0, \tag{5.13}$$

which is a particular consequence of (5.10). By insering in (5.13) the generic expansion:

$$R^{ab} = \frac{1}{2} V^s \wedge V^t \wedge R^{ab}{}_{st} + V^t \wedge (\bar{\psi} T^{ab}{}_t) + V^t \bar{T}^{ab}{}_t \psi +$$
$$+ \text{terms in } \bar{\psi}, \psi,$$

one sees that the space-time part must vanish. The theory contains therefore a trivial gravity, and this means that $R^A = 0$; so we must set B = 0. In general, in building actions similar to (5.7) according to A), B), C), we must include forms which have the same dimension as the standard Einstein term:

$$R^{a_1 a_2} \wedge V^{a_3} \wedge \ldots \wedge V^{a_n} \varepsilon_{a_1 \ldots a_n}$$

under the scaling law (5.11), or of the de-Sitter cosmological term:

$$\Lambda V^{a_1} \wedge \ldots \wedge V^{a_n},$$

where Λ must be given a dimension $(k/\tau)^2$. Although this rule has not been proved rigorously, it seems to be well-respected in all the cases examined so far.

In place of (5.7) we have now:

$$S = \int (R^{ab} \wedge V^c \wedge V^d \varepsilon_{abcd} + 4 \bar{\psi} \gamma_5 \gamma_a \rho V^a). \tag{5.14}$$

The absence of the Hodge dual operator in this action has the effect that the resulting variational equations can be implemented either directly on space-time, or on any large manifold containing space-time. By going up to the group manifold, we obtain a set of algebraic equations between curvature components (inner) along space-time, and outer curvature components along normal directions. Then it is possible to extend a solution given initially on space-time to the whole group manifold. This bears some analogies to the standard Cauchy problem: the space-time configuration is a set of initial data for a system of differential equations on the group manifold. This possibility, and the fact that the lifted solution is unique, is called rheonomy. A trivial rheonomy is already provided by factorization, where the outer components along the SO(3,1) direction are all vanishing and the lifting amounts to a standard gauge transformation.

The rheonomic lifting can be seen as a space-time transformation of the fields. Indeed once we reach another space-time imbedded in the group manifold, we can operate on it with a suitable diffeomorphism and bring it on the initial space-time. The old solution is thus mapped into a new one. Related to this discussion is the role of Bianchi identities, which pose constraints on the curvatures as calculated from a set of potential one-forms ω^A. The rheonomic conditions appear as an additional set of constraints on the curvatures and as a set of differential equations on the potentials, when considered on the whole group manifold. If we insert the rheonomic conditions into the Bianchi identities, we see that they are compatible only if the inner equations hold. Therefore it is in general impossible to lift rheonomically an arbitrary field configuration: only field solutions on space-time can be lifted to the whole group manifold.

This corresponds, in the conventional approach to supersymmetry, to the fact that supersymmetry transformations admit an infinitesimal algebra which closes only on-shell. Should we want to lift an arbitrary field configuration, we have to relax the rheonomy conditions so that they do not imply any more the field equations. This is done by inserting extra fields into the parametrization of the curvatures; the Bianchi identities then merely state the supersymmetry transformation rules and determine the outer derivatives of the auxiliary fields. From the discussion and from other concrete examples to be discussed, we add a fourth principle:

D) The non-trivial solutions of the variational equations must satisfy the rheonomy condition, i.e., the outer curvature components must be completely determined by the inner components.

Then conditions A),..,D) guarantee that the set of classical solutions on a space-time manifold will have a closed algebra of supersymmetric transformations. Infinitesimal transformations can be reached as follows. For an infinitesimal diffeomorphism, through the Lie derivative, we have:

$$\begin{aligned}\delta_\varepsilon \omega^A &= \ell(\varepsilon)\,\omega^A = (\iota_\varepsilon d + d\,\iota_\varepsilon)\,\omega^A = \\ &= \mathcal{D}\varepsilon^A - 2\varepsilon^B dx^C \wedge R^A_{\ CB}\ ;\quad R^A = -dx^B \wedge dx^C R^A_{\ CB}\ .\end{aligned} \tag{5.15}$$

Here all components of curvature appear explicitly; as stated, one has to replace all outer components of the curvature as functions of the inner ones, using the rheonomic conditions. In this way the infinitesimal change is given entirely in terms of the space-time components of the field and a properly defined transformation is found. (5.15) and these substitutions form an algebra which closes on-shell. In fact, if we define the Lie derivatives through (5.15), we see that closure implies the Bianchi identities. The rheonomy principle gives a direct construction of on-shell symmetries on space-time.

But is it possible to extend these symmetries to off-shell configurations or, better, to symmetries of the action? There is no clear-cut extension to off-shell configurations, or rather this extension is given only modulo the field equations. However, in the cases where this extension has been found, it is seen to imply that the derivative of the Lagrangian form $d\mathcal{L}$ vanishes on G. Obviously, this is equivalent to say that the action, as computed on any space-time manifold, does not depend on it. This discussion can be exemplified in N = 1 supergravity (5.14). Here (5.10) read:

$$\begin{aligned} &\varepsilon_{ijkl} R^k \wedge V^l = 0 , \\ &R^{ab} \wedge V^k \varepsilon_{abkl} + 2\bar{\psi}\gamma_5\gamma_l \rho = 0 , \\ &2\gamma_5\gamma_m \rho \wedge V^m - \gamma_5\gamma_m \psi \wedge R^m = 0 . \end{aligned} \tag{5.16}$$

By taking into account rheonomy, for the outer equations one has:

$$\begin{aligned} &R^a = R^a{}_{mn} V^m \wedge V^n , \\ &\rho = \tfrac{1}{2}\rho_{mn} V^m \wedge V^n - \tfrac{1}{2}(\gamma_m\gamma_k R^k{}_{mn} - \tfrac{i}{3}\gamma_m \Sigma^{rs}\gamma_k R^k{}_{rs})\psi \wedge V^m , \\ &R^{ab} = \tfrac{1}{2} R^{ab}{}_{mn} V^m \wedge V^n - \tfrac{1}{2}\varepsilon^{abrs}\bar{\psi}\gamma_5\gamma_m \rho_{rs} \wedge V^m - \\ &\qquad - \tfrac{1}{2}\varepsilon^{trsa}\bar{\psi}\gamma_5\gamma_t \rho_{rs} V^b . \end{aligned} \tag{5.17}$$

The inner field equations are:

$$\begin{aligned} &R^k{}_{mn} = 0, \quad R^{km}{}_{lm} - \frac{1}{2}\delta^k_l R^{mn}{}_{mn} = 0, \\ &\gamma^m \rho_{mn} = 0, \quad \varepsilon^{abcd}\gamma_b \rho_{cd} = 0 . \end{aligned} \tag{5.18}$$

Besides these field equations, one has also the Bianchi identities:

$$\begin{aligned} &dR^{ab} - \omega^{at} \wedge R_t{}^b + \omega_t{}^b \wedge R^{at} = 0 , \\ &dR^a - \omega^{at} \wedge R_t - i\bar{\psi}\gamma^a \rho + R^{ab} \wedge V_b = 0 , \\ &d\rho + \tfrac{i}{2}\Sigma_{ab}\omega^{ab}\rho - \tfrac{i}{2}\Sigma_{ab} R^{ab}\psi = 0 , \end{aligned} \tag{5.19}$$

which imply:

$$\rho^{*ab} = \frac{1}{2}\varepsilon^{abrs}\rho_{rs} = i\gamma_5 \rho^{ab} , \tag{5.20}$$

so that the Rarita-Schwinger spinor ρ^{ab} is self-dual on-shell. The expression for the curvatures is then much simpler if all on-shell conditions and Bianchi identies are taken into account:

$$R^a = 0, \quad \rho = \frac{1}{2} \rho_{mn} V^m \wedge V^n , \tag{5.21}$$
$$R^{ab} = \frac{1}{2} R^{ab}{}_{mn} V^m \wedge V^n + i \bar{\psi} \gamma_m \rho^{ab} V^m .$$

From these one obtains the familiar transformation rules:

$$\delta_\varepsilon V^a = -i \bar{\psi} \gamma^a \varepsilon, \quad \delta_\varepsilon \omega^{ab} = i \bar{\varepsilon} \gamma_m \rho^{ab} V^m,$$
$$\delta_\varepsilon \psi = d\varepsilon - \frac{i}{2} \omega^{ab} \Sigma_{ab} \varepsilon . \tag{5.22}$$

Their closure on-shell is guaranteed by the Bianchi identities. For a complete discussion of the off-shell supersymmetries, the reader may consult Ref. 3), where this kind of problem is very clearly discussed; múch of the material contained in this section has been borrowed by said reference.

The same techniques can be applied also to other theories. A non trivial example is provided by supergravity in five dimensions [8]. Here we have a four-component spinor ψ besides the fünfbein V^a and the spin connection ω^{ab}. The curvatures are given by:

$$R^{ab} = d\omega^{ab} - \omega^{at} \wedge \omega_t{}^b, \quad R^a = DV^a - \frac{i}{2} \bar{\xi} \wedge \Gamma^a \xi,$$
$$R^{\otimes} = dB - i \bar{\xi} \wedge \xi, \quad \rho = \mathcal{D} \xi, \quad \bar{\rho} = \mathcal{D} \bar{\xi}, \tag{5.23}$$

where ξ is not a Majorana spinor, according to the general rules. The manifold M is now imbedded in the group SU(2,2/1). A detailed study (see Ref. 8) yields the first order action:

$$\mathcal{L} = \frac{1}{3} R^{ab} \wedge V^i \wedge V^j \wedge V^k \varepsilon_{abijk} + R^{ab} \wedge V_a \wedge V_b \wedge B + \frac{i}{2} R^{\otimes} \wedge \bar{\xi} \wedge \xi \wedge B +$$
$$+ i R^a \wedge \bar{\xi} \Gamma_a \xi \wedge B - 2i R^{\otimes} \wedge \bar{\xi} \wedge \Gamma_a \xi \wedge V^a - \frac{1}{2} (\bar{\xi} \wedge \xi) \wedge (\bar{\xi} \wedge \xi) \wedge B +$$
$$+ \bar{\xi} \wedge \xi \wedge (\bar{\xi} \wedge \Gamma^a \xi) \wedge V_a + \frac{1}{4} R^{\otimes} \wedge R^{\otimes} \wedge B + R^a \wedge R_a \wedge B + \tag{5.24}$$
$$+ 4 \bar{\xi} \wedge \Sigma_{ab} \xi \wedge V^a \wedge V^b - (1-\eta) R^a \wedge R^{\otimes} \wedge V_a ,$$

and the variational equations:

Einstein equations: (5.25.1)

$$\varepsilon_{abijk} R^{bi} \wedge V^j \wedge V^k + 2i (\bar{\xi}\wedge\xi) \wedge R_a - i\left(\frac{3}{2} - \frac{n}{2}\right) R^{\otimes} \wedge \bar{\xi}\Gamma_a \xi - 4(\bar{\rho}\wedge\Sigma_{ab}\xi + \bar{\xi}\wedge\Sigma_{ab}\rho)\wedge V^b + i(1-\eta)(\bar{\rho}\wedge\xi - \bar{\xi}\wedge\rho)\wedge V^a + 2\eta R_a \wedge R^{\otimes} = 0$$

Maxwell equations: (5.25.2)

$$\eta R^{ab} \wedge V_a \wedge V_b - \frac{i}{2}(3-\eta) R^a \wedge \bar{\xi}\wedge\Gamma_a\xi + \frac{3i}{2} R^{\otimes}\wedge\bar{\xi}\wedge\xi + \eta R^a \wedge R_a - i\,\frac{3+n}{2}(\bar{\rho}\wedge\Gamma_a\xi - \bar{\xi}\wedge\Gamma_a\rho)\wedge V^a - \frac{3}{4} R^{\otimes}\wedge R^{\otimes} = 0$$

Rarita-Schwinger equations: (5.25.3)

$$4\,\Sigma_{ab}\,\rho\wedge V^a \wedge V^b - 4\,\Sigma_{ab}\,\xi\wedge R^a \wedge V^b + i(1-\eta)\xi\wedge R^a \wedge V_a - \frac{i}{2}(\eta+3)\Gamma_a\xi\wedge R^{\otimes}\wedge V^a = 0.$$

The Bose sector of this theory has been discussed in detail; the theory contains an electromagnetic field. The Fermi sector yields two additional neutral gravitinos; these may acquire a charge if we deal with a d = 5 theory in a de Sitter space. The d = 5 theory is itself the dimensional reduction of a d = 6 theory [10]. This is the simplest one containing a non-trivial (i.e. not all potentials are 1-forms) free differential algebra given by:

$$R^{ab} = d\omega^{ab} - \omega^{at}\wedge\omega_t^{\ b} = 0, \quad R^a = DV^a - \frac{i}{2}\bar{\psi}\wedge\Gamma^a\psi = 0$$

$$\rho = D\psi = 0, \quad R^{\otimes} = dB - \frac{i}{2}\bar{\psi}\wedge\Gamma^a\psi\wedge V_a = 0, \quad (a,b = 1,\ldots,6) \tag{5.26}$$

where the connection ω^{ab}, the sechsbein V^a and the gravitino ψ are gauge potentials for the six-dimensional super-Poincaré group. Here the signature of the metric is: (1,−1,−1,−1,−1,−1). The Clifford algebra is given by:

$$\{\Gamma_a,\Gamma_b\} = 2\eta_{ab}, \quad \Gamma_1^{\dagger} = \Gamma_1, \quad \Gamma_i^{\dagger} = -\Gamma_i \quad i = 2,\ldots,6, \quad \Gamma_7 = \Gamma_1\Gamma_2\ldots\Gamma_6,$$

$$\Gamma_7^{\dagger} = \Gamma_7, \quad \Gamma_7^2 = \mathbb{1}, \quad \varepsilon_{1\ldots6} = 1, \quad \Gamma_a^T = C\Gamma_a C^{-1}, \quad C = C^T = C^{-1},$$

$$\Gamma^{b_1\ldots b_l}\varepsilon_{a_1\ldots a_k b_1\ldots b_l} = (-1)^{\frac{l(l-1)}{2}}\,\Gamma_7\, l!\,\Gamma_{a_1\ldots a_k} \tag{5.27}$$

Requiring that the gravitino is Weyl:

$$\Gamma_7 \psi = \psi , \tag{5.28}$$

leads to the Fierz identity:

$$\Gamma_a \psi \wedge \bar{\psi} \Gamma^a \psi = 0 , \tag{5.29}$$

and the only non-vanishing currents are:

$$\bar{\psi} \wedge \Gamma^a \psi , \quad \bar{\psi} \wedge \Gamma^{abc} \psi . \tag{5.30}$$

The three-index current is self-dual:

$$\bar{\psi} \wedge \Gamma_{abc} \psi = \frac{1}{6} \varepsilon_{abcpqr} \bar{\psi} \wedge \Gamma^{pqr} \psi . \tag{5.31}$$

The action must remain invariant under the scaling properties:

$$\begin{aligned} &\omega^{ab} \to \omega^{ab} , \quad R^{ab} \to R^{ab} , \quad V^a \to e V^a , \\ &R^a \to e R^a , \quad \psi \to \sqrt{e}\, \psi , \quad B \to e^2 B , \text{ etc.} \end{aligned} \tag{5.32}$$

Furthermore we require exact gauge invariance under the gauge transformation:

$$B \to B + d\varphi . \tag{5.33}$$

Rheonomy and all the other conditions leave us with two choices:

$$\begin{aligned} S = \int \{ & R^{ab} \wedge \tfrac{1}{4} \varepsilon_{abc_1 .. c_4} V^{c_1} \wedge .. \wedge V^{c_4} + 3i R^a \bar{\psi} \Gamma^b \psi \wedge V_b \wedge V_a - \\ & - 3i R^{\otimes} \bar{\psi} \wedge \Gamma^a \psi \wedge V_a - i \bar{\rho} \Gamma_{abc} \wedge \psi \wedge V^a \wedge V^b \wedge V^c + (\text{herm.} \\ & \text{conj.}) \pm 2\sqrt{3}\, R^a \wedge R^{\otimes} \wedge V_a \} . \end{aligned} \tag{5.34}$$

The variational equations are then:

Torsion equation

$$\varepsilon_{abijkl} R^i \wedge V^j \wedge V^k \wedge V^l - \xi R^{\otimes} \wedge V_a \wedge V_b = 0 \tag{5.35}$$

Einstein equation

$$\begin{aligned} & \varepsilon_{abcdef} R^{ab} \wedge V^c \wedge V^d \wedge V^e - 6i R^a \wedge V_a \wedge \bar{\psi} \wedge \Gamma_f \psi \text{-} 6i \bar{\psi} \Gamma^b \wedge \psi R_f \wedge V_b + \\ & + 3i (\bar{\rho} \wedge \Gamma^b \psi - \bar{\psi} \wedge \Gamma^b \rho) \wedge V_b \wedge V_f - 3i (\bar{\rho} \Gamma_{abf} \wedge \psi + \bar{\psi} \Gamma_{abf} \wedge \rho) V^a \wedge V^b + \\ & + 2\xi R_f \wedge R^{\otimes} = 0 \end{aligned} \tag{5.36}$$

Maxwell equation:

$$\xi R^{ab}\wedge V_a\wedge V_b - i\left(3+\frac{\xi}{2}\right)(\bar{\rho}\wedge\Gamma^a\psi-\bar{\psi}\wedge\Gamma^a\rho)\wedge V_a - 3i\, R^a\wedge\bar{\psi}\wedge\Gamma_a\psi +$$

$$+\xi\, R^a\wedge R_a = 0 \qquad (5.37)$$

Gravitino equation:

$$\left(-3+\frac{\xi}{2}\right) R^a\wedge\Gamma^b\psi\wedge V_a\wedge V_b + \left(3+\frac{\xi}{2}\right) R^{\otimes}\wedge\Gamma_a\psi\wedge V^a - 2\,\Gamma_{abc}\rho\wedge V^a\wedge V^b\wedge V^c +$$

$$+ 3\,\Gamma_{abc}\psi\wedge R^a\wedge V^b\wedge V^c = 0. \qquad (5.38)$$

Moreover one gets:

$$R^a = -\frac{1}{12}\xi\,\varepsilon^{abcpqr}F_{pqr}\,V_b\wedge V_c$$

$$R^{\otimes} = F_{abc}\,V^a\wedge V^b\wedge V^c,\ \xi=\pm 2\sqrt{3}\,. \qquad (5.39)$$

Had we chosen an anti-Weyl spinor, then we would have an anti-self-dual current. By analogy one expects that the strengths F_{abc}, that is the components of the curvature of B, are themselves self-dual on the shell. If we compute the degrees of freedom of the fields we find that they are nine in the sechsbein, twelve in the gravitinos. The strengths of B must provide the matching three degrees of freedom since it is well known from general rules that any realistic theory must have the same number of Fermi and Bose degrees of freedom. We find three degrees of freedom in the F_{abc} only assuming self-duality:

$$F^{abc} = F_{(+)}{}^{abc} = \frac{1}{6}\varepsilon^{abcpqr}F_{pqr}\,. \qquad (5.40)$$

The following type of identity is then very useful in the computations:

$$\varepsilon_{abcpqr}A^{abc}B^{pqr} = 0 \qquad (5.41)$$

where A_{abc}, B_{abc} have the same duality type. In particular:

$$F_{apq}\,\bar{\psi}\wedge\Gamma^{ars}\psi\,\varepsilon^{pq}{}_{rslm} = 0. \qquad (5.42)$$

Then it is possible to construct a Lagrangian satisfying the axioms of the theory. The Bianchi identities are given by:

$$DR^{ab} = 0,\ DR^a + R^{ab}\wedge V_b + \frac{i}{2}(\bar{\rho}\wedge\Gamma^a\psi-\bar{\psi}\wedge\Gamma^a\rho) = 0,\ D\rho + \frac{1}{4}\Gamma_{ab}R^{ab}\psi = 0,$$

$$dR^{\square} + \frac{i}{2}(\bar{\rho}\wedge\Gamma^{a}\psi - \bar{\psi}\wedge\Gamma^{a}\rho)\wedge V_a + \frac{i}{2}\bar{\psi}\wedge\Gamma^{a}\psi\wedge R_a = 0 \,. \tag{5.43}$$

It can be checked that duality follows only if we take into account the normal components of the equations. On space-time the theory is consistent with a generic strength, but admits no supersymmetry unless we assume self-duality. This raises a number of interesting questions. On space-time the supersymmetry holds only if we restrict the manifold of the solutions with a condition which does not follow from variation of the space-time fields; we need the gravitino components. This implies that the set of theorems which can be described by the present scheme is somewhat larger than the usual set of supergravities. The group-manifold approach regards all variational equations on the same footing and therefore also the normal components have the same states as the space-time ones.

Finally, let us consider the supergravity which is most important from the physical point of view, namely the CJS theory in eleven dimensions. It is based on the FDA:

$$\begin{aligned} R^{ab} &= d\omega^{ab} - \omega^{ac}\wedge\omega_c{}^{b}\,, \quad R^{a} = \mathcal{D}V^{a} - \frac{i}{2}\bar{\psi}\wedge\Gamma^{a}\psi\,, \\ \rho &= \mathcal{D}\psi\,, \quad R^{\square} = dA - \frac{1}{2}\bar{\psi}\wedge\Gamma^{a_1a_2}\psi\wedge V_{a_1}\wedge V_{a_2}\,, \end{aligned} \tag{5.44}$$

where the V^{a} and the ω^{ab} are the conventional elfbein and spin connection for the Poincaré group, the ψ are Majorana and generate the supertranslations. The potential A is a three-form. We have therefore strengths with four indices F^{abcd}. The action is then given by:

$$\begin{aligned} \mathcal{L} = &-\frac{1}{9}R^{a_1a_2}\wedge V^{a_3}\wedge..\wedge V^{a_{11}}\varepsilon_{a_1\dots a_{11}} - 840\,R^{\square}\wedge R^{\square}\wedge A + \\ &+\frac{7i}{30}R^{a}\wedge V_a\wedge\bar{\psi}\wedge\Gamma^{b_1..b_5}\psi\wedge V^{b_6}\wedge..\wedge V^{b_{11}}\varepsilon_{b_1\dots b_{11}} + \\ &+2\bar{\rho}\wedge\Gamma_{a_1..a_8}\psi\wedge V^{a_1}\wedge\dots\wedge V^{a_8} - \frac{1}{330}F_{a_1..a_4}F^{a_1..a_4}V^{c_1}..V^{c_{11}}\varepsilon_{c_1..c_{11}} - \\ &-84\,R^{\square}\wedge(i\bar{\psi}\wedge\Gamma_{a_1..a_5}\psi\wedge V^{a_1}..V^{a_5} - \Lambda\wedge\psi\wedge\Gamma^{a_1a_2}\psi\wedge V_{a_1}\wedge V_{a_2} + \\ &+\frac{1}{4}\bar{\psi}\wedge\Gamma^{a_1a_2}\psi\wedge\bar{\psi}\wedge\Gamma^{a_3a_4}\psi\wedge V^{a_5}\wedge..\wedge V^{a_{11}}\varepsilon_{a_1..a_{11}} - \\ &-210\,\bar{\psi}\wedge\Gamma^{a_1a_2}\psi\wedge\bar{\psi}\wedge\Gamma^{a_3a_4}\psi\wedge V_{a_1}\wedge..\wedge V_{a_4}\wedge A + \\ &+2\,F_{a_1..a_4}R^{\square}\wedge V_{a_3}\wedge..\wedge V_{a_{11}}\varepsilon^{a_1\dots a_{11}} \end{aligned} \tag{5.45}$$

The normal components are seen to satisfy rheonomy as a consequence of the choice of the coefficients in the action. The resulting equations are:

$$R^a = 0,\; R^{\Box} = F_{m_1 \dots m_4} V^{m_1} \wedge \dots \wedge V^{m_4};\; \rho = \rho_{ab} V^a \wedge V^b + \rho^{(\text{on-shell})}_{(\psi V)} .$$

Quite likely the scalar components can be eliminated and the theory written entirely in terms of p-forms. I do not know any way to do this at the moment, nor do we have any information as to whether or not the CJS theory is the ultimate one. Certainly it contains a number of very interesting dimensional reductions. The actual algebraic manipulations involved in dealing with higher FDA's is enormous and have discouraged attempts towards the construction of more general theories.

In (5.45) we see that the last term violates the scheme which we have followed so far, as it contains a set of space-time scalars which ultimately represent the strengths and allow us to form the dual of the form $R^{\Box}$. We can in fact regard the scalars as zero forms in the FDA; the duality is actually achieved in the tangent space and is not dangerous. We are allowed to use the general method explained before. In fact the same procedure allowes us to discuss the electromagnetic field coupled to gravity in any dimension. The Bianchi identities appear as:

$$\nabla R^{ab} \equiv \mathfrak{D} R^{ab} = 0 ,$$

$$\nabla R^a \equiv \mathfrak{D} R^a + R^{ab} \wedge V_b - i\bar{\psi} \wedge \Gamma^a \rho = 0 ,$$

$$\nabla \rho \equiv \mathfrak{D} \rho + \frac{1}{4} \Gamma_{ab} \psi \wedge R^{ab} = 0,$$

$$\nabla R^{\Box} = dR^{\Box} - \bar{\psi} \wedge \Gamma^{a_1 a_2} \rho \wedge V_{a_1} \wedge V_{a_2} + \bar{\psi} \wedge \Gamma^{a_1 a_2} \psi \wedge R_{a_1} \wedge V_{a_2} = 0 .$$

References.

1 Y. Ne'eman and T. Regge, Rivista Nuovo Cimento 1,(1978)5.
2 For a general review and formal conventions see: R. D'Auria, P. Fré and T. Regge, Group-Manifold Approach to Gravity and Supergravity Theories, IC/81/54, Trieste, Lectures given at the Spring School on Supergravity, Trieste (April 1981).
3 A. D'Adda, R. D'Auria, P. Fré and T. Regge, Rivista del Nuovo Cimento 6 (1980).
4 E. Cartan, Comptes Rendus 174, (1922) 593.

5 P. van Nieuwenhuizen, CERN Preprint TH.3499 (1982).

6 R. D'Auria, P. Fré, E. Maina and T. Regge, Geometrical First Order Supergravity in Five Space-Time Dimensions, IFTT 404 (1981), to appear in Annals of Physics.

7 R. Coquereaux, Phys. Lett. 115B (1982) 389.

8 P. van Nieuwenhuizen, Stony Brook Preprint ITP-SB-81-67 (1981).

9 R. D'Auria, P. Fré, and T. Regge, Rivista del Nuovo Cimento 3 (1980)12.

10 R. D'Auria, P. Fré and T. Regge, CERN Preprint TH.3563(1983).

GRAVITY ON A LATTICE

Tullio Regge
Istituto di Fisica Teorica
dell'Università degli Studi di Torino, Italy.

1. Introduction

The current revival of interest on field theories on a lattice has focused the attention on the problems presented by gravity on piecewise linear spaces. In the original paper by the author gravity on a lattice was intended primarily as an approximation procedure particularly suited for numerical work. It has turned out however that most of the interesting features of the continuum limit are preserved on the lattice; in particular space-time remains a manifold at all stages of the approximation. The metric tensor, the curvature, the Bianchi identites and the Einstein-Hilbert action can be all given a natural definition on a gravitational lattice. The use of this technique in quantum gravity appears quite advantageous in that one works with a lattice which preserves the structure of space-time.

The simplest example of a piecewise linear manifold (PLM) is provided by a polyhedron imbedded in a 3 dimensional euclidean space. We expect to be able to approximate (locally) any differentiable surface with a polyhedron whose faces are sufficently small. Higher dimensional PLM's are the natural generalization of this concept to generic manifolds.

It is essential to understand the nature of a 2-dimensional PLM and the way it is related to a differentiable surface. The gaussian curvature of a surface is itself (for C^{∞} surfaces) a differentiable function. On a PLM it appears as a distribution with support on the vertices of the polyhedron. We can check this fact by looking at a geodesic triangle T on the surface.

Let α, β, γ the internal angles of the triangle. We have the relation:

$$\alpha + \beta + \gamma - \pi = \int_T K \, d\sigma \tag{1}$$

between angles and the integrated gaussian curvature. This formula reduces to a well-known trigonometric formula in the case of a sphere. The value of the gaussian curvature K appears as a parameter measuring the violation of Euclid's axiom of the parallels.

On a PLM the value of the quantity $\alpha + \beta + \gamma - \pi$ depends only on the vertices lying inside the triangle. A triangle which contains no vertices differs in no way, in its intrinsic geometry, from one built in the euclidean plane. One can easily convince oneself by looking at a triangle drawn on two contiguous face of a cube. One can always think of cutting away the faces, folding them along the common edge and flattening the pair on a plane. The resulting triangle is euclidean and contains no gaussian curvature. Therefore the edges of a polyhedron contain no curvature.

On the contrary it is possible to draw on a cube a geodesic triangle which is trirectangular. We have then $\alpha+\beta+\gamma=\frac{3\pi}{2}$

This triangle encloses exactly one vertex of the cube. In general one can prove, through elementary geometrical methods, that one has:

$$\alpha+\beta+\gamma-\pi = \sum_i \epsilon_i \tag{2}$$

where ϵ_i is an angle associated with the i-th vertex V_i. This angle can be calculated from the formula:

$$\epsilon_i = 2\pi - \sum_K \theta_{Ki} \tag{3}$$

where the sum is carried on all angles having V_i as vertex.

It is always possible to add an apparent vertex to any face by joining an internal point to the vertices; this vertex does not carry any gaussian curvature and indeed it does not contribute to the sum (3). We can understand (2) by replacing the function K by the Dirac function:

$$K(x) = \sum_{\forall i} \epsilon_i \, \delta^2(x-x_i) \tag{4}$$

We see in this way that curvature has support on a subset of codimension 2. This is also true in higher dimensions.

The intrinsic geometry of the neighbourhood of a vertex V can be easily identified with that of an ϵ-cone $C(\epsilon)$ obtained as follows. We introduce polar coordinates in the plane: ρ, φ with origin in V.

We consider then the subset $0 \leq \varphi \leq 2\pi-\epsilon$ with the rays $\varphi = 0$ and $\varphi = 2\pi-\epsilon$ identified. The resulting manifold is a PLM and is everywhere flat with the exception of V. Our construction hols actually only for $\epsilon > 0$ but can be easily supplemented for $\epsilon < 0$.

Parallel transportation along a closed loop has no effect if the loop does not enclose V but results in a anticlockwise rotation of n ϵ if the loop encircles V n times (anticlockwise).

This result is a straightforward extension of known facts about Levi--Civita transportation.

Higher dimensional ϵ-cones can be constructed essentially as pro-

ducts $\mathbb{R}^{n-2} \times C(\epsilon)$ One can take similarly in $\mathbb{R}^3$ cilindrical coordinates and identify the semiplanes $\varphi = 0$ and $\varphi = 2\pi - \epsilon$
The resulting PLM carries now curvature along the z axis.
The construction can be readily adapted to n dimension and the support of curvature will be a n-2 dimensional axis.
The most general PLM will be obtained by fitting together several cone subsets. This is most commonly done through a simplicial decomposition. The manifold is assumed to remain flat inside the simplex, curvature arises only when we fit together the simplexes.
The simplicial decomposition consists of a collection of flat simplexes which are fitted together by suitably identifying common faces (or n-1 dimensional boundary cells). This implies also identifying corresponding edges in the boundaries.
Any given edge may belong to several simplexes, up to subsimplexes of dimension n-2. The familiar operation of assigning a metric to a differentiable manifold reduces here to the setting of all lengths of all edges in the lattice. Once these are given the geometry of all simplexes is completely determined. Once all edges are known it is possible to calculate through elementary formulas all internal angles of all subsimplexes. In particular it is possible to calculate the internal dihedral angles θ between two boundary n-1 dimensional cells hinging on a common n-2 dimensional subsimplex V. The sum:

$$\epsilon = 2\pi - \sum_k \vartheta_k \tag{5}$$

extended over all simplexes sharing V is called the defect angle of V. This quantity is a natural extension of the sum (3). If we are dealing with a simplicial decomposition of an euclidean flat space then this sum must vanish. ϵ measures therefore the curvature of the decomposition. In this sense it represents here the analog of the Riemann tensor. It should then be obvious that one can identify a suitable neighbourhood of the hinge with a suitable neighbourhood of an ϵ- cone. The hinge can be identified with a compact subset of the axis of the cone. This appears a new feature not present in a 2-dimensional PLM where hinges are just points and obviously do not intersect or have boundaries.
In 3 dimensions hinges are segments which meet at vertices.
What happens and what significance have the vertices (or the n-3 simplexes)?. Just as hinges are the carriers of the Riemann tensor (n-3)--simplexes are the carriers of the Bianchi identities. In order to see it we must analize the Levi-Civita parallel transportation in a PLM.
If we carry a vector around the axis of a ϵ -cone we find it rotated by the angle ϵ around the axis.

We see therefore that transportation acts on the vector through an element of the orthogonal (or pseudoorthogonal) group O(n,m) which characterizes the flat metric. This action will depend only on a certain equivalence class of the loop used for transportation. It is indee possible to deform the loop without changing the result provided we never cross the hinge. In 2 dimensions this means in fact that we never allow a vertex to cross the loop. If we have a PLM M and we call $\mathbb{B}$ the subset of all hinges then we see that the final element in O(n,m) will depend only on the element in $\pi_1(M-\mathbb{B})$, the fundamental group. Our construction has given us a representation $\pi_1 \rightarrow$ O(n,m), this representation is more properly the analog of the Riemann tensor. This makes it obvious that there is a relations among all elements of O(n,m) which correspond to hinges meeting at a common n-3 simplex. It can be indeed show that $\pi_1(M-\mathbb{B})$ can be presented as a collection of generators, one for each hinge, with a relation at each n-3-subsimplex. If l labels all hinges meeting in this subsimplex and α_ℓ is the corresponding element in π_1 the there is a relation of the kind:

$$\alpha_1 \alpha_2 \cdots \alpha_n = \mathbb{1} \tag{6}$$

which implies a relation in the images in O(n,m). This relation is the analog of the Bianchi identities. Notice that π_1 is not in general abelian and that the element α_ℓ is then defined only up to some conjugation in the group. This introduces some arbitrariness in the writing of (6), this can be removed by a suitable convention in the definition of the generators of $\pi_1(M-\mathbb{B})$. Hinges never meet in 2 dimensions, this corresponds to the known fact that Bianchi identites are trivially satisfied in this case.

2. The geometry of the PLM.

We see therefore that curvature is concentrated exclusively on a subset $\mathbb{B}$ of codimension 2. Obviously this dimensionality cannot be reduced further, if the consider the group $\pi_1(M-\mathbb{B})$ we would find it to be trivial if the codimension of $\mathbb{B}$ is larger than 2. The gravitational field is therefore concentrated on the hinges in $\mathbb{B}$ and vanishes outside. This structure has some analogy with the behaviour of superconductors of type II. Here the magnetic lines of force can penetrate the superconductor, for sufficently high external fields. This does not signal the transition to the normal state.
The magnetic field is in fact confined into flux tubes which are quantized and which represent topological singularities. The material reverts to the normal state inside the tube and the size of this is dictated by the simple condition that the energy per unit length should

be a minimum. It is interesting to notice that there is an analogy here as summarized by the following table:

Object	gravity	Superconductor
Potential	Frames	A_μ
Stenght	Curvatures	Magnetic field
Group	$SO(n,m)$	$U(1)$
Support	$\mathbb{B}$	flux tubes

If we give a thickness to the hinges than this is obviously equivalent to the rounding off of the sharp corners in the PLM and this becomes once again a differentiable manifold. And of course in the gravitational lattice one does not see any particular reason for the elements of O(n,m) to be quantized.

Finally it is not clear at this stage what the hypothetical gravitational superconductor might be. We leave these speculations to some future work and discuss instead the physics of the gravitational lattice. Since the days of general relativity it has been often stated that physics is geometry. To me physics is geometry plus an action principle. Without it we could not select among all lattices those which carry some physical meaning. The action I proposed in my first paper was more in the nature of a conjecture than a direct offspring of the Einstein--Hilbert action.

In order to write it we need some notation. I indicate with L_K the measure of the K-th hinge. Apart from a costant the action is then given by:

$$S = \sum_K L_K \epsilon_K \tag{8}$$

where the sum is carried over all hinges. The measure L_K is calculated according to the metric. This means that some provision must be made for hinges which are time-like versus hinges which are space like. There is indeed a factor i between both cases which is compensated by another factor i in the corresponding ϵ_K so that the result is an way real.

The dimensionaly of (8) is in agreement with that of the conventional action:

$$S = \text{const.} \int R \sqrt{-g}\, d^n x \tag{9}$$

where R is indeed an inverse area so that the total action has dimension ℓ^{n-2} as in the (8). The real proof that (8) has (9) as a limit when the lattice approximates a differentiable manifold has been given

by Cheeger et al. (18). They show that we are not dealing with a pointwise convergence (8) by rather with a convergence only when we are considering summs over many contiguous simplexes. The problem has been also considered by Feinberg et al. with a particular attention to the possible applications to discrete mechanics. (See (19-31-32) for a complete discussion on this subject. The interesting fact is that the action (8) has been tested in a variety of cases and found to be in most of them an acceptable if not eccellent approximation to countinous gravity. I cannot summarize here all cases which have been so far examined and which mimic known exact solutions of the field equations.
An obvious application deals with a toy Friedmann univers where the homogeneous t=const. section is replaced with a regular polytope (there are 6 regular 3-polytopes). Already with the simplest polytope, the five points 4-simplex, one finds a reasonable agreement with the standard model. (One should be remainded here that the actual time slice of the space time is the boundary of the simplex).
The discrete action (8) has some formal features in common with the countinuous one. We have already seen that the lengths of the edges of the lattice, whatever the dimension n+m, are the basis field variables and play a role similar to that of the metric tensor. The angles instead are throught to be rather complicated inverse trigonometric functions of the lengths, just as the Riemann tensor is after all a rather lenghty expression in the derivatives of the metric tensor. The problem of finding the stationary configurations of the field for the action (8) looks very difficult. Luckily there occurs an important cancellation which greatly simplifies the treatment.
It can be in fact proved that for triangulated manifolds one has always:

$$\sum_K \frac{\partial \epsilon_K}{\partial \ell_i} L_K = 0 \tag{10}$$

that is one can calculated the variation of the action as if the angles were constant. The field equations are then:

$$\frac{\partial S}{\partial \ell_i} = \sum_K \frac{\partial L_K}{\partial \ell_i} \epsilon_K = 0 \tag{11}$$

If the manifold has a boundary then there are corresponding surface terms. In n=2 the property is trivial, the action reduces to the known Gauss-Bonnet formula for the Euler characteristics and this does not change for compact manifolds under small changes. One has a similar property for the continuum case. The action (9) is given more explicitly as:

$$S = \int g^{\mu\nu} R_{\mu\nu} \sqrt{-g}\, d^4x \tag{12}$$

Upon variation we find in this action two contributions, one from the metric tensor and related g and the other from the contracted Riemann tensor $R_{\mu\nu}$. It turns out that this one is always a divergence and that it can be integrated away. This property finds an exact discrete analog in the above mentioned property of the angles ϵ_K.
The field equation (11) are particularly simple, if not trivial, in 3 dimensions. Here the measures L_K themselves are the lengths of the edges. Therefore the field equations are most simply:

$$\epsilon_K = 0 \tag{13}$$

i.e. space is flat. This result is again a discrete analog of a well known theorem in the continuous theory.
In 4 dimensions the measures of the hinges are areas. The area of a triangle of sides a,b,c is given by the formula:

$$A^2 = \frac{1}{16}\left(2a^2b^2 + 2b^2c^2 + 2c^2a^2 - a^4 - b^4 - c^4\right) \tag{14}$$

and in general the measure of a n dimensional simplex can be given as a determinant as follows. The vertices are numbered as $P \ldots P_{n+1}$. The distance between P_i, P_k is then ℓ_{ik} The square hypervolume is then:

$$2^n (n!)^2 L^2 = - \begin{vmatrix} 0 & \ell_{12}^2 & \cdots & \ell_{1,n+1}^2 & 1 \\ \ell_{21}^2 & 0 & & \cdots & 1 \\ & & & \vdots & \\ 1 & 1 & \cdots & \cdot & 0 \end{vmatrix} \tag{15}$$

It is possible to express all relevant trigonometric functions of all angles in the simples in terms of these determinants, their minors, and their derivatives with respect to the edges. Our discussion would not be complete without pointing out some of the not so pleasant features of the discrete theory. Because of its very nature it cannot possess a true coordinate invariance and it is unclear at this stage how coordinate transformation fit at all into the picture.

Coordinate invariance may be related to property (16) which seems so far to be distinctive of the discrete Einstein-Hilbert theory.
Related to this discussion is the role of the Dirac constraints $\mathcal{H}_\perp$, $\mathcal{H}_i$ in the continuous theory and the possible existence of similar constraints on the lattice. The $\mathcal{H}$'s have commutators which are again linear in the $\mathcal{H}$'s that is they form what is named, by abuse of language, a Lie algebra with variable structure constants. If the $\mathcal{H}_i$ are replaced by a discrete group then it is conceivable that also $\mathcal{H}_\perp$ must be discretized. This makes it unlikely that space only will remain discrete while time may run continuously. In some of the approximations developed for instance by Williams and Rocek one has in fact a discrete time slice for the Friedmann universe while time suns continously. It is to be noted however that their choice really fixes the lapse and shift functions. It follows that their time is really the analog of the cosmological time and it cannot made to run differently at different point by using the coordinate gauge or any discrete analog of it. The question of the true nature of the coordinate transformation on the lattice remains open and intriguing. In their paper Williams and Rocek [illegible] that the number of degrees of freedom in the propagation of gravitons in the lattice, in the weak field approximation, corresponds to that of the continuous theory, perhaps this result is related to (10).

3. - The quantum theory in 3 dimensions.

Both the discrete and the continuous theory are trivial in 3 dimensions at the classical level, the only solution is locally flat space. It is interesting however to notice that there is a surprising formal connection between the theory of angular momentum and quantum field theory on a 3 dimensional lattice.
The idea is to replace classical vectors, the edges of the tetrahedra, with quantum angular momenta. I recall here a few essential definition in the quantum theory of angular momenta. Those who have done work in nuclear physics are already quite familiar with them.
The first object is the 3j-Symbol. If we compose two angular momenta $\vec{a}$, $\vec{b}$ we may write their wave functions as Ψ_a^α, Φ_b^β where a,b label the corresponding representations in SU(2), and α, β are the azimuthal quantum numbers. The states of the total angular momentum $\vec{c}$ are the given by:

$$\Theta_c^\gamma = \sum_{\alpha,\beta} C^{a\ b\ c}_{\alpha\ \beta\ \gamma}\ \Psi_a^\alpha\ \Phi_b^\beta \tag{16}$$

where a+b $\leq$ c $\leq$ a-b othewise the coefficent vanishes.
According to standard practice one defines then a symmetrical symbol:

$$\begin{pmatrix} a & b & c \\ \alpha & \beta & \gamma \end{pmatrix} = (-1)^{a-b-\gamma} (2c+1)^{-\frac{1}{2}} C^{abc}_{\alpha\beta-\gamma} \tag{17}$$

This 3j-symbol depends on the choice for the basis of the representations of SU(2), it has therefore a certain amount of convention built into it. If we have 3 angular momenta a,b,d then we may compose them in different order either as;

$$\vec{c} = \vec{a} + \vec{b} \qquad \vec{e} = \vec{c} + \vec{d} \tag{18}$$

or as:

$$\vec{f} = \vec{b} + \vec{d} \qquad \vec{e} = \vec{a} + \vec{f} \tag{19}$$

given representation in the total sum may then be present more than nce in the final decomposition. (Notice that it appears at most once n the product of 2 representations).

This accounts for the different decompositions appearing in (17) and (18) resp. These correspond to different mixings in the final set of representations sharing the same value of e. According to Racah one has then a relation of the kind:

$$\sum_{\alpha\beta\gamma} C^{cde}_{\gamma\delta\epsilon} C^{abc}_{\alpha\beta\gamma} \psi^{\alpha}_{a} \phi^{\beta}_{b} \xi^{\delta}_{d} = \sqrt{(2c+1)(2f+1)} \cdot \sum C^{afe}_{\alpha\varphi\eta} C^{bdf}_{\beta\delta\varphi} W(abed;cf) \psi^{\alpha}_{a} \phi^{\beta}_{b} \xi^{\delta}_{d} \tag{20}$$

where the 6j-symbol W achieves the recoupling of the 3 representations. It can be written as:

$$W(a\,b\,e\,d : c f) = (-1)^{a+b+d+e} \begin{Bmatrix} a & b & c \\ d & e & f \end{Bmatrix} \tag{21}$$

and has the symmetry:

$$\begin{Bmatrix} a & b & c \\ d & e & f \end{Bmatrix} = \begin{Bmatrix} b & c & a \\ e & f & d \end{Bmatrix} = \begin{Bmatrix} d & e & c \\ a & b & f \end{Bmatrix} \quad \text{etc.} \tag{22}$$

From (20) it can be easily seen to be independent of the basis of the representations of SU(2). The 6j-symbol is the prototype of an infinite set of recoupling coefficents which can be used when several angular momenta are added together. According to Wigner we introduce the symbol:

$$\begin{pmatrix} a \\ \alpha\ \alpha' \end{pmatrix} = \delta_{\alpha,-\alpha'} (-1)^{a-\alpha} \tag{23}$$

If we concentrate on a particular summation index, say α we see that it appears in the form:

$$\cdots \begin{pmatrix} b & c & a \\ \beta & \gamma & \alpha \end{pmatrix} \begin{pmatrix} a \\ \alpha\ \alpha' \end{pmatrix} \begin{pmatrix} a & e & f \\ \alpha' & \epsilon & \varphi \end{pmatrix} \cdots \tag{24}$$

which guarantees invariance under the choice of basis. The generalized 3nj-symbols can be all generated by writing the product of 2n 3j-symbols and by contracting all azimuthal quantum numbers with the help of Wigner's symbol. There are 3n summations (since $\alpha = -\alpha'$ etc.). This number can be further decreased by noticing that in fact the sum of all 3 lower indices in any 3j must vanish. This reduced drastically the amount of summs which need to carried out. The final result is still extremely complicated. The 6j-symbol admits a beautiful and compact formula, due to Racah, which is used extensively in the literature.

I am not reporting here this celebrated formula because it is of limited validity in our discussion. I am interested in the recoupling of a very large number of angular momenta and when the value of the individual momenta is also large.

In other words I am interested in the semiclassical limit of the theory. Here the formula of Racah becomes inadequate because it appears as the sum of a large number of factorials with alternating signs, the errors involved in the computation go quickly out of hand if we persist in the general practice of rounding off, as done in most computers. It is much better to use the semiclassical approximation. There are 6 angular momenta in the symbol, they can be considered as the 6 sides of a tetrahedron. Similarly the higher 3nj-symbols can be considered as polyhedra with 2n faces and 3j edges. The genus of the polyhedron need not to be that of the sphere, and this has raised a number of very interesing question in statistical mechanics, in which unfortunately I have no time to dwell.

If we choose at random six edges we have no guarantee that a corresponding tetrahedron will exist. We must satisfy the triangular inequality.

In each face the sum of two sides must be larger than the remaining one. But even if this is true it may happen that a global tetrahedron does not exist although each of its four faces can be constructed.
We need a stronger inequality, that is that the square volume should be positive, as calculated from (14).
If we are indeed in this case then it can be shown that the 6j-symbol has the asymptotic value:

$$\begin{Bmatrix} a & b & c \\ d & e & f \end{Bmatrix} \sim \frac{1}{\sqrt{12\pi V}} \cos\left(\sum J_{hk} \theta_{hk} + \frac{\pi}{4} \right) \tag{25}$$

where V is the volume, θ_{hk} is the internal dihedral angle hinging on the side J_{hk}. It is intended that the value of that J which corresponds to a, say J_{12} is $J_{12} = a+\frac{1}{2}$. This extra $\frac{1}{2}$ is needed because of the quantum fluctuations of the angular momentum and is well-known in the semiclassical theory.
If we let one side vary gradually till V becomes negative than we enter the quantum region of the 6j. The values of the 6j are no longer oscillating and decrease exponentially when we move away from the classical region.
Consider now the generic 3nj-symbol. According to the general theory it can be expressed directly in terms of the 6j-symbol, thus bypassing the 3j-symbol.
This can be understood as follows. For simplicity I assume that the generic 3nj-symbol T triangulates a topological sphere S^2.
I introduce a 3-dimensional triangulation of the interior of the S^2 whose boundary is just the original 3nj-symbol.
In this way I have to introduce a collection of tetrahedra filling the interior of S^2, as well as new edges of these simplexes. I use the label ℓ for the external edges, x for the internal ones. Once the topology, the x's and the l's are given the triangulation is completely determined and we can write the collective amplitude for it as defined by;

$$A = \prod_{\forall \text{ tetrahedra}} \{ \ell, x \} \tag{26}$$

where the product is extended to all simplexes. It can the be show that there is a formula of the kind:

$$T(\ell) = \sum_{\text{all } x\text{'s}} A(\ell, x) (-1)^{\Phi} \tag{27}$$

yielding the global 3nj-symbols. Here the (-1) is a well defined phase, linear in the l's and the x's, whose exact definition depends on the topology of the lattice.
If we expand asymptotically the formula (27) we see therefore that it appears as the product of terms of the form:

$$\cos\left(\sum j_{hk}^{\alpha}\,\theta_{hk}^{\alpha}+\frac{\pi}{4}\right) \qquad j_{hk}=l,x \tag{28}$$

α labels simplexes

times some slowly varying volume factor (which is anyway essential for a complete discussion). Let us concentrate on a given internal edge x. It will belong to a subset of simplexes and it will appear in the total sum (27) as:

$$\prod_{\alpha} \cos\left(\left(x+\tfrac{1}{2}\right)\theta_{12}^{\alpha}+\text{etc}\right)(-1)^{\phi} \tag{29}$$

If we expand the cosines into positive and negative frequencies we see that the coefficent of, in the exponent is given by:

$$\sum_{\alpha} \theta_{12}^{\alpha}\,(\pm 1) \tag{30}$$

The term with all positive frequencies is the just the sum:

$$\sum_{\alpha} \theta_{12}^{\alpha} \tag{31}$$

which differ from 2π by the defect angle, and this represents the curvature. It can be shown that the extra phase accounts for the 2π term. Moreover it is con eivable that the different choices of signs really correspond to gluing of simplexes with different orientations and that they come naturally into play is we really want to sum on all lattices. The final result really look like a summation on all x's of the quantity:

$$e^{\,i\sum_{k} x_k\,\epsilon_k} \tag{32}$$

which is the discrete Einstein-Hilbert action.
Therefore the problem of analizing the behaviour of the generic 3nj-symbol is closely related to the sum (32) and to the quantum theory of the

gravitational field in 3 dimensions.
The analogy is unsatisfactory in many ways. It tells us nothing interestin for the more cogent 4-dimensional case. Finally the summation (32) over histories is really a euclidean theory and we expect a real Boltzmann factor at the exponent. Therefore the summation (32) really deals with imaginary temperatures or coupling constants, should we go back to the 2+1 theory.
In spite of this the formal structure of the summation in (32) is quite interesting. There is no need to impose the triangular inequalities on the angular momenta, nor the volume inequality. The corresponding symbols vanishes anyway automatically.
The restriction to imaginary coupling constant may not be tragical if we succed to control enough of the analytical domain to be able to carry analytic continuations.
Also it hints at the possibility to define a 4 dimensional symbol, function of lenghts (or perhaps areas) defining the analog of a 4 metric. The symbol would be a sort of amplitude for a domain in 4 dimensional space, just as the 6j-symbol is the amplitude for a quantum tetrahedron. An attempts to move into 4 dimensions hare met formidable difficulties. The 6j-symbol is a "selfquantizing" amplitude.
If we fix 5 angular momenta and move the sixth outside the classical range we have seen that the corresponding amplitude decreases exponentially us we move into the quantum region.
This decrease holds only if the 5 momenta satsfy the ordinary selection rules for the angular momentum, in particular they must be quantized in integer or half integer units of $\hbar$.
This really implies again the quantization of all angular momenta, else the amplitude would be meaningless.
A similar reasoning done on the 4 dimensional action leads to infamous difficulties. The defect angles ϵ_k are given only modulo 2π and in order that the exponential $\sum L_k \epsilon_k$ be uniquely defined we see that the values of the areas L_k of the hinges must be quantized; we expect therefore a result in many ways analog to the 3 dimensional theory, where now areas are quantized instead of edges. But then these areas should be the independent variables defining the metric instead of the edges. At first this seems to be appealing, a 4 dimensional simplex has indeed 10 edges and as many areas. The algebraic relation between these two description is far from trivial and probably very intricated. Already the problem of expressing the square measure of the simplex in terms of areas only appears to be very difficult.
Finally the fundamental "brick" from which we should build the 4 dimensional geometry may not be a simplex at all. If we look for a 4 dimensional Einstein metric interpolating the boundary of the simplex into the interior we see clearly that this metric is trivially the flat one,

the reason is that any 4-simplex can be embedded into a flat euclidean or pseudoeuclidean space. This need not to be true for an object having many more vertices n edges and faces. Such an object will need at least a 6 dimensional space to be imbedded and the corresponding interpolating Einstein metric would not be trivial. Only if we attempt to construct the quantum amplitude for such an object we have any possibility to see any non trivial metric.

References

(1) - S.W. Hawking, in "General Relativity - An Einstein centenary survey", edited by S.W. Hawking and W. Israel, Cambridge University press (1979).

(2) - T. Regge, Nuovo Cimento 19 (1961) 558.

(3) - M. Rocek and R.M. Williams, Phys. Lett. 104B (1981) 31 and Z. Phys. C21 (1984) 371, and in "Quantum Structure of Space and Time", edited by M.J. Duff and C.J. Isham (Cambridge University press 1982).

(4) - J.A. Wheeler, in "Relativity, Groups and Topology", edited by C. De Witt and B. De Witt (Gordon and Breach, New York 1964); C.Y. Wong, J. Math. Phys. 12 (1971) 70.

(5) - A. Das, M. Kaku and P.K. Townsend, Phys. Lett. 81B (1979) 11

(6) - L. Smolin, Nucl. Phys. B148 (1979) 333.

(7) - C. Mannion and J.G. Taylor, Phys. Lett. 100B (1981) 261.

(8) - J. Cheeger, J. Diff. Geom. 18 (1983) 575.

(9) - J. Cheeger, W. Muller and R. Schrader, Comm. Math. Phys. 92 (1984) 405.

(10)- H. Coxeter, "Regular Polytopes", (Mathuen & Co. Ltd, London 1948).

(11)- R. Friedberg and T.D. Lee, Columbia University preprint CU-TP-277 (March 1984).

(12)- T.D. Lee, "Discrete Mechanics", (1983 Erice Lecture Notes), Columbia University preprint CU-TP-267 (March 1984), and references therin.

Aspects of Supergravity Theories.

G. W. Gibbons.
D. A. M. T. P., Silver Street,
Cambridge, CB3 9EW, United Kingdom.

Contents.

Lecture 1: An Introduction to Anti-de Sitter Space

Contents:

1) Motivation for studying AdS_4.

2) The Anti-de Sitter Group SO(3,2) and its contraction to the Poincare Group

3) The Supersymmetric Extension of SO(3,2)-Osp(4|N).

4) The necessity of a non-positive cosmological constant from spontaneous compactification.

5) A description of AdS_4.

6) The Stability and Uniqueness of AdS_4.

7) Conclusion.

1) Motivation

There have been 3 principal motivations for studying Anti-de Sitter spacetime (AdS_4).

a) The historical reason was an attempt to generalize the familiar Poincare covariant quantum field theory to curved spacetime while at the same time retaining the crutch of a spacetime symmetry group. Since the only groups which contract (in the Wigner Ioniu [1] sense) to the Poincare group are O(3,2) and O(4,1) and since only O(3,2) has "positive energy" representations this - at least as Fronsdal [2] was concerned - singled out O(3,2). Thus Dirac [3] first studied the electron wave equation in de Sitter space and this was later clarified by Lee and Gursey [4]. Fronsdal has developed, over the years, a considerable body of work relating to elementary particles in AdS_4 [5] and particular aspects such as time ordering [6] and Goldstone's theorem [7] have been described by Castell.

b) With the realization that gauged extended supergravity required that the cosmological constant Λ and the gauge coupling constant g by related to [8]

$$\Lambda = -3g^2/4\pi G$$

it became clear that the appropriate background for these theories was (AdS_4) and Avis, Isham and Storey [9] embarked on a systematic study of quantum field theory in AdS_4. The definition of energy in $(AdS)_4$ was given by Abbott and Deser [10] and applied to stability problems in gauged extended supergravity by Breitenlohner and Freedman [11, 12] and Gibbons, Hull and Warner [13].

c) Current interest in spontaneous compactification lead Freund and Rubin [14] to find a solution of the 11-dimensional supergravity theory of the form

$$M = M_4 \times B_7$$

where B_7 is a compact internal 7-manifold and M_4 turned out to be

AdS_4. This last fact is no accident I shall point out in section 4) that if one assumes certain reasonable energy conditions on the stress tensor in higher dimensions a negative cosmological constant almost <u>inevitably</u> results.

2) <u>O(3,2) and its contraction to Poincare</u>

One usually defines O(3,2) as the group leaving invariant the quadractic form

$$\eta_{AB} X^A X^B$$

$A = 0,1,2,3,4,$ $\quad \eta_{AB} = \text{diag}\,[+1,-1,-1,-1,+1]$. The 10-Killing vectors.

$$-K_{AB} = X_A \partial_B - X_B \partial_A$$

satisfy the Lie Algebra:

$$[K_{AB}, K_{CD}] = K_{AC}\eta_{BD} + K_{BC}\eta_{AD} - K_{AD}\eta_{BC} - K_{BD}\eta_{AC}$$

In quantum theory we prefer to think of the quantum generators

$$M_{AB} = -i\,K_{AB} \qquad\qquad = -M_{BA}$$

which satisfy

$$+\,i[M_{AB}, M_{CD}] = M_{AC}\eta_{BD} + M_{BC}\eta_{AD} - M_{AD}\eta_{BC} - M_{BD}\eta_{AC}$$

The quadratic casimir $\frac{1}{2}M_{AB}M^{AB}$ defines a metric on the Lie Algebra of the group which may be transferred to the group by right translation to give the bi-invariant Killing metric (see e.g. Helgasson). This metric is non-degenerate, so O(3,2) is semi <u>simple</u>, and has signature ++++ ------ since the "compact" generators, (time translation and rotations)

$$M_{04} \quad \text{and} \quad M_{ij}, \qquad i,j = 1,2,3$$

appear with positive sign and the "non-compact" generators, M_{04} and M_{4i}, with a negative sign.

As quantum operators on a Hilbert space with positive norm M_{AB} should be Hermitean. In a finite dimensional representation however the compact generators are Hermitean and the non-compact generators skew Hermitean with negative sign. [In fact O(3,2) is not merely semi-simple but simple]. The connected component of the group, $O_o(3,2)$, thus had topology SO(2) x SO(3) x $\mathbb{R}^6$ [15], the maximal compact subgroup being SO(2) x SO(3). Since time translations, M_{04}, are compact generators we encounter the phenomenon of periodic time unless we pass to a covering space. This will become clearer in a moment when we consider AdS_4 itself.

To recover the Poincare group we must perform a group contraction by rescaling the generators appropriately setting: $M_{\mu 4} = aP_\mu$, where a has the dimension of length (or mass $^{-1}$) the algebra becomes:-

$$-i[P_\mu, P_\nu] = \eta_{44} M_{\mu\nu}/a$$

$$-i[M_{\alpha\beta}, M_{\gamma\delta}] = M_{\alpha\gamma}\eta_{\beta\delta} + M_{\beta\gamma}\eta_{\alpha\delta} - M_{\alpha\delta}\eta_{\beta\nu}M_{\beta\delta}\eta_{\alpha\gamma}$$

$$-i[M_{\gamma\delta}, P_\alpha] = \eta_{\alpha\delta}P_\gamma - \eta_{\alpha\gamma}P_\delta$$

Now let $a \to \infty$. We recover the familiar Poincare group. Further we see that since $M_{04} = aP_0$ the indentification of M_{04} with time translations is reasonable.

Everything we have said so far applies also to the de Sitter group - O(4,1) except η_{44} changes sign so M_{04} becomes a non-compact generator while the M_{4i} become compact generators and the group has topology SO(4) x $\mathbb{R}^4$ with maximal compact subgroup SO(4). According to Levy-Nahas [18] these are the only two groups which contract to the Poincare group.

What does distinguish the groups is the nature of the spectrum of the energy operator M_{04}. For O(4,1) the map: U:

$$\begin{array}{ccc} X^0 & \to & X^0 \\ X^4 & \to & -X^4 \\ X^i & \to & -X^i \end{array}$$

is in the connected component of the identity $O_o(4,1)$.

Since the M_{AB} transforms under the adjoint representation the transformed M_{04} is:

$$U M_{04} U^{-1} = -M_{04}$$

Thus if $|E\rangle$ is a state in the Hilbert space of states with energy E

$$M_{04} \mid E\rangle = E|E\rangle$$

$U \mid E\rangle$ is a state with energy $-E$. Thus the notion of positive energy makes no sense for $O(4,1)$ as first pointed out by Wigner and Phillips [16]. This lack of positivity is intimately connected with the existence of an <u>event</u> <u>horizon</u> in deSitter spacetime and the associated Hawking radiation. AdS_4 has, as we shall see, no event horizon and a zero Hawking temperature. For AdS_4 U is <u>not</u> in the connected component of the identity $(U\epsilon O(3,2); U \neq O_o(3,2))$. The representations of $O(3,2)$ have been discussed by a number of authors [17]. A good pedagogic discussion is contained in the last paper of ref [17]. It is convenient to decompose an irrep of $O_o(3,2)$ into a sum of irreps of the maximal compact subgroup $O(2) \times SO(3)$. Reps of $SO(2) \times SO(3)$ are given by 3 labels, a positive integer and a <u>sign</u> for $SO(2)$ (i.e. whether the "frequency", or in our case the energy, is positive or negative) and the spin, j, of the $SO(3)$ rep. One may now label the reps of $O_o(3,2)$ by the lowest weights, giving the so called $D^{\pm}(E_o,s)$ representations when E_o is the lowest energy (necessarily an integer) and s the lowest spin For example if $s = 0$, the angular momentum operator $\frac{1}{2}M_{ij}M^{ij}$ and the energy M_{04} for the remaining $SO(2) \times SO(3)$ reps take the following values:

$D^{\pm}(E_o,0)$ $\qquad j = 1,2,3,\ldots\ldots$

$$E = E_o + 2n + j, n = 0,1,2,3$$

The "massless" representations $D(s+1,s)$ are of special interest and have

$$E = 1 + n + j, \quad n = 0,1,2,\ldots\ldots$$

$$D(s \pm 1,s) \quad j = s, s+1, \ldots\ldots\ldots$$

If one considers the full group $O(3,2)$, rather than its connected components $O_o(3,2)$ one encounters another operator the <u>antipodal</u> <u>map</u> J, which sends $X^A \to -X^B$

$$J \ : \ X^A \to -X^A$$

Clearly J commutes with all elements of $SO(3,2)$ and had eigenvalues ± 1 (since $J^2 = 1$). Thus representations of $O(3,2)$ are labelled by this extra "J -parity" which has no analogue in flatspacetime. One can in addition introduce time reversal T and space parity P.

$$T: \quad \begin{array}{ccc} X^o & \to & -X^o \\ X^4 & \to & x^4 \\ X^i & \to & X^i \end{array}$$

$$P: \quad \begin{array}{ccc} X^o & \to & X^o \\ X^4 & \to & X^4 \\ X^i & \to & -X^i \end{array}$$

However P and J are related by

$$P = V J V^{-1}$$

where $V \epsilon SO_o(3,2)$ is given by

$$V: \quad \begin{array}{ccc} X^o & \to & -X^o \\ X^4 & \to & -X^4 \\ X^i & \to & X^i \end{array}$$

So O(3,2) has just 4 connected components just like the Poincare group i.e. $SO_o(3,2)$ is an invariant subgroup of O(3,2) and

$$O(3,2)/SO_o(3,2)$$

is Klein's "viergruppe" generated by P and T. [For O(4,1) the situation is similar but T and J are in the same connected component.]

As mentioned earlier M_{04} is a compact generator which is why energies are quantized in integer units. It also has the consequence that time is periodic. To overcome this one may pass to a covering space. Then the eigenvalues of M_{04} need not be integers, merely spaced by integers. The fractional part of the energy gives us a new "constant of motion" analogous to the fermion number in Poincare physics as was first pointed out by Wigner [15]. This covering space is not the universal cover since although we have "unwrapped" the SO(2) factor in the group we have not unwrapped the SO(3) factor. To do this we turn to spinors and O(3,2) which is intimately connected with supersymmetry.

3) Supersymmetry and O(3,2)

Using the γ-matrices $\Gamma^A = (i\gamma^\mu\gamma^5, \gamma^5)$ which satisfy

$$\Gamma^A\Gamma^B + \Gamma^B\Gamma^A = \eta^{AB}$$

we discover that the matrices

$$\sigma^{AB} = \tfrac{1}{4}[\Gamma^A, \Gamma^B]$$

satisfy the O(3,2) algebra. We may use the standard 4-dimensional charge conjugation matrix $C = -C^t$ as a charge conjugation matrix for O(3,2) since

$$C\,\Gamma_A\,C^{-1} = -\Gamma_A^t$$

If follows that O(3,2) allows us to define Majorana spinors which, by definition satisfy

$$i\,\bar{\Psi} = \Psi^t\, C$$

where the left hand side is the Dirac adjoint ($= \Psi^+\gamma^o$) and the right hand side the Majorana conjugate. If we think of Ψ as having an index "upstairs" then

$$\Psi \equiv \Psi^a$$
$$\bar{\Psi} \equiv (\bar{\Psi})_a$$

it is convenient to adopt the convention for raising and lowering indices using the charge conjugation matrix, $C_{ab} = -C_{ba}$, which is to contract from top left to bottom right. The Majorana conjugate $(\Psi M)_a = \Psi^b C_{ba}$. If we lower the last spinor index on $\sigma_{AB}{}^a{}_b$, we find that

$$\sigma_{ABab} = \sigma_{AB\ ba}$$

Further in a Majorana representation the γ_α 's are pure imaginary. It follows that σ_{ABab} are real symmetric matrices which establishes that there is a homomorphism between $O_o(3,2)$ and the real <u>symplectic</u> group $Sp(4,R)$. $Sp(4,R)$ consists for real 4×4 matrices preserving a real skew form, i.e. C_{ab}, and its Lie Algebra consists (when indices are lowered using C_{ab}) of symmetric matrices. In fact $Sp(4,R)$ is a 2-fold cover of $O_o(3,2)$. It is convenient to regard $Sp(4,R)$ as the real transformation leaving invariant the skew form in <u>Grassmannian</u> variables

$$C_{ab}\ \theta^a\ \theta^b$$

The <u>orthosymplectic</u> supergroups Osp (4|N) are obtained by demanding that the mixed form obtained by adjoining N commuting variables X^i

$$C_{ab}\theta^a\theta^b + \delta_{ij}\ X^i\ X^j$$

in left invariant. The algebra is thus augmented by generators of $O(N)$, T_{ij}, which rotate the X^i into themselves and some anticommuting generators Q^{ia} which mix Φ^a with X^i. These Majorana spinor generators satisfy the fundamental anticommutation relation

$$\left\{Q^i, \bar{Q}^j\right\} = i\,\delta^{ij}(\sigma_{AB}M^{AB}) + i\,T^{ij}$$

An important consequence is a Positive Energy Theorem which follows by multiplying on the right side γ^0 and tracing over spinor and internal, i,j indices and taking a quantum mechanical expectation value in some state $|S\rangle$. It follows that in that state

$$M_{04} \geq 0$$

with equality if and only if $|S\rangle$ admits the maximal number of supersymmetries. In fact a rather stronger inequality holds, bounding the energy by the angular momentum [13, 30].

It is of some interest to ask what happens in the case of $O(4,1)$. The relevant γ - matrices are

$$\Gamma^A = \left\{\gamma^{\mu}, \gamma^5\right\}$$

but now the charge conjugation matrix becomes

$$C_5 = C\,\gamma_5$$

which satisfies

$$C\,\Gamma_A C_5^{-1} = \Gamma_A{}^t$$

One can convince oneself that the Majorana condition would now lead, an iteration, to the contradictory relation:

$$\bar{\Psi} = -\Psi$$

Thus, in the naive sense of the word, $O(4,1)$ does not admit Majorana spinors. However, this is no obstacle to introducing supersymmetry since we may follow Cremmer [19] and introduce symplectic spinors satisfying

$$\bar{Q}^i = \Omega^{ij} Q^{jt} C_5$$

where $\Omega^{ij} = -\Omega^{ji}, \Omega^{ij}\,\Omega^{jk} = -\delta^{jk}$ is an N x N skew matrix (N must be even). Superalgebras containing O(4,1) as a bosonic parts exist and have recently been discussed by Lukierski and Novickni [20]. The problem is that they cannot be realized on a Hilbert space with positive norm. Suppose that they could. Using the symplectic condition one finds an anticommutator of the form:

$$\left\{Q^{ia}, \bar{Q}^j_b\right\} = \Omega^{ij}(\sigma_{AB} J^{AB})^a_b + \delta^a_b\, T\, \Omega^{ij}$$

If one now multiplies on the right by γ^o and traces the left hand side is positive semi-definite and the right hand side vanishes. Thus either

$$Q^{ia} \mid S\rangle = 0$$

for all states and hence $J_{AB} = 0$ or we have an indefinite norm in Hilbert space. The recent paper of Pilch et. al. [21] which appeared after this school, confirms that this latter possibility actually occurs in the N = 2 case.

Representations of supersymmetry in AdS were first worked out by Heidrenreich for N = 1. The higher N case is dealt with by Freedman and Nicolai. The last reference of [17] contains a nice account with full references.

4) Spontaneous Compactification and AdS_4

Let us now turn to the question of spontaneous compactification. We seek a solution of some higher dimensional field equations which is topologically a product

$$M_n = M_4 \times B$$

where B is a compact "internal" space and M_4 is a locally Lorentz invariant homogeneous, spacetime i.e. Anti de Sitter space, de Sitter space or preferably Minkowski spacetime. Local lorentz invariance and homogeneity now require that the metric on M_n is that of a "warped product", i.e.

$$dS^2 = g_{mn}(y)\, dy^m dy_n + W^2(y) g_{\alpha\beta} dX^\alpha dX^\beta$$

where $W \neq 0$, y are the internal coordinates and X the spacetime coordinates – $\alpha,\beta = 0,1,2,3,$; $m,n = 5,\ldots..n$. This form of the metric is more general than the usual product metric for which $W = 1$, but is equally compatible with fundamental physical principles. The non-vanishing components of the Ricci tensor are then

$${}^{(n)}R_{\alpha\beta} = {}^{(4)}R_{\alpha\beta} - \frac{1}{4} W^{\frac{1}{4}} \nabla_m \nabla^m (W^{-\frac{1}{4}}) g_{\alpha\beta}$$

$${}^{n}R_{mn} = R_{mn} - \frac{4}{W} \nabla_m \nabla_n W$$

The Einstein equations in the higher dimensional space are

$$R_{AB} = 8\pi G \left\{ T_{AB} - \frac{1}{n-2} g_{AB} T^C_C \right\}$$

$A = 0,1,2,3,5,\ldots n$, where T_{AB} is the stress tensor in higher dimensions which we expect to obey the following Energy conditions:

1) <u>The Weak Energy Condition</u>: $T_{AB} t^A t^B \geqslant 0$ for all non-spacelike t_A, i.e. $T_{o\dot{o}} \geqslant 0$ in all frames. This is the obvious positive energy density condition. 2) <u>The Dominant Energy Condition</u>: $T_{AB} t^B$ is future pointing and non-spacelike for all future pointing non-spacelike t^B. This was introduced by Hawking [22] and is equivalent to the statement that

$$T_{oo} \geqslant | T_{AB} |$$

in an orthonormal frame. It basically says that mater moves subluminally and is used in the positive mass theorem.

3) <u>The Strong Energy Condition</u>: $(T_A - \frac{1}{n-2} g_{AB} T^C_C)\, t^A t^B \geqslant 0$ for all non-spacelike t^A. This basically says that gravity is attractive and is

used in the singularity theorems.

Whereas the weak and the dominant energy conditions hold for all known classical stress tensors with satisfactory physical properties the strong energy condition breaks down for scalars for example, since

$$T_{AB} - \frac{1}{n-2} g_{AB} T^C_C = \phi_{,A}\phi_{,B} - \frac{2}{n-2} V(\phi) g_{AB}$$

so if $V(\phi) > 0$ it may be violated. Even a standard mass term $V = \frac{1}{2}m^2\phi^2$ can violate it.

All these conditions may be further strengthened by the requirement that equality requires $T_{AB} = 0$ and the field configurations to be trivial. We call such conditions "strict". An immediate consequence of the strong energy condition is obtained by integrating the ${}^{(n)}R_{oo}$ equation over the compact manifold B.

Since

$$^{n}W^{-\frac{1}{4}} R_{oo} = {}^{(4)}W^{-\frac{1}{4}} R_{oo} - g_{oo} \nabla^2 (W^{-\frac{1}{4}})$$

we get

$$-\int W^{-\frac{1}{4}} = \int W^{-\frac{1}{4}} (T_{oo} - \frac{1}{n-2} T^C_C)$$

so $\Lambda \leqslant 0$.

Thus Λ is negative or zero. Zero is only allowed if the strict strong energy condition fails to hold. One can consider as an example p- form potentials.

$$L = \frac{(-1)^{p-1}}{2p!} F_{AB....S} F^{AB....S} \surd g$$

$$T_{AB} - \frac{1}{n-2} g_{AB} T^C_C = \frac{(-1)^{p-1}}{(p-1)!}$$

$$F_{A....S} F_B^{....S} + (p-1)\frac{(-1)^p}{p!}(n-2) g_{AB} F_{m....S} F^{m....S}$$

$$T_{oo} - \frac{1}{n-2} g_{oo} T^C_C$$

contains the dangerous term

$$(-1)^{p-1}\frac{p}{p!} F_{o....S} F^{o....S}$$

$$- \frac{p(p-1)}{p!(n-2)} F_{o...S} F^{o...S} (-1)^{p-1}$$

Thus the strong energy condition holds unless $n > p + 1$.
Thus if

1) $n = p$ which by duality is equivalent to <u>scalars</u>

or

2) $n = p + 1$ which is the case considered by Aurilia-Nicolai and Townsend [23] in which

$$F_{AB.....S} = f\, \epsilon_{AB.....S}$$

$$T_{AB} = \left(\frac{n-2}{2n}\right) f^2\, g_{AB}$$

An interesting case of equality has been pointed out by Gell-Mann and Zwibach [24]. They chose a non-linear sigma model lagrangian

$$L = ½\, g_{AB}\, \partial_A \phi^i \partial_B \phi^j G_{ij}(\phi^k)$$

where $G_{ij}(\phi)$ is the metric on an Riemannian space C with coordinates $\phi^i(x)$. The Einstein equations are

$$R_{AB} = 2\kappa^2\, \partial_A \phi^i\, \partial_B \phi^j\, G_{ij}(\phi)$$

The trick is now to identify B and C. This is accomplished by setting

$$\phi^i = y^m$$

so

$$R_{mn} = 2\kappa^2 G_{mn}$$

$$R_{\alpha\beta} = 0$$

This gives a flat Minkowski spacetime and the internal space is an Einstein space. This model is not realistic but it does illustrate what is required to avoid having $n < 0$. In fact allowing $T_{oo} - \frac{1}{n-2} g_{oo} T^c_c = 0$ is, in effect, allowing antigravity (i.e. no gravity; not negative gravity), since such sources would generate no gravitational field in spacetime. No supergravity model has this property and we are almost inevitably led to a negative cosmological term.

5. (AdS_4) and its Uniqueness

In the last section we saw that Λ is generally negative in Kaluza-Klein theories. Thus AdS_4 is the natural background or ground state. I shall begin by describing $(AdS)_4$ and then discussing its stability and uniqueness. AdS_4 has constant curvature

$$R_{\alpha\beta\gamma\delta} = -\frac{\Lambda}{3} (g_{\alpha\gamma} g_{\beta\delta} - g_{\alpha\delta} g_{\beta\gamma})$$

which follows from the local lorentz invariance about each point and isometry group SO(3,2). We may thus think of AdS_4 as the coset space SO(3,2)/SO(3,1) or more concretely as the hyperboloid

$$(X^o)^2 + (X^4)^2 - (X^1)^2 - (X^2)^2 - (X^3)^2 = 1$$

(in units such that $\Lambda = -3$). The generator M_{04} in SO(3,2) generates rotations in the $X^o - X^4$ plane and so time will be periodic with the consequent closed timelike loops unless we pass to the universal covering space $(CAdS)_4$. Since time translations act freely and continuously on AdS_4 we can define a positive direction of time - thus AdS_4 is time orientable and in fact space orientable (since it clearly inherits an orientation from 5-dimensions). Using the angle $t = \tan^{-1} (X^o/X^4)$ we can still define a "time ordering" by deeming the later of two events to be that which has a smaller t value. If differs from the usual concept in not being transitive. If A is earlier than B and B earlier than AC A may nevertheless be later than C! However it seems to be what is required in quantum field theory in AdS_4 [6,9]. We could, if we wished identify points under the antipodal map J defined earlier, $X^A \to -X^A$. The resultant space would still be time orientable (since J commutes with the action of M_{04}) but not space orientable. Incidentally the conformal group of AdS_4 is SO(4,2), but the extra conformal motions do not commute with J so AdS_4/J has no conformal motions. One way of viewing the boundary conditions necessary for supersymmetry on AdS [11,12] is to regard the fields as being defined on AdS_4/J.

We may coordinalize AdS_4 in a number of ways. The static frame:

$$X^0 = (1 + r^2)^{1/2} \quad \sin t$$

$$X^4 = (1 + r^2)^{1/2} \quad \cos t$$

$$X^i = rn^i, \qquad n^i n^i = 1$$

which leads to

$$ds^2 = (1 + r^2)\, dt^2 - \frac{dr^2}{1 + r^2} - r^2(d\theta^2 + \sin^2\theta d\phi^2)$$

with $0 \leqslant t \leqslant 2\pi$ on $(AdS)_4$

or we may choose not to identify time, $-\infty < t < \infty$ and work on the universal covering space $CAdS_4$. The action of J is now given by $t \to t + \pi$, $\theta \to \pi - \theta$, $\phi \to \phi + \pi$. Since $g_{00} = (1 + r^2) \neq 0$ AdS_4 is globally static and contains no horizons. Thus $\frac{\partial}{\partial t}$ is everywhere timelike. This should be contrasted with the de Sitter space for which $g_{00} = (1 - r^2)$. The Killing vector $\frac{\partial}{\partial t}$ is timelike and future directed inside the Cosmological Horizon at $r = 1$ but spacelike or timelike past directed outside it [see 25]. The antipodal map J would interchange the regions where $\frac{\partial}{\partial t}$ is future directed with those where it is past directed in that case. That is geometrically why positivo energy is not possible in the de Sitter case. Incidentally the spatial compactness in the de Sitter case also precludes non zero mass or supercharge for the spacetime since these are 2-surface integrals at infinity and no such infinity exists in that case.

Another useful set of coordinates are the horospheric or quasi euclidean coordinates

$$X^4 = t/x$$

$$X^3 = z/x$$

$$X^2 = y/x$$

$$X^o + X^1 = \frac{1}{x}$$

$$X^o - X^1 = x + (y^2 + z^2 - t^2)/x$$

in which

$$ds^2 = \frac{1}{x^2}(dt^2 - dx^2 - dy^2 - dz^2)$$

The action of J is to send $x \to -x$. The surface $x = 0$ is a coordinate singularity. One may use $t =$ constant as a time surface provided one takes either $x > 0$ or $x < 0$. These are useful if one wants to use fourier series representations [26]. Geometrically the flat Minkowski planes $x =$ constant are rather interesting and are called horospheres. These coordinates make manifest the 3-dimensional Poincare subgroup of the isometry group SO(3,2). The remaining elements of SO(3,2) are conformal motions of 3-dimensional Minkowski space whose conformal group is SO(3,2).

Finally we may regard $(AdS)_4$ as an (unphysical) cosmological model in Robertson-Walker coordinates

$$X^o = \cos t \quad , \quad X^4 = \sin t \cosh \chi$$

$$X^i = \sin t \sin \chi \, n^i$$

in which

$$ds^2 = dt^2 - \sin^2 t(d\chi^2 + \sin^2\chi(d\theta^2 + \sin^2\theta d\phi^2))$$

The apparent singularities at $t = 0$ are entirely fictional and due to focussing of the geodisic timelines by the attractive cosmological term. Addition of a spatially uniform matter distribution would in general however lead to physical spacetime singularities. Thus to work in AdS_4 we must impose boundary conditions which require fields to fall off at large distances [see 11, 12, 27, 28]. This might be a problem if one wished to indulge in Kaluza-Klein Cosmology.

To understand infinity in AdS_4 it is useful to compactify conformally by setting $R = \tan\rho$, $\Omega = \frac{1}{\cos}\rho$. The metric becomes

$$ds^2 = \Omega^2(dt^2 - d\rho^2 - \sin^2\rho(d\theta^2 + \sin^2\theta d\phi^2))$$

where the metric in the brackets is that of the Einstein Static Universe – i.e. the product metric on $\mathbb{R} \times S^3$. Since $\Omega = \infty$ at $\rho = \frac{\pi}{2}$, AdS_4 maps conformally onto (one half of S^3) x R. The boundary $\rho = \frac{\pi}{2}$ is called I and has the topology $S^2 \times \mathbb{R}$, the S^2 being the equatorial 2-sphere of S^3.

We can now consider metrics which tend to AdS_4 at infinity. The metric is conformal to a regular metric with smooth timelike boundary such that

$$\Omega = 0$$

and

$$\partial_\alpha \Omega \neq 0 \qquad \text{at } I$$

We must also require that $O(3,2)$ act on the boundary at infinity as its conformal group, and that the matter tensor vanishes at least as fast $\frac{1}{r^3}$ near I, where r is an area coordinate. An example of such an asymptotically anti-de Sitter metric is Schwarzchild – Anti-de Sitter:

$$ds^2 = (1 - \frac{2GM}{r} - \frac{\Lambda r^2}{3})dt^2$$

$$(1 - \frac{2GM}{r} - \frac{\Lambda r^2}{3})^{-1} dr^2 - r^2(d\theta^2 + \sin^2\theta d\phi^2)$$

which satisfied $R_{\alpha\beta} = -\Lambda g_{\alpha\beta}$. The boundary conditions at infinity have been discussed in detail in [11, 12, 27, 28].

6) The Stability and Uniqueness of AdS_4.

The quantity M in the Schwarzschild anti-de Sitter metric may be identified with the Abbott Deser mass [10]. It enjoys positivity properties [10, 29, 13] which follow from supersymmetry and may be established by generalizing the Witten argument for positive energy in the asymptotically flat case [see 30' on Freedman's article in this volume for a review]. To establish the posivity one must assume that any matter tensor satisfies the dominant energy condition. If black holes are present one must impose suitable boundary conditions on the apparent horizon [29]. The condition that the mass should vanish requires the existence of 4 Killing spinors satisfying

$$\nabla_\alpha \epsilon + \tfrac{1}{2} \sqrt{(-\tfrac{\Lambda}{3})}\, i\gamma_\alpha \epsilon = 0$$

which implies that the metric is that of $(AdS)_4$. In fact there exist generalized plane wave metrics with killing spinors but they are not asymptotically flat [31]. The positivity of the energy and the uniqueness of the supersymmetric vacuum state indicate that this vacuum is stable both classically and quantum mechanically since there is no other zero energy state for the vacuum to tunnel to.

Another question is whether there are classical "lumps" or "solitons" which tend to the supersymmetric vacuum at large distances but which have some non-trivial structure in the interior. In fact unless one allows horizons the answer is no. There are no classical solitons without horizons which are time independent - i.e. possess an everywhere timelike Killing vector [32]. To prove this we consider the static case where in addition there is time- reversal invariance. The postulated metric is

$$ds^2 = -Vdt^2 + h_{ij}dx^i dx^j$$

where $V \neq 0$ and both V and h_{ij} do not depend on t. The field equations are (in the absence of matter)

$$\nabla_i \nabla^i V = -\Lambda V$$

$$R_{ij} = V^{-1} \nabla_i \nabla_j V + \Lambda h_{ij}$$

where ∇_i is covariant derivative with respect to the Riemannian metric h_{ij} whose Ricci tensor is R_{ij}.
One defines

$$W = \nabla_i V \nabla^i V$$

$$W_o = \tfrac{1}{3}\Lambda (1 - V^2)$$

and the Bach-Cotton tensor R_{ijk} by

$$R_{ijk} = \nabla_k R_{ij} - \nabla_j R_{ik} + \tfrac{1}{4}(h_{ik} \nabla_j R^m_m - h_{ij} \nabla_k R^m_m$$

R_{ijk} vanishes if and only if h_{ij} is conformally flat. W_o is the term of W in the exact AdS_4 case. Thus $W = W_o$, $R_{ijk} = 0$ implies that the metric is (AdS_4). Lindblom has given the following identity:

$$\nabla^i (V^{-1} \nabla_i (W - W^o))$$

$$= \tfrac{1}{4} V^3 W^{-1} R_{ijk} R^{ijk} + \tfrac{3}{4} V^{-1} W^{-1} \nabla_i (W - W^o) \nabla^i (W - W^o)$$

If we integrate this over the surfaces t = constant we get

$$\int_\infty V^{-1} \nabla_i (W - W^o) d\sigma^i \geq 0$$

where the integral is over the 2 surface on I at infinity. Equality of course implies the metric is AdS_4.
Now near infinity

$$W - W_o = 4 \frac{GM}{r} \frac{|\Lambda|}{3}$$

and so the boundary term is

$$-4\pi G|\frac{\Lambda}{3}|\ M$$

Thus unless M actually vanishes, in which case the metric is AdS_4, it must be negative which contradicts the positive mass theorem. Thus there can be no solitons. To complete the proof one must show that if the metric were merely stationary (time independent but not time-reversal invariant) it must be static. This is done by generalizing a well-known Lemma of Lichnerowicz in the asymptotically flat case.

The conclusion is (as is true in ungauged supergravity) that there are no solitons without horizons. In fact such black holes exist in AdS_4 backgrounds but they are unstable because of Hawking radiation. Some aspects of their thermodynamics have been discussed in [33]. It would be interesting to know whether this uniqueness argument generalizes to the case of matter sources.

7) <u>Conclusion</u>.

I have outlined some of the simpler properties of $(AdS)_4$. Clearly it cannot describe our present universe and it does not seem to be directly relevant for the Very Early Universe in contrast to de Sitter space. Its properties are perhaps unfamiliar but not completely unreasonable and it admits a consistent physical interpretation. If gauged extended supergravity models have relevance to physics it should perhaps be regarded as a symmetric phase of the theory that does not happen to be realized, the true phase being a spontaneously broken one. If affords a useful testing ground for various ideas in supergravity. By exploring its properties further we may hope to understand whether the appropriate symmetry breaking may take place. Associated with that may come an understanding, which is required even if one does not expect supergravity to be important, of why the presently observed value of the cosmological constant is so small. Therefore an examination of quantum corrections in AdS_4 is of some importance. Some work on this may be found in [34]

References.

[1] E. Inoniu and E.P Wigner, Proc. Natl. Acad. Sci. U.S., 39, 510 (1953).

[2] C. Frondsal, Rev. Mod. Phys., 37, 221 (1965).

[3] P.A.M. Dirac, Ann. Math., 36, 657 (1935).

[4] F. Gursey and T.D. Lee, Proc. Natl. Acad. Sci. U.S., 49, 179 (1969).

[5] C. Frondsel, Phys. Rev. D., 10, 589 (1974)

C. Frondsel, Phys. Rev. D., 12, 3819 (1975).

C. Frondsel and Hangen, Phys. Rev. D., 12, 3810 (1975).

[6] L. Castell, Nucl. Phys. B., 5, 601 (1968).

[7] L. Castell, Nuovo Cimento, 61A, 585 (1969).

erratum, Nuovo Cimento, 62A, No 2.

[8] D.Z. Freedman and A. Das, Nucl. Phys. B., 120, 221 (1977).

[9] S. Avis, C.J. Isham and D. Storey, Phys. Rev. D., 18, 3568 (1978).

[10] L. Abbott and S. Deser, Nucl. Phys. B., 195, 76 (1982).

[11] P. Breitenlohner and D.Z. Freedman, Phys. Letts. B. 115, 197 (1982).

[12] P. Breitenlohner and D.Z. Freedman, Ann. Phys. (N.Y.), 144, 249 (1982)

[13] G.W. Gibbons, C.M. Hull and N.P. Warner, Nucl. Phys. B., 28, 173 (1983).

[14] P. O. Freund and R. A. Rubin, Phys. Lett. B., 97, 233 (1980).

[15] E. P. Wigner, Proc. Natl. Acad. Sci. U. S., 36, 184 (1950).

[16] E. P. Wigner in "Group Theory and its Applications" ed. T. U. Philips, New York (1968).

[17] J. B. Ehrman, Proc. Camb. Phil. Soc., 53, 290 (1957).

P. A. M. Dirac, J. M. P., 4, 901 (1963).

N. T. Evans, J. M. P., 8, 170 (1967).

H. Nicolai, "Representation of Supersymmetry in Anti-de Sitter Space", CERN preprint, Th 3882.

[18] M. Levy-Nahas, J. M. P., 8, 1211 (1967).

[19] E. Cremmer in "Superspace and Supergravity", eds. S. W. Hawking and M. Rocek, publ. by Cambridge University Press.

[20] J. Lukierski and A. Novicki, Wroclaw preprint, 609 (1984).

[21] K. Pilch, P. Van Nieuwenhuizen and M. F. Sohnius, preprint I. T. P., SB-84-46.

[22] S. W. Hawking, Comm. Math. Phys.

[23] A. Aurilia, H. Nicolai and P. K. Townsend, in "Superspace and Supergravity", eds. S. W. Hawking and M. Rocek, publ. Cambridge University Press.

[24] M. Gell-Mann and B. Zwiebach, Phys. Letts. B., 141, 333 (1984).

[25] G. W. Gibbons and S. W. Hawking, Phys. Rev. D., 15, 2738 (1977).

[26] Castell, J. Math. Phys., 11, 1669 (1969).

[27] S. W. Hawking, Phys. Letts. B., 126, 175 (1983).

[28] A. Ashtekar and A. Magnon, Class.Q.Grav., 1, L39 (1984).

[29] G.W. Gibbons, S.W. Hawking, G. Horowitz and M.J. Perry, Comm.Math.Phys., 88, 295 (1983).

[30] G.W. Gibbons in Proceedings 4th Silarg Symposium, ed. C. Aragone, World Scientific (1984).

[31] W. Boucher, G.W. Gibbons and G. Horowitz, to be published.

[32] G.W. Gibbons and S.T.C. Siklos, in preparation.

[33] S.W. Hawking and D.N. Page, Comm.Maths.Phys., 87, 577 (1983).

[34] B. Allen and S. Davies, Phys.Lett., 124B, 353 (1983)

G.W. Gibbons and H. Nicolai, CERN preprint TH. 3185.

T. Inami and K. Yamagishi, Kyoto preprint, UT KOMABA-84-8.

N. Sakai and Y. Tanii Tokyo preprint TIT/HEP-80

Lecture 2: Black Hole Solitons in Ungauged Extended Supergravity

Contents

1: <u>The soliton concept for gravity.</u>

In flat space physics we expect of a soliton or in Coleman's famous phrase a "classical lump" that it be a

a) time independent (static or stationary)

b) finite (positive) energy

c) non-singular

d) classically and semi-classically stable
solution of the classical equations of motion which frequently

e) spatially interpolates between different vacua.

The "vacua", since they contain nothing, must be space-time homogeneous solutions of the classical equations which may or may not be stable. If the soliton spatially interpolates between one vacuum in the interior and one at infinity the vacuum at infinity should be stable. Energy is of course measured relative to the vacuum at infinity. Since the soliton is an excitation above that vacuum or ground state it should have positive energy and in order to have predictable dynamics it should be non-singular. If the soliton is to persist in time and possess particle-like properties it should itself be stable both classically and quantum mechanically.

Solitons frequently possess conserved, sometimes topological, charges and their stability arises because they minimize the energy for a fixed value of the charge. The fact that they are stationary usually implies that they extremize the energy. In a supersymmetric theory the charge is often a central charge and the configurations extremizing it a partially supersymmetric one. The solitons then fit into supermultiplets. Because of the partial supersymmetry these multiplets are "shortened". They are smaller than the usual massive multiplets. The archetypal example is that in 2-dimensions discussed by Witten and Olive [1]. The most interesting 4-dimensional example are the monopoles which are reviewed in [2].

If we apply these ideas to a gravity theory it is natural to demand of a soliton that it be a complete and non-singular stationary spacetime which is asymptotically flat and solves the relevant field equations. Actually it is too strong a restriction to demand non-singularity. Because of the well-known singularity theorems of Penrose and Hawking almost all interesting solutions will contain singularities. However provided these singularities lie inside event horizons the dynamics of the spacetime outside the horizon will be predictable. Thus we will be forced to assume the truth of the Cosmic Censorship Hypothesis [3].

An important role in our considerations will be played by the Positive Mass Theorem [4, 5] which states that for an asymptotically flat spacetime with a complete Cauchy surface and whose stress tensor satisfies the dominant energy condition that the mass M is non-negative and vanishes if and only if the spacetime is flat. This result can be extended to spacetimes with singularities provided they contain a Cauchy surface which is complete outside an inner boundary which is an apparent horizon [6]. An apparent horizon on a spacelike 3-surface is compact 2-surface whose outward null l^{α} normals have everywhere vanishing convergence, $l^{\alpha}{}_{;\alpha} = 0$.

We shall see in the next section that there are at least in the ungauged supergravity models, no solitons without horizons. The existence of horizons raises the question of the quantum mechanical stability of our solutions since the work of Hawking [7] shows that horizons will, in general, emit thermally because of pair creation by the gravitational field. This will in general render unstable quantum mechanically solutions which are stable classically. Thus the requirement of quantum mechanical stability is additional to the purely classical one. An analogy might be to the dyons in Yang-Mills theory which may lose charge by pair creation.

An interesting discussion of the soliton concept in gravity theory in which these issues are discussed is [7]

2: Absence of solitons without horizons.

The ungauged extended supergravity theories all possess an important symmetry, the classical equations are invariant under a spacetime independent dilation. That is the metric $g_{\alpha\beta} \to \lambda^2 g_{\alpha\beta}$, the spin 1 field strengths $F^i_{\alpha\beta} \to \lambda F^i_{\alpha\beta}$ and the scalar fields $\phi^A \to \phi^A$. The classical Lagrangian density $L \to \lambda^2 L$. This symmetry arises because the particles are massless. It is broken if one gauges the theory, i.e. the field strengths become non-Abelian since the unit of charge introduces a scale. (Remember classically the coupling constant has a dimension $\sqrt{\hbar}$.) Such a rescaling takes a classical configuration of mass M to one of mass λM, since by suitable rescaling the coordinates one can obtain a new asymptotically flat metric which tends to the flat metric $\eta_{\alpha\beta}$ at infinity. Now any stationary solution extremizes the total energy amongst all possible initial data. This energy amongst all possible initial data. This is a general statement proved by Schutz and Sorkin [8]. Thus the deformation produced by rescaling the metric either $M = \infty$, which is not interesting, or $M = 0$. One may readily check that all ungauges extended supergravity bosonic matter stress tensors satisfy the dominant energy condition. Therefore by the Positive mass theorem the solution must be flat space – the trivial vacuum. We summarize this statement as "No solitons without Horizons". For pure gravity, or Einstein Maxwell theory the result may be proven directly from the stationary field equations [9]. Note that in principle the argument should work in any spacetime dimension.

It is clear from this General Relativistic version of Derrick's Theorem [10] that in order to get an interesting solution of the equations we must break the scale invariance. This maybe done by means of boundary conditions which break scale invariance. This is, in effect, how the Schwarzschild solution

$$ds^2 = (1 - 2\frac{GM}{r})dt^2 - (1 - 2\frac{GM}{r})^{-1}dr^2 - r^2(d\theta^2 + \sin^2\theta d\phi^2)$$

escapes this theorem. This is static and finite, positive mass. Clearly the mass is not extremized since given one solution there exists another with mass $M + \delta M$ say. However such a perturbation will change the area $A = 16\pi G^2 M^2$ of the event horizon. The mass of the black hole data set will only extremize the mass M if one holds fixed the area A of the outermost apparent horizon. Formally one finds that when one tries to prove that the mass is extremized one encounters boundary terms which spoil the argument. Consider, for simplicity the time symmetric vacuum case. The initial data is a 3-metric h_{ij} subject to the constraint

$$h^{ij}\ {}^3R_{ij} = 0$$

The integration being over a spacelike surface with inner boundary the apparent horizon (a minimal 2-sphere in this case). The mass M is read off from the asymptotic metric $h_{ij} \approx \delta_{ij}(1 + \frac{2M}{r}) + O(\frac{1}{r^2})$. Varying M subject to the constraint yields if <u>integration by parts were permissible</u>.

$$\int V h^{ij}\,{}^3R_{ij}\sqrt{h}\, d^3x$$

$${}^3R = 0$$

$${}^3R_{ij} = \nabla_i \nabla_j V$$

when ∇_i is the covariant derivative with respect to the metric h_{ij}. But these are just the static field equations for the 4-metric:

$$ds^2 = V^2 dt^2 - h_{ij} dx^i dx^j$$

Thus as stated above, static solutions extremize the mass provided boundary terms may be discarded. This will be possible if we keep the apparent area of the horizon fixed. In fact there are reasons for believing that an isoperimetric inequality holds for black holes and their apparent horizons, i.e.

$$M \geq \sqrt{\left(\frac{A}{16\pi G^2}\right)}$$

with equality only in the Schwarzschild case. The evidence for this isoperimetric inequality is reviewed in [11]. In any event by Israel's Theorem [12], the Schwarzschild metric is the unique static metric.

regular outside a single event horizon.

One might ask whether black holes should be regarded as classical solitons. They are classically stable, however the mass may grow with time as matter or gravitational radiation falls into the horizon. Indeed classically a theorem of Hawking states that the area of the event horizon (which lies outside the apparent horizon) cannot decrease. Similar remarks apply to the angular momentum of rotating Kerr black holes. An initial data set consisting of gravitational waves can disperse to infinity and collapse to form a black hole. One might ask whether it can do anything else? In particular are the solutions of the classical solutions which are periodic in time? Bound states of gravitational radiation or black holes in periodic orbits perhaps? The answer to this question is no. Because any such system must inevitably radiate gravitational radiation and hence lose mass to infinity periodic motions, other than those with an infinite period are not allowed. this result has recently been shown for pure gravity [13] and it almost certainly holds in the ungauged extended supergravity models. The import of this result is that there are no solutions analogous to the "breathers" of Sine Gordon theory.

In my view it is questionable whether one regards Schwarzschild black holes as classical solitons but it is clear that at the quantum level they are not, as will be explained in the next section.

3: Black Holes and the breakdown of supersymmetry at finite temperature.

In what follows I shall speak only of the Schwarzschild black hole. Similar remarks apply to the rotating Kerr black holes but the formulae are significantly more complicated without any significantly new qualitative features arising. As is well-known by now a black hole of mass **M** acts like a hot body of temperature

$$T = \frac{1}{8\pi GM}$$

This means that it will emit massless particles at a rate inversely proportional to the square of the mass. This will lower the mass and hence raise the temperature leading to an even larger emission rate.

Presumably, although this is controversial, it will ultimately disappear altogether.

The temperature is most easily calculated by noting that if one introduces Kruskal coordinates

$$T = \left(1 - \frac{2GM}{r}\right)^{1/2} \exp\left(\frac{r}{4GM}\right) \sinh\left(\frac{t}{4GM}\right)$$

$$X = \left(1 - \frac{2GM}{r}\right)^{1/2} \exp\left(\frac{r}{4GM}\right) \cosh\left(\frac{t}{4GM}\right)$$

to remove the coordinate singularities at $r = 2GM$ the metric is periodic in the imaginary time coordinate $t = i\tau$ with period

$$\beta = \frac{1}{T} = 8\pi GM$$

Thus a functional integral around this background is one at finite temperature. In fact any boson field which is regular on the horizon must be periodic with β while any fermion field must anti-periodic with period β [14].

This suggests that, like a heat bath in flat spacetime, [15] a black hole should violate supersymmetry, since the constant parameter of global supersymmetry, which should apply at infinity, cannot depend upon imaginary time. To make this idea more precise recall that in $N = 1$ the supersymmetry transformations are

$$\delta e^{a}_{\mu} = -iK\bar{\epsilon}\gamma^{0}\psi_{\mu}$$

$$\delta\psi_{\rho} = \frac{1}{K}\nabla_{\rho}\epsilon$$

where e^{a}_{μ} are the vierbein fields and ψ_{μ} the Rarita-Schwinger field. A purely bosonic background possesses a supersymmetry if there exists a Killing spinor field ϵ such that

$$\delta_{\epsilon}\psi_{\rho} = 0$$

i.e.

$$\nabla_\rho \epsilon = 0$$

There are no such Killing spinors in the Schwarzschild background since the only asymptotically flat spacetime with Killing spinors is flat space.

Thus the Schwarzschild solution has no exact residual supersymmetry. However what is of interest here are global supergauge transformations. We want ψ_μ's which fall of at infinity. Thus the ϵ's should at most tend to constants at infinity. We want to factor the space of all such supergauge transformations ($\epsilon \to \epsilon_o$ at ∞) by those which tend to zero at infinity ($\epsilon \to 0$ at ∞). The resulting equivalence class of supergauge transformations is the global supergauge group and it should be parameterized by giving a constant spinor at infinity. To effect this factorization it is convenient to impose a gauge condition - e.g.

$$\gamma^\alpha \psi_\alpha = 0$$

This fixes ϵ up to solutions of the massless Dirac equation.

$$\gamma^\alpha \nabla_\alpha \epsilon = 0$$

If $\gamma^\alpha \phi_\alpha = 0$ is a good gauge condition there should be no regular solutions of the Dirac equation which tend to zero at infinity and 4 real Majorana solutions which tend to constants at infinity. These 4 real constant solutions - or 2 complex - allow us to make global gauge transformations of our background if the background is static the solutions should be time-independent. The gauged transformation solutions should also be static. These gauged transformed superpartners will fit into supermultiplets multiplets. In the quantum theory for example one would associate with these solutions a spin $\frac{3}{2}$ solutions

$$\psi_\mu^i = \nabla_\mu \epsilon^i$$

where

$$\gamma^\alpha \nabla_\alpha \epsilon^i = 0$$

$$\epsilon^i \to \epsilon^i_o$$

with the $4\psi^i_\mu$ one would have 4 creation operators a^+_i and the original state $|s\rangle$ would be generate with the 16 states:

$a^+_i \, | \, s\rangle$ $\qquad a^+_i a^+_j \, |s\rangle$ $\qquad a^+_i a^+_j a^+_k \, |s\rangle$

$a^+_l a^+_i a^+_j a^+_k \, |s\rangle$

If ϵ^i where covariantly constant, as it is in flat space, where $|s\rangle$ represents the flat vacuum the 15 extra states vanish, i.e.

$$a^+_i \, |s\rangle = 0$$

For the Schwarzschild solution one might have anticipated 4 zero modes and 15 superpartners of the original black holes (cf. [16]). However this is wrong. The time independent solutions of the Dirac equation may be found, in isotropic coordinates defined by

$$r = \rho(1 + \frac{GM}{2\rho})^2$$

so that

$$ds^2 = \frac{(1 - \frac{GM}{2\rho})^2}{(1 + \frac{GM}{2\rho})^2}dt^2 - (1 + \frac{GM}{2\rho})^4 \; (d\rho^2 + \rho^2(d\theta^2 + \sin^2 d\phi^2))$$

the solutions are

$$\epsilon = \frac{1}{(1 - \frac{2G^2M^2}{4\rho^2})}\epsilon_o$$

Clearly these solutions are singular on the horizon $r = 2GM$ or $\rho = GM/2$.

Thus it is not possible to perform a supergauge transformation of the black hole in this way.

Cordero and Teitleboim [17] have suggested that

$$\epsilon = \frac{s}{16\pi GM}$$

is a suitable spinor with which to perform on supergauge transformation, where s is a constant spinor in the obvious spinor frame. But this spinor frame is singular on the horizon, since the tetrad is. One must pass to a non-singular tetrad, related to the present one by a boost which becomes infinite as the horizon is approached. Alternatively note that

$$\bar{\epsilon}\,\gamma_{\mu}\epsilon = \left[\frac{1 - \frac{GM}{2\rho}}{1 + \frac{GM}{2\rho}}\right]^{-1} \delta^{o}_{\mu}$$

if we choose $\gamma^{o}\,\epsilon = \epsilon$, but the right hand side is singular as $\rho \to GM/2$.

Thus it seems that it is just not possible to perform a time independent global supersymmetry transformation of a black hole. Of course nothing prevents one performing time dependent supergauge transformations.

A similar thing happens at finite temperature. The finite temperature state is not supersymmetric – for example the euclidean metric or $\mathbb{R}^3 \times S^1$ admits no spinors which are both antiperiodic in imaginary time and covariantly constant. This is not surprising since the thermal distribution of fermions and bosons is different. In fact it would take an infinite amount of energy to perform a supersymmetry transformation of a heat bath since the total energy of bosons differs from that of the fermions by an infinite amount.

In the quantum theory black holes cannot correspond to stationary states, since they are unstable. If one prevents them from evaporating by sending in a thermal flux of particles from infinity they can be kept time independent. One can then use the classical Schwarzschild solution to approximate this mixed thermal state. However this is different from using solitons to approximate the corresponding pure one-soliton state in the quantum theory. There just does not seem to be a pure state corresponding to a Schwarzschild black hole.

One can try to relate the absence of solitons in pure gravity to the existence of vacua. There are 3 homogeneous solutions of the vacuum Einstein Equations [18].

A) Flat space, invariant under the Poincare group.

B) A special case of the pp-wave metrics invariant under a 6-dimensional isometry group.

C) A solution due to Petrov invariant under a 4-dimensional isometry group.

Flat space has maximal supersymmetry, (i.e. 4 Majorana covariantly constant spinors). all the pp-waves have half the maximal supersymmetry and this may be taken as a characterization of them. Thus the special case B) has half the maximal supersymmetry. Case C) has no supersymmetry. There appear to be no time independent solutions interpolating between flat space and vacuum B) or C). We shall see that the situation is different in N=2 supergravity.

4: Extended supersymmetry: central charges and multiplet shortening.

The extended supersymmetry algebras allow the possibility of a qualitatively new feature - the existence of central charges [19]. These are central in that they commute with every other generator in the algebra. In particular they commute with time translations and Lorentz rotations. Thus they are conserved invariant charges. They have the dimension of mass and are thus only non zero in a massless theory of the dilation invariance is broken in the presence of solitons. Since the massless fields can carry off energy but not the central charges the configurations with least energy for fixed central charge are likely to be stable. These least energy configurations are partially supersymmetric [1].

Specifically the basic anticommunicator becomes, in a Weyl basis

$$\{Q_A^i \ , \ \overline{Q}_{\dot{A}}^{\,j}\} = \sqrt{2} \ P_{A\dot{A}} \delta^{ij}$$

$$\{Q_A^i \ , \ Q_A^j\} = \epsilon_{AB} Z^{ij}$$

where ϵ_{AB} is the SL(2,C) invariant skew form and $Z^{ij} = -Z^{ji}$ and $1 = 1, z, \text{--}N$ where N is the number of supercharges.

If we move to the rest frame (assuming $P_\alpha \rho^\alpha \neq 0$) we have

$$\{Q_1^i \ , \ Q_1^{j\dagger}\} = \delta^{ij} M$$

$$\{Q_2^i \ , \ Q_2^{j\dagger}\} = \delta^{ij} M$$

where $Q_1^{i\dagger}$ is say spin up and $Q_2^{i\dagger}$ is spin down.

while

$$\{Q_1^i \ , \ Q_2^j\} = Z^{ij}$$

Consider the N = 2 case for simplicity. The last anticommunication relation becomes

$$\{Q_1^1 \ , \ Q_2^2\} = U + iV$$

$$\{Q_1^2 \ , \ Q_2^1\} = -U - iV$$

If $U + iV = |\lambda| e^{i\theta}$ we can multiply Q_A^i by $e^{+i\theta/2}$ to make the last two equations become

$$\{Q_1^1 \ , \ Q_1^2\} = |\lambda|$$

$$\{Q_1^2 \ , \ Q_2^1\} = -|\lambda|$$

Thus

$$\{Q_1^1 \pm Q_1^{2\dagger}, \ Q_1^{1\dagger} \pm Q_1^2\} = 2(M \mp |\lambda|)$$

$$\{Q_1^1 \pm Q_1^{2+} \quad , \quad Q_1^1 \mp Q_1^{2\dagger}\} = 0$$

These are the anticommulation relations of a pair of creation and annihilation operators creating states of spin ½. There is a similar set with lower index 1 changed to 2 which create states of spin −½. They yield immediately the inequality

$$M \geqslant |\lambda|$$

Further if $M = |\lambda|$ we get multiplets of dimension 4 rather than 16 since the operators $Q_1^1 - Q_1^2$ and $Q_2^2 - Q_2^2$ act trivially on the states. We shall refer to the inequality as the Bogomolny inequality though the supersymmetric version is due to Olive and Witten [1]. One can carry out a similar analysis for higher N. One finds that the skew matrix Z^{ij} has $\left[\frac{N}{2}\right]$ skew eigen-values λ_n and the mass cannot be less than the modulus of any λ_n , $M \geqslant |\lambda_n|$ for $n = 1, 2, \ldots \left[\frac{N}{2}\right]$

5: <u>Black Holes in N = 2 ungauged extended Supergravity: Charge without Charge.</u>

The physical fields of N = 2 ungauged supergravity [20] are a graviton, 2 majorana gravitini and a photon. They are all massless and all electrically neutral with respect to the U(1) charge of the theory. Nevertheless classical solutions exist for which this charge is non-zero. In fact the theory possesses a global "dual-chiral" invariance if the gravitino is chirally rotated $\psi_\mu^i \to \exp i\,\theta\frac{\gamma_5}{2}\,\psi_\mu^i$ and the Maxwell field $F_{\alpha\beta}$ dually rotated $F_{\alpha\beta} \to \exp i\theta *\, F_{\alpha\beta}$ the equations of motion are unaltered. This corresponds to the invariance of the algebra under $Q_A \to e^{i\frac{\theta}{2}} Q_A$, $Z_{ij} \to e^{i\theta} Z_{ij}$. Using the invariance we can perform a dual transformation to make the charge of a soliton purely electric which we shall assume, for simplicity of exposition, has been done from now on. However one should bear in mind that both electric and magnetic charges are possible. These are the Reissner-Nordstrom black holes or their rotating versions the Kerr-Newman solutions. As in section 3 we shall discuss only the non-rotating holes for simplicity. The metric is

$$ds^2 = (1 - \frac{2GM}{r} + G\frac{Q^2}{4\pi r^2})dt^2$$

$$- (1 - \frac{2GM}{r} + G\frac{Q^2}{4\pi r^2})^{-1}dr^2 + r^2(d\theta^2 + \sin^2\theta d\phi^2)$$

and in order that the singularity at $r = 0$ be hidden by an event horizon g_{oo} must change sign between $r = 0$ and $r = \infty$ which leads to the inequality

$$M \geqslant \frac{|Q|}{\kappa}$$

The temperature of a Reissner-Nordstrom black hole is easily worked out using the method described in section 3 and one obtains:

$$T = \frac{1}{4\pi G}\,\frac{\sqrt{(M^2 - Q^2/\kappa^2)}}{(M + \sqrt{(M^2 - Q^2/\kappa^2)})^2}$$

we see that for the limiting or extreme value for the charge that the temperature vanishes.

Consider now the quantum evolution of a black holes in this theory. Its mass will be reduced by the thermal emission of gravitons, photons and gravitini. However its charge Q cannot change since these particles do not carry this charge. The result is that the holes will evolve towards the state with least mass for fixed Q, i.e. the extreme zero temperature Reissner-Nordstrom black hole. It will in fact take an infinite amount of time to do so, since the specific heat (at constant charge) changes sign for $Q/\kappa > \frac{3}{4}M$ and the evolution slows down as $Q/\kappa \to M$. Thus the stable end points of gravitational collapse in this theory are zero temperature black holes, rather than flat space as in the $N = 1$ theory.

Thus the geometry of the 3-surface metric is if two asymptotically flat sheets connected by an Einstein-Rosen Bridge or "throat". The electric flux lines pass from one asymptotically flat region to the other without ending on any sources. This is the phenomenon of "charge without charge". (Note however that the full spacetime contains singularities at $r = 0$ which do not lie on these t = constant surfaces.)

It is interesting to note that the metrics for $\frac{Q}{\kappa} < M$ are qualitatively different from those with $\frac{Q}{K} = M$. This is most easily seen from the Penrose diagrams. Another way is to look at the t = constant surfaces. Introducing isotropic coordinates the Reissner– Nordstrom metric restricted to the surface t = constant becomes

$$-\left[(1 + \frac{GM}{2\rho})^2 - (\frac{GQ}{8\pi\rho})^2\right]^2 \left[d\rho^2 + \rho^2(d\theta^2 + \sin^2\theta d\phi^2)\right]$$

where $0 < \rho < \infty$. The area of a sphere ρ = constant is

$$4\pi\left[\rho + GM + \frac{1}{4} G^2(M^2 - Q^2/\kappa^2)\frac{1}{\rho}\right]^2$$

If $M^2 > Q^2/\kappa^2$ this expression is symmetrical under the transformation:

$$\rho \rightarrow \frac{1}{\rho}(M^2 - \frac{Q^2}{\kappa^2})\frac{1}{4}$$

of the 11-dimensional supergravity theory. It describes a spontaneous compactification from 4 dimensions to 2 dimensions similar to that studied in [22].

On the other hand if $M^2 = Q^2/K^2$ there is no other asymptotically flat region. In fact as $M \rightarrow |Q|/K$ the throat becomes infinitely long. The horizon may then be thought of as an internal infinity.

The geometry of this throat approximates that of the Robinson-Bertotti metric. This is the product metric of the form $(AdS)_2 \times S^2$ where the radii of curvature of the two factors are equal in magnitude to $|Q|/\kappa$ [24]. This is of considerable interest because the Robinson-Bertotti metric may be thought of as the analogue of the Freund-Rubin solution.

In fact apart from certain plane waves the only homogeneous solution of the Einstein Maxwell equations apart from flat spacetime is the Robinson-Bertotti metric [2]. Thus from the point of view of homogeneity we might claim that there are 2 vacua in Einstein-Maxwell theory and the extreme Reissner-Nordstrom black holes spatially interpolate between them. We shall see in the next section that this interpretation is in harmony with the supersymmetric interpretation.

6: The Bogomolny Inequality.

The striking similarity between the Bogolmolny inequality of supersymmetry and the inequality from black hole physics that $M \geq |Q|/K$ suggests that the positive energy argument may be extended to cover this case. This is in fact true [24, 6]. The relevent supersymmetry transformations for $N = 2$ is

$$\delta\psi^{i}_{\mu} = \frac{1}{\kappa}(\nabla_{\mu}\epsilon^{i} - \epsilon^{ij}\kappa F_{\alpha\beta}\gamma^{\alpha}\gamma^{\beta}\gamma_{\beta})\epsilon^{i}$$

$$= \frac{1}{\kappa} \quad \hat{\nabla}^{\,ij}\epsilon^{j}$$

where the second equation maybe taken as a definition of the "hatted derivative". If one uses the hatted derivative in the Witten positive energy argument one arrives at the statement that

$$M \geq \sqrt{\left(\frac{Q^2}{\kappa^2} + \frac{P^2}{\kappa^2}\right)}$$

with equality if and only if the spacetime admits solutions of the Killing spinor condition

$$\hat{\nabla}^{\,ij}_{\mu}\,\epsilon^{j} = 0$$

One can show [23, 24] that the only such spacetimes other than pp waves are the Israel-Wilson metrics [25] well known to classical relativists. It was shown by Hartle and Hawking [26] that the only regular black hole solutions in this class are the Papapetrou-Majumdar metrics [27] which describe an arbitrary number of extreme Reissner-Nordstrom black holes in neutral equilibrium, the gravitational attraction being just balanced by the electrostatic repulsion. This balance was named (perhaps misleadingly) antigravity by Scherk [28] who pointed out that matter multiplets with maximum central charge coupled to supergravity exhibit this phenomenon.

The Israel–Wilson metrics admit, in general, 4 Majorana or 2 Dirac Killing spinors – i.e. ½ the maximum number of supersymmetries for $N = 2$. However in one special case – the Robinson–Bertotti metric – they admit 8 Majorana Killing spinors – i.e. the maximum allowed by supersymmetry. As I mentioned in section 5 the Robinson–Bertotti metrics are homogeneous product metrics with a non-vanishing Maxwell field. In the purely magnetic case the Maxwell field is proportional to the alternating tensor on the S^2 factor. This is precisely analogous to the Freund–Rubin soluton. Thus the Robinson–Bertotti metric qualifies as an alternative, maximally supersymmetric, vacuum state to the usual flat vacuum. The Extreme Reissner–Nordstrom metrics spatially interpolate between them as one expects of a soliton.

7: <u>The zero modes.</u>

As explained in section 3 we are interested in pure gauge solutions of the spin $\frac{3}{2}$ equations. We may again imposed the gauge condition

$$\gamma^{\mu}\psi^{i}_{\mu} = 0$$

and find that if

$$\psi_{\mu} = \frac{1}{\kappa}\hat{\nabla}_{\mu}\,\epsilon$$

then ϵ satisfies the massless Dirac equation. Now the following lemma is useful: If the metric takes the form:

$$ds^2 = A^2dt^2 - B^2d\underline{x}^2$$

where A and B depend only on $\underline{x}$ then the time independent solutions of the Dirac equation have the form

$$\epsilon = \frac{\epsilon_o}{AB^2}$$

where ϵ_o is a time-independent solution of the massless Dirac equation in flat space. Note that any spherically symmetric static metric can be cast in this "isotropic" from. It is clear that if $A = 0$, $B \neq \infty$, then the spinor field is singular on the horizon. This is true for the entire Reissner-Nordstrom family except in the limiting case $M = |Q|/\kappa$ in which case

$$AB = 1$$

Thus the spinor field ϵ vanishes at the horizon rather than blowing up. If we take for ϵ_o a constant spinor we will get 4 spinor zero modes which tend to constants at infinity. Of these 2 are in fact Killing spinors. They give a vanishing spin $\frac{3}{2}$ field

$$\psi_\mu = \frac{1}{\kappa}\hat{\nabla}_\mu \epsilon = 0$$

The remaining 2 zero modes can be used to perform supersymmetry transformations on the purely bosonic background. Thus the extreme Reissner- Nordstrom holes fit into a supermultiplet of 4 states. The multiplet is shortened exactly as we would expect for a system saturating the Bogomolny bound.

8: Conclusion.

We have seen that the extreme Reissner-Nordstrom Black Holes behave much like solitons. They are stable against Hawking radiation. They have zero temperature. They have the least mass consistent with their conserved central charge and are (partially) supersymmetric. They also fit into shortened multiplets. The geometry spatially interpolates between two maximally supersymmetric vacua of the ungauged $N = 2$ theory.

However there is no way known to me of fixing their mass. Because of the scale invariance of the theory and the lack of a coupling constant in the theory the mass, or equivalently the charge, is arbitrary. Usually one expects the mass of a soliton to be a fixed quantity. In Yang-Mills theory the monopole mass is given in terms of the expectation value of the Higgs field. In the next lecture I will show how this defect can be remedied by using the Kaluza-Klein idea.

References.

[1] E. Witten and D. Olive, Phys.Lett., 99B 229 (1981).

[2] "Monopoles in Quantum Field Theory" ed. N. Craigie, P. Goddard and W. Nahm, publ. World Scientific, (1982).

[3] R. Penrose, Ann.N.Y.Acad.Sci., 224 125 (1973).

[4] S.T. Yau and R. Schoen, Commun.Math.Phys., 65 45 (1979). Commun.Math.Phys., 79 231 (1981).

[5] E. Witten, Commun.Math.Phys., 80 381 (1981).

[6] G.W. Gibbons, S.W. Hawking, G. Horowitz and M.J. Perry, Comm.Math.Phys., 88, 295, (1983).

[7] P. Hajicek, Nucl.Phys., B185 254 (1981).

[8] B.F. Schutz and R. Sorkin, Ann.Phys., 107 1 (1977).

[9] R. Serini, Acad.Na.Lincei.Mem.Cl.Sci.Mat.Nal., 27 235 (1918).

A. Einstein and W. Pauli, Ann.Math., 44 131 (1943).

A. Lichnerowicz., C.R.Acad.Sci. Paris, 222 432 (1975).

[10] C.H. Derrick, J.Math.Phys., 5 1252 (1962).

[11] G.W. Gibbons in "Global Riemannian Geometry", ed. T. Wilmore and N.J. Hitchin: Ellis Horwood, 1984.

[12] W. Israel, Phys.Rev., 164 1776 (1967).

[13] G.W. Gibbons and J.M. Stewart in "Classical General Relativity", ed. W.B. Bonnor and J. Islam, C.U.P., 1984.

[14] G.W. Gibbons and M.J. Perry, Phys.Rev.Letts., 36 985 (1976).

[15] L. Girardello, M.T. Grisaru and P. Salomonson, Proc. Roy. Soc., A358 467 (1978).

Nucl. Phys., B178 331 (1981).

[16] T. Yoneya, Phys. Rev., D17 2567 (1978)

[17] P. Cordero and C. Teitleboim, Phys. Letts., 78B 80 (1978).

[18] D. Kramer, H. Stephani, M. MacCallum and E. Herlt, "Exact Solution of Einstein's Field Equations", C.U.P. and V.E.B., Deutsches Verlag der Wissenschaften.

[19] R. Haag, J.T. Lopouszanshi and M.F. Sohnius, Nucl. Phys., B88 257 (1975).

[20] S. Ferrara and P. van Nieuwenhuizen, Phys. Rev. Letts., 37 1669 (1976)

[21] B. Carter, "Black Holes", eds. C. DeWitt and B.S. DeWitt, Gordon and Breach, New York (1973).

[22] D.Z. Freedman and G.W. Gibbons, Nucl. Phys., B233 24 (1984).

[23] P. Tod, Phys. Letts., B121 241 (1983).

[24] G.W. Gibbons and C.M. Hull, Phys. Letts., 109B 190 (1982).

[25] W. Israel and G.A. Wilson, J. Math. Phys., 13 865 (1972).

[26] J.B. Hartle and S.W. Hawking, Commun. Math. Phys., 26 87 (1982).

[27] A. Papapetrou, Proc. Roy. Irish Acad., A51 191 (1947).

S.D. Majumdas, Phys. Rev., 72 390 (1947).

[28] J. Scherk, Phys. Letts., 88B 265 (1979).

Lecture 3 Supersymmetric Monopoles in Kaluza–Klein Theory.

Contents

1: Kaluza-Klein Theory.

In this lecture we shall deal with the simplest Kaluza-Klein theory - pure gravity in 5-dimensions and its (N = 8) supersymmetric extension. The main reason for this is that, as discussed in lecture 1, more sophisticated spontaneous compactifications entail dealing with an AdS_4 background in which the concept of a soliton is, as yet, rather obscure. As is well-known if we assume that the 5-dimensional spacetime is independent of one coordinate - X^5 we may parameterize the metric as:

$$ds^2 = e^{\frac{4\sigma\kappa}{\sqrt{3}}}(dx^5 + 2\kappa A_\alpha dx^\alpha)^2 + e^{\frac{-2\sigma\kappa}{\sqrt{3}}} g_{\alpha\beta} dx^\alpha dx^\beta$$

where the scalar field σ, the vector field A_α and metric $g_{\alpha\beta}$ are independent of X^5. The 5-dimensional Einstein lagrangian becomes, up to a total derivative,

$$\sqrt{g}\left(\frac{R}{4\kappa^2} - \frac{1}{4}e^{2\sigma\kappa\sqrt{3}} F_{\alpha\beta}F^{\alpha\beta} + \tfrac{1}{2} g^{\alpha\beta}\partial_\alpha\sigma\partial_\beta\sigma\right.$$

where R is the Ricci scalar of $g_{\alpha\beta}$ and $F_{\alpha\beta} = \partial_\alpha A_\beta - \partial_\beta A_\alpha$. The unusual coupling of scalars to the vectors means that in general an isolated, charged, system will possess a scalar charge Σ, which we may define by the requirement that

$$\sigma \to \frac{1}{4\pi}\frac{\Sigma}{r}$$

at infinity. This is because if $F_{\alpha\beta}F^{\alpha\beta} \neq 0$, it is inconsistent to set $\sigma =$ constant. Two scalar sources then experience in the "Born approximation" an attractive force $\frac{\Sigma_1 \Sigma_2}{r^2}$. One point should be made about regularity requirements. We have parameterized the metric in this way so as to put the action in standard Einstein conformal gauge. Starting from this 4-dimensional form of the theory it is natural to demand that the metric $g_{\alpha\beta}$ be complete and non-singular, the scalar field σ be everywhere bounded and the Maxwell tensor $F_{\alpha\beta}$ be regular. These 4-dimensional regularity conditions, which guarantee a regular 5-metric, are too strong as far as the 5-dimensional metric is concerned since there are regular 5-metrics which violate them as we shall see later. In particular since a 5-metric may be regular even though $g_{55} = 0$ regular 5-metrics may give rise to scalar fields with logarithmic singularities. In what follows we shall take the liberal view that what matters is regularity in 5-dimensions rather than 4.

The vacuum of Kaluza-Klein theory is the flat metric on $S^1 \times$ Minkowski space. Killing spinors will exist and hence the vacuum be supersymmetric and thus stable if we demand that spinor fields be periodic in X^5 rather than antiperiodic. This will also rule out the bounce solution of Witten [1] which would (in the antiperiodic case) render the vacuum unstable. The perturbative excitations around the vacuum consist of the massless, electrically neutral sector containing the graviton, $g_{\alpha\beta}$, graviphoton A_α, and dilaton σ. The dilaton maybe thought of as the goldstone particle arising from the broken dilation invariance of the vacuum. In addition we obtain the tower or ladder of charged massive states with charge e and mass m given by

$$\frac{e^2}{K^2} = 4\frac{m^2}{K}$$

$$m = \frac{n}{R_K}$$

where $n = \pm 1, \pm 2, \ldots, \; 0 < X^5 < R_K$. The lowest rung of the ladder $n^2 = 1$, are absolutely stable particles. They cannot decay into massless states because that would not violate the conservation of the Noetherian charge e arising from the U(1) invariance of the vacuum [2]. Of course these particles, which following [2] we call "pyrgons" can annihilate with their antiparticles.

2) Gross-Perry-Sorkin Monopoles.

We now seek solutions of the 5-dimensional vacuum Einstein Equations which are independent of X^5 and also a time coordinate t. One class of such solutions has the ultrastatic form

$$ds^2 = dt^2 - h_{\alpha\beta}dx^\alpha dx^\beta$$

when $h_{\alpha\beta}$ is a complete riemannian 4-metric which has an additional U(1) isometry. That is, $h_{\alpha\beta}$ is the metric of a gravitational instanton with a Killing vector. To be asymptotic to the Kaluza-Klein vacuum, $h_{\alpha\beta}$ must be asymptotically locally flat (ALF) i.e. tend to a twisted product of S^1 with $S \times \mathbb{R}$. If the bundle is twisted the resulting objects will have a magnetic charge. In the language of [3] the NUT charge is now reinterpreted as magnetic charge. The self-dual Taub-NUT metrics provide an example [4,5,6]. The Rieman tensor $R_{\alpha\beta\gamma\delta}$ of $h_{\alpha\beta}$ satisfies

$$R_{\alpha\beta\gamma\delta} = \pm \tfrac{1}{2}\, \epsilon_{\alpha\beta}^{\ \ \mu\nu}\, R_{\mu\nu\gamma\delta}$$

where $\epsilon_{\alpha\beta}^{\ \ \mu\nu}$ is the alternating tensor Explicitly we have:

$$h_{\alpha\beta}dx^{\alpha}dx^{\beta} = W^2 d\underline{x}^2 - \frac{1}{W^2}(dx^5 + 2\kappa\underline{A}.d\underline{x})^2$$

$$\mathrm{grad}W = -\frac{\kappa}{W}\mathrm{curl}A$$

$$W = 1 + 4GM\Sigma_{p=1}^{p=n} \frac{1}{|\underline{x} - \underline{x}_p}$$

The coordinate singularities at $\underline{X} = X_p$ can be removed [3] provided X^5 is identified

$$0 \leqslant X^5 \leqslant 4\kappa^2 M$$

One readily reads off the magnetic moment of each monopole as

$$\frac{P}{\kappa} = 2M$$

Now since the identification of X^5 in Kaluza-Klein gives the unit of electric charge:

$$0 \leqslant X^5 \leqslant \frac{4\pi\kappa}{e}$$

we deduce that the Dirac quantization condition holds:

$$eP = 2\pi$$

Since $P = 2\frac{\pi}{e}$ and the ADM mass is M we have

$$M = \frac{\pi}{\kappa e}$$

so indeed we have obtained a quantization rate for the mass of our solitons. Notice, as alluded to in section 1, the points $\underline{X} = \underline{X}_p$ are 5-dimensional coordinate singularities. From the conservative point of view of 4-dimensions they appear as singularities. One may read off the scalar charge Σ, it is

$$\Sigma = \sqrt{3}GM$$

Thus the antigravity condition

$$M^2 = \Sigma^2 / G - \frac{P^2}{\kappa^2} - Q^2 / \kappa^2 = 0$$

is satisfied.

3) N=8 Supergravity in 5-dimensions.

We shall now consider the supersymmetry algebra in 5-dimensions and then the supersymmetric extension of Einstein's theory in 5-dimensions given by Cremmer[7]. Remarks on the 5-dimensional algebra and central charges may also be found in [8]. The present analysis is essentially that given in [9]. We consider the $N = 8$ algebra in 4-dimensions for which, in 2 component form:

$$\{Q^i_A, \bar{Q}^j_{B'}\} = \sqrt{2}P_{AB'}\ \delta^{ij}$$

$$\{Q^i_A, Q^j_B\} = Z^{ij}E_{AB}$$

where $Z^{ij} = U^{ij} + i\ V^{ij}$
are the central charges, U^{ij} being P.T. even is magnetic and V^{ij} being PT even is odd is electric.

In the most symmetric case

$$V^{ij} = V\Omega^{ij} , \quad U^{ij} = U\Omega^{ij}$$

where $\Omega^{ij}\ \Omega^{jk} = -\delta^{ik}$

Thus the algebra is Usp(8) invariant, Ω^{ij} being the symplectic form. 5-dimensional gamma matrices Γ^A, A = 0, 1, 2, 3, 5 may be chosen as

$$\Gamma^A = (\gamma^\alpha, \quad i\gamma^5)$$

The 5-dimensional charge conjugation matrix C_5 is given by

$$C_5 = C\gamma^5$$

where C is the 4-dimensional charge conjugation matrix and satisfies

$$\Gamma^t_A = C_5 \Gamma_A\ C_5^{-1}$$

One cannot have Majorana 4-spinors in 5-dimensions, however one can have symplectic spinors.

These satisfy

$$(\overline{S}^i)_a = \Omega^{ij}(S^{jt}C_5)_a$$

In terms of the Weyl spinors we find that

$$S^{ia} = \begin{bmatrix} Q^{iA} \\ \Omega^{ij}\overline{Q}^j{}_A{}^i \end{bmatrix}$$

One may now rewrite the algebra in the form

$$\{S^{ia}, \overline{S}^b_b\} = -U\delta^{ij}\ \delta^a_i + \delta^{ij}(\gamma^1 P_i - i\gamma^5 V)^a$$

This looks more 5-dimensional if we identify the electric central charge,

V with the 5th component of momentum

$$P^5 = V$$

$$\{S^{ia}, \bar{S}^j{}_b\} = -U\delta^{ij}\delta^a_b + \delta^{ij}(\Gamma^A P_A)^a_b$$

Thus the magnetic charge is identified with a central charge in 5-dimensions.

The Bogomolny inequality now becomes

$$M \geq \sqrt{(U^2 + (P^5)^2)}$$

but according to the usual Kaluza-Klein interpretation the 5th component of momentum, P^5, and electric charge, Q, are related by

$$P^5 = \frac{Q}{2K}$$

whence

$$M \geq \sqrt{(U^2 + \frac{Q^2}{4K^2})}$$

Note the factor of 4 in the denominator of the last term inside the square-root sign. This differs from the inequality in Einstein-Maxwell theory given in lecture 2.

Now consider 5-dimensional, $N = 8$, supergravity. This was constructed by Cremmer [*]. The lagrangian has an $E_{6(+6)}$ global and Usp(8) local symmetry with fields as follows:

graviton	g_{AB}	$\underline{1}$	$\underline{1}$
gravitino	ψ^i_A	$\underline{1}$	$\underline{8}$
vector	$A_A^{\alpha\beta}$	$\underline{27}$	$\underline{1}$
spin½	χ^{ijk}	$\underline{1}$	$\underline{48}$
scalar	$V^{ij}_{\alpha\beta}$	$\underline{27}$	$\underline{27}$

$A, B = 0, 1, 2, 3, 5$	are spacetime indices
$i, j = 1, 2, \ldots\ldots..8$	are Usp(8) indices
$\alpha, \beta = 1, 2, \ldots\ldots..8$	are $E_{6(+6)}$ indices

The spinors are symplectic

$$\bar{\psi}^i = \Omega^{ij} \psi^{it} C_5$$

$$\bar{\chi}^{ijk} = \Omega^{il} \Omega^{jm} \Omega^{kn} \chi^{lmn\,t} C_5$$

$$\bar{\chi}^{ijk} \Omega_{ij} = 0$$

The matrix-valued scalars $V^{ij}_{\alpha\beta}$ provide a harmonic map into $E_{6(+6)}/Usp(8)$. The bosonic action is

$$-\frac{1}{4\kappa^2}R - \frac{1}{8}G_{\alpha\beta,\gamma\delta} F^{\alpha\beta}_{AB} F^{AB\gamma\delta}$$

$$+ \frac{1}{24\kappa^2} P_{Aijkl} P^{Aijkl} - \frac{\kappa}{12}\eta^{ABCDE} F^{\alpha\beta}_{AB} F_{CD\beta\gamma} A^{\gamma}_{E\alpha}$$

where

$$G_{\alpha\beta,\gamma\delta} = V^{ij}_{\alpha\beta} \Omega_{ik} \Omega_{jl} V^{kl}_{\gamma\delta}$$

is the metric on $E_{6(+6)}$ and

$$P^{ij}_{A\,kl} = \left[(V^{-1})_{kl}{}^{\alpha\beta} \partial_A V^{ij}_{\alpha\beta}\right]_{|-}$$

For backgrounds for which the scalars are constant and the vector fields vanish the supersymmetry transformations reduce to

$$\delta \psi^i_A = \frac{1}{\kappa} \nabla_A \epsilon^i$$

4: <u>Supersymmetries</u> <u>and</u> <u>Zero-Modes.</u>

The vacuum of Kaluza-Klein theory, $S^1 \times R^3 \times$ Time, has 32 real or 16 complex Killing spinors constructed from the 4 covariantly constant Dirac spinors. These are naturally independent of time. The solitons possesses 2 covariantly constant time-independent Dirac fields which are eigenstates of γ^o with eigenvalue +1. This is because the spatial curvature is self-dual (or anti-self dual) and "chiral" eigenstates with negative (respectively positive) chirality don't feel the curvature. Now what one usually thinks of γ_5 is γ^o (i.e. " γ^{extra} "). If one looks for solutions of the Dirac equation (which would be zero modes) one finds that those which tend to constants at infinity have the form

$$\tfrac{1}{2}(1 + \gamma^o)\xi$$

or

$$\frac{1}{2W}(1 - \gamma^o)\xi$$

where ξ is a constant spinor. In the first case we get Killing spinors and in the second case regular zero modes. Thus we get half the maximum number of supersymmetries and the maximum amount of multiplet shortening allowed in $N = 8$.

One may go through the Witten identity to derive the Bogomolny inequality. This also serves to fix the relation between the central charge U and the magnetic moment P. One obtains

$$U = \frac{P}{2\kappa}$$

whence

$$M \geq \sqrt{\left(\frac{P^2}{4\kappa^2} + \frac{Q^2}{4\kappa^2}\right)}$$

5: Electric-Magnetic Duality.

The standard Kaluza-Klein theory has a descrete, electric magnetic duality which is a symmetry of the equation of motion but not the lagrangian. The equations of motion for the vectors are

$$\nabla_\alpha(e^{2\sqrt{3}\kappa\sigma}F^{\alpha\beta} = 0$$

The Bianchi identity is

$$\nabla_\alpha(*F^{\alpha\beta}) = 0$$

The substitution

$$e^{2\sqrt{\kappa}\sigma}\ F^{\alpha\beta} \rightarrow *F^{\alpha\beta}$$

$$\sigma \rightarrow -\sigma$$

leaves both and the σ equation of motion:

$$-\nabla^2\sigma + \frac{\sqrt{3\kappa}}{2}\ F_{\alpha\beta}F^{\alpha\beta}e^{2\sqrt{3}\kappa\sigma}$$

unchanged.

For static metrics one can introduce electric and magnetic potentials ψ and χ. If the metric takes the form

$$ds^2 = e^{-2\sqrt{3}\kappa\sigma}(e^{2\eta}\ dt^2 - e^{-2\eta}\ \gamma_{ij}dx^i dx^j$$

$$-e^{4\sqrt{3}\kappa\sigma}(dx^5 - 2\kappa\psi\ dt - 2\kappa\ A_i dx^i)^2$$

where

$$\text{grad}\chi = -e^{2(\eta - \kappa\sigma)}\ \ \text{curl}A$$

the static equations of motion can be derived from the effective 3-action:

$$\frac{3R}{\kappa^2} - 2\frac{\gamma^{ij}}{\kappa^2}\partial_i\eta\partial_j\eta - 2\partial_i\sigma\partial_j\sigma\gamma^{ij}$$

$$+ 2e^{-2\eta}\ \gamma^{ij}\ (\partial_i\psi\partial_j\psi e^{2\sqrt{3}\sigma} + \partial_i\chi\partial_j\chi e^{-2\sqrt{3}\sigma})$$

The duality:

$$\eta \to \eta$$
$$\gamma_{ij} \to \gamma_{ij}$$
$$\psi \to \chi$$
$$\chi \to \psi$$
$$\sigma \to -\sigma$$

taking electric fields into magnetic fields is now manifest.

One may use the duality to show that the monopoles are dual to some singular limits of black holes which antigravitate because they move at the speed of light in 5-dimensions.

The 4-dimensional black hole solutions were discussed in [17]. If one insists on regularity in 4-dimensions (c.f. section 1) one finds that the spherically symmetric holes are parameterized by their mass M and magnetic and electric charges Q and P. The scalar charge Σ is a function of these, not an independent parameter. Regularity, i.e. Cosmic Censorship then requires

$$M^2 + \frac{4\pi\Sigma^2}{\kappa^2} \geq \frac{Q^2}{\kappa^2} + \frac{P^2}{\kappa^2}$$

In the case of equality we further require

$$QP \neq 0$$

Note that the inequalities are not, in this case, identical with the Bogomolny inequality.

Clearly the duality interchanges the electric and magnetic charges and reverses the sign of Σ. The only supersymmetric objects are the monopoles:

$$Q = 0$$
$$P = 2\kappa M$$
$$\Sigma = 3\sqrt{3}GM$$

or their dual partners with

$$Q = \sqrt{2\kappa M}$$
$$P = 0$$
$$\Sigma = \sqrt{3}GM$$

In fact the regular black holes with zero magnetic charge may all be obtained by boosting a Schwarzschild black hole in the 5-th direction. As the velocity approaches that of light one must take a suitable limit and one gets precisely the dual partner of the monopoles. The singular limiting geometry is singular even in the 5-dimensional sense. Corresponding to the multi-monopole solutions one also has multi-object solutions with purely electric charge. They antigravitate (since $M^2 + \frac{4\pi}{\kappa^2}\Sigma - \frac{Q^2}{\kappa^2} = 0$) because they are moving at the speed of light in 5-dimensions. The multi- object-metrics have the form

$$ds^2 = -W'(dx^5)^2 + 2dtdx^5 - d\underline{x}^2$$

where $W' = 1 + \Sigma_{p=1}^{p=n} \frac{4GM_p}{|\underline{x} - \underline{x}p|}$.

The "masses" M_p are in general arbitrary (though those obtained by duality are of course all equal). The metric is in fact a pp wave in 5-dimensions and so is naturally supersymmetric. The pp – waves represent classical gravitational waves in 5-dimensions. Thus we see that the duality takes gravitational waves into monopoles. In the next section we shall discuss this duality in more detail.

6) Monopole-Pygron Duality.

In the last section we wish to extend the concept of duality discussed in section 5). There we saw that the monopoles are dual to gravitational waves in 5-dimensions. In 5-dimensions these gravitational waves fall into massless supermultiplets . Viewed in 4 dimensions these "waves" are massive multiplets which have 256 members. This is the same multiplet as that of the "pygrons". This suggests that the following duality may hold.

monopole <-> pygron
massless sector <-> sector sector
antimonopole <-> anti-pyrgon

The noetherian central charge would then be dual to the topological NUT charge. The monopoles and the pyrgons would then be stable for essentially the same reason.

More specifically I am suggesting that there exist operators creating monopoles and antimonopoles and that these together with the operators creating the massless states satisfy an effective field theory which is identical to Cremmer's 5-dimensional supergravity.

Suppose this is true. Then the β -function for the electrically charged pyrgons and that for the magnetically charged monopoles must have the same form.

$$\mu\frac{dP}{d\mu} = \beta(P)$$

$$\mu\frac{de}{d\mu} = \beta(e)$$

to maintain the Dirac relation

$$eP = 2\pi$$

we must have

$$\beta(e) = \beta(P) = 0$$

Now to one-loop this is trivially true since Kaluza-Klein is one loop finite in 5-dimensions. What is being suggested here is that this is true to all orders. In fact $\beta(e)$ vanishes to one loop not only when the entire tower of massive states is taken into account but "floor by floor" by spin sum rules. This may indicate that the divergences of the 5-dimensional theory are less severe than appears at first sight.

The duality being conjectured here would have other consequences which could be checked. For example the pyrgons all have magnetic dipole moments depending upon their spin. The states in the monopole multiplets with non-vanishing spin should have electric dipole moments. These are calculable and the gyroelectric ratios should equal the gyromagnetic ratios.

The situation with respect to monopoles in N = 8 Kaluza-Klein theory appears to be very analogous to the Yang-Mills monopoles in N = 4 Yang-Mills theory [11]. The present lecture is just a beginning of what promises to be an interesting subject.

References.

[1] E. Witten, Nucl. Phys., B195 481 (1982).

[2] E.W. Kolb and R. Slansky, Phys. Lett., 135B 378 (1984).

[3] G.W. Gibbons and S.W. Hawking, Commun. Math. Phys., 66 291 (1979).

[4] S.W. Hawking, Phys. Letts., 60A 81 (1977).

[5] D. Gross and M.J. Perry, Nucl. Phys., B226 29 (1983).

[6] R. Sorkin, Phys. Rev. Lett., 51 87 (1983).

[7] E. Cremmer, " Supergravities in 5 Dimensions" in `Superspace and Supergravity' ed. S.W. Hawking and M. Rocek, C.U.P., (1981).

[8] J. Lukierski and L. Rytel, Phys. Rev. D., 27 2354 (1983).

[9] G.W. Gibbons and M.J. Perry, "Soliton-Supermultiplets and Kaluza-Klein Theory", Princton Preprint (1984).

[10] G.W. Gibbons, Nucl. Phys., B207 337 (1982).

[11] C. Montonen and D. Olive, Phys. Lett., 72B 117 (1977)

D. Olive in "Monopoles in Quantum Field Theory", eds. N. Craigie, P. Goddard and W. Nahm, World Scientific (1981).

$d = 4$ SUPERGRAVITY FROM $d = 11$: CONJECTURES REVISITED

M.J. Duff[1]
Institute for Theoretical Physics
University of California
Santa Barbara, California, 93106

ABSTRACT

We examine the relationship between the massless sector of the theory obtained by Kaluza-Klein compactification of $d = 11$ supergravity on the round S^7 and the $N = 8$ gauged $SO(8)$ supergravity in $d = 4$. Some time ago it was conjectured that (I) the two theories are the same and (II) non-symmetric extrema of the $d = 4$ effective potential correspond to other solutions in $d = 11$ with the same S^7 topology; non-zero VEV's for scalars and/or pseudoscalars corresponding to distortion of and/or "torsion" on S^7. As yet neither conjecture is rigorously proved. Here we review the evidence in their favor, which is now overwhelming. As an interesting spin-off, one discovers the criterion for the consistency of the Kaluza-Klein ansatz.

1. d = 11 Supergravity

1.1 The Lagrangian, its symmetries, and the equations of motion

The Lagrangian of $N = 1$ supergravity in $d = 11$ is given with signature $(-++++++++++)$ by [1]

$$
\begin{aligned}
\mathcal{L} = & \frac{1}{4} e\, e_B{}^N e_A{}^M R_{MN}{}^{AB}(\omega) \\
& - \frac{i}{2} e \bar{\Psi}_M \Gamma^{MNP} D_N[\frac{1}{2}(\omega + \tilde{\omega})]\Psi_P \\
& - \frac{1}{48} e F_{MNPQ} F^{MNPQ} + \frac{2}{(12)^4} e \epsilon^{M_1 \ldots M_{11}} F_{M_1 \ldots M_4} F_{M_5 \ldots M_8} A_{M_9 \ldots M_{11}} \\
& + \frac{3}{4(12)^4} e[\bar{\Psi}_M \Gamma^{MNWXYZ} \Psi_N + 12 \bar{\Psi}^W \Gamma^{XY} \Psi^Z](F_{WXYZ} + \tilde{F}_{WXYZ})
\end{aligned}
\quad (1.1.1)
$$

[1]On leave from the Theory Division, CERN, Geneva and from the Blackett Laboratory,-

Imperial College, London.

where $e = \det e_M{}^A$, $\bar{\Psi} = \Psi^+\Gamma_0$, $D_M(\omega) = \partial_M - \frac{1}{4}\omega_M{}^{AB}\Gamma_{AB}$ and

$$\omega_{MAB} = \frac{1}{2}(-\Omega_{MAB} + \Omega_{ABM} - \Omega_{BMA}) + K_{MAB}, \tag{1.1.2}$$

$$K_{MAB} = \frac{i}{4}[-\bar{\Psi}_N\Gamma_{MAB}{}^{NP}\Psi_P + 2(\bar{\Psi}_M\Gamma_B\Psi_A - \bar{\Psi}_M\Gamma_A\Psi_B + \bar{\Psi}_B\Gamma_M\Psi_A)], \tag{1.1.3}$$

$$\Omega_{MN}{}^A = 2\partial_{[N}e^A_{M]} \tag{1.1.4}$$

$$\tilde{\omega}_{MAB} = \omega_{MAB} + \frac{i}{4}\bar{\Psi}_N\Gamma_{MAB}{}^{NP}\Psi_P \tag{1.1.5}$$

$$F_{MNPQ} = 4\partial_{[M}A_{NPQ]} \tag{1.1.6}$$

$$\tilde{F}_{MNPQ} = F_{MNPQ} - 3\bar{\Psi}_{[M}\Gamma_{NP}\Psi_{Q]} \tag{1.1.7}$$

We are using the convention that $M, N, P...$ refer to $d = 11$ world indices and $A, B, C...$ refer to $d = 11$ tangent space indices. $\epsilon^{MNP...}$ is a tensor, rather than a tensor density, and

$$\epsilon_{12...11} = e \tag{1.1.8}$$

The Γ-matrices satisfy

$$\{\Gamma_A, ,\Gamma_B\} = -2\eta_{AB} \tag{1.1.9}$$

where η_{AB} is the metric in the tangent space and $\Gamma_{A_1...A_p} \equiv \Gamma_{[A_1}...\Gamma_{A_p]}$. The spinors appearing in (1.1.1) are anticommuting and satisfy the Majorana condition

$$\bar{\Psi} = \Psi^T C \tag{1.1.10}$$

where the charge conjugation matrix C is antisymmetric and defined by

$$C^{-1}\Gamma_A C = -\Gamma_A{}^T \tag{1.1.11}$$

Upon variation with respect to $e_M{}^A$, Ψ_M, and A_{MNP} we obtain the field equations [2]

$$R_{MN}(\tilde{\omega}) - \frac{1}{2} g_{MN} R(\tilde{\omega})$$
$$= \frac{1}{3}[\tilde{F}_{MNPQ}\tilde{F}_N{}^{PQR} - \frac{1}{8} g_{MN}\tilde{F}_{PQRS}\tilde{F}^{PQRS}] \tag{1.1.12}$$

$$\Gamma^{MNP}\tilde{D}_N(\tilde{\omega})\Psi_P = 0 \tag{1.1.13}$$

$$\nabla_M(\tilde{\omega})\tilde{F}^{MPQR} = -\frac{1}{576}\epsilon^{PQRM_1\ldots M_4 M_5\ldots M_8}\tilde{F}_{M_1\ldots M_4}\tilde{F}_{M_5\ldots M_8} \tag{1.1.14}$$

where

$$\tilde{D}_M(\tilde{\omega})\Psi_N = D_M(\tilde{\omega})\Psi_N + T_M{}^{PQRS}\tilde{F}_{PQRS}\Psi_N \tag{1.1.15}$$

and

$$T^{SMNPQ} = \frac{1}{144}(\Gamma^{SMNPQ} - 8\Gamma^{[MNP}g^{Q]S}) \tag{1.1.16}$$

We see that the appearance of $\frac{1}{2}(\omega + \tilde{\omega})$ and $\frac{1}{2}(F + \tilde{F})$ in the Lagrangian ensures that only the supercovariant $\tilde{\omega}$ and $\tilde{F}$ enter the field equations.

The action and equations of motion are invariant under the following symmetries

a) $d = 11$ general covariance, with parameter ξ^M

$$\delta e_M{}^A = e_N{}^A \partial_M \xi^N + \xi^N \partial_N e_M{}^A \tag{1.1.17}$$
$$\delta \Psi_M = \Psi_N \partial_M \xi^N + \xi^N \partial_N \Psi_M \tag{1.1.18}$$
$$\delta A_{MNP} = 3A_{Q[NP}\partial_{M]}\xi^Q + \xi^Q \partial_Q A_{MNP} \tag{1.1.19}$$

b) Local $SO(1,10)$ Lorentz transformations with parameter $\alpha_{AB} = -\alpha_{BA}$

$$\delta e_M{}^A = -e_M{}^B \alpha_B{}^A \tag{1.1.20}$$
$$\delta \Psi_M = -\frac{1}{4}\alpha_{AB}\Gamma^{AB}\Psi_M \tag{1.1.21}$$
$$\delta A_{MNP} = 0 \tag{1.1.22}$$

c) $N = 1$ supersymmetry transformations with anticommuting parameter ϵ

$$\delta e_M{}^A = -i\bar{\epsilon}\Gamma^A\Psi_M \tag{1.1.23}$$

$$\delta\Psi_M = \tilde{D}_M(\tilde{\omega})\epsilon \tag{1.1.24}$$

$$\delta A_{MNP} = \frac{3}{2}\bar{\epsilon}\Gamma_{[MN}\Psi_{P]} \tag{1.1.25}$$

d) abelian gauge transformations with parameter $\Lambda_{MN} = -\Lambda_{NM}$

$$\delta e_M{}^A = 0 \tag{1.1.26}$$

$$\delta\Psi_M = 0 \tag{1.1.27}$$

$$\delta A_{MNP} = \partial_{[M}\Lambda_{NP]} \tag{1.1.28}$$

e) odd number of space or time reflections together with

$$A_{MNP} \rightarrow -A_{MNP} \tag{1.1.29}$$

This summarizes the Lagrangian and transformation rules to be used in subsequent sections. For possible Chern-Simons modifications, see [3]. The ramifications of adding such a term and its effect on the $d = 4$ theory are very interesting but will not be pursued here.

1.2 The Freund-Rubin ansatz: $M_4 \times M_7$ ground state.

We now see how $d = 11$ supergravity admits spontaneous compactification to $d = 4$. We start by looking for solutions of (1.1.12) to (1.1.14) which might be candidates for a ground state, at least at the tree level.

We are eventually interested in obtaining a four dimensional theory which admits maximal spacetime symmetry. With signature $(-+++)$ this means that the vacuum should be invariant under $SO(1,4)$, Poincaré, or $SO(2,3)$ according as the cosmological constant is positive, zero, or negative, corresponding to de Sitter, Minkowski or anti-de Sitter space (AdS), respectively.

The first requirement of maximal symmetry is that the VEV of any fermion field should vanish and accordingly we set

$$< \psi_M >= 0\,, \tag{1.2.1}$$

and focus our attention on the equations for g_{MN} and A_{MNP} which from (1.1.12) and (1.1.14) are

$$R_{MN} - \frac{1}{2} g_{MN} R = \frac{1}{3} \left[F_{MNQP} F_N{}^{PQR} - \frac{1}{8} g_{MN} F_{PQRS} F^{PQRS} \right], \tag{1.2.2}$$

$$\nabla_M F^{MNPQ} = -\frac{1}{576} \epsilon^{M_1 \ldots M_8 NPQ} F_{M_1 \ldots M_4} F_{M_5 \ldots M_8} \tag{1.2.3}$$

We look for solutions of the direct product form $M_4 \times M_7$, compatible with maximal spacetime symmetry. Thus we set

$$\begin{aligned} &< g_{\mu\nu} > = \overset{\circ}{g}_{\mu\nu}(x) \quad , \quad < F_{\mu\nu\rho\sigma} >= \overset{\circ}{F}_{\mu\nu\rho\sigma}(x), \\ &< g_{mn} > = \overset{\circ}{g}_{mn}(y) \quad , \quad < F_{mnpq} >= \overset{\circ}{F}_{mnpq}(y) \end{aligned} \tag{1.2.4}$$

but

$$< g_{\mu n} >= 0\,, < F_{\mu\nu\rho q} >=< F_{\mu\nu p q} >=< F_{\mu n p q} >= 0, \tag{1.2.5}$$

where x^μ are the spacetime coordinates and y^m the extra coordinates. The y independence of $\overset{\circ}{g}_{\mu\nu}$ and the x independence of $\overset{\circ}{g}_{mn}$ in (1.2.4) are necessary if $\overset{\circ}{g}_{MN}$ is to be a product metric, whereas the y independence of $F_{\mu\nu\rho\sigma}$ and the x independence of F_{mnp}, are consequences of the Bianchi identity

$$\partial_{[M} F_{NPQR]} = 0 \tag{1.2.6}$$

and (1.2.5).

It is important to note that maximal spacetime symmetry alone would not rule out a "warped-product" ansatz for which $< g_{\mu\nu}(x,y) >= f(y) g_{\mu\nu}(x)$.

Note however that since the Bianchi identity does not involve the metric, the y-independence of $F_{\mu\nu\rho\sigma}$ is still guaranteed. In what follows we shall confine our attention to "warp-factor one" *i.e.* $f(y) = 1$. The case $f(y) \neq 1$ will be postponed until Section 3.

The ansatz of Freund and Rubin [4] is to set

$$\overset{\circ}{F}_{\mu\nu\rho\sigma} = 3m\overset{\circ}{\epsilon}_{\mu\nu\rho\sigma}, \tag{1.2.7}$$
$$\overset{\circ}{F}_{mnpq} = 0, \tag{1.2.8}$$

where m is a real constant and the factor of 3 is chosen for future convenience. The superscript zero on the Levi-Cevita tensor $\epsilon_{\mu\nu\rho\sigma}$ means that $\epsilon_{0123} = \sqrt{-\overset{\circ}{g}}$. Substituting (1.2.7) and (1.2.8) into the field equaitons we find that (1.2.3) is trivially satisfied while (1.2.2) yields the product of a four-dimensional Einstein spacetime

$$\overset{\circ}{R}_{\mu\nu} = -12m^2\overset{\circ}{g}_{\mu\nu} \tag{1.2.9}$$

with Minkowski signature $(- + ++)$ and a seven-dimensional Einstein space

$$\overset{\circ}{R}_{mn} = 6m^2\overset{\circ}{g}_{mn} \tag{1.2.10}$$

with Euclidean signature $(+ + + + + + +)$.

For future reference we also record the form taken by the supercovariant derivative $\tilde{D}_M$ appearing in (1.1.15), when evaluated in the Freund-Rubin background geometry. First we decompose the $d = 11$ $\hat{\Gamma}$ matrices

$$\hat{\Gamma}_A = (\gamma_\alpha \otimes \mathbf{1}\,, \gamma_5 \otimes \Gamma_a) \tag{1.2.11}$$

where

$$\{\gamma_\alpha, \gamma_\beta\} = -2\eta_{\alpha\beta} \tag{1.2.12}$$
$$\{\Gamma_a, \Gamma_b\} = -2\delta_{ab} \tag{1.2.13}$$

and where $\alpha, \beta ...$ are spacetime indices for the tangent space group $SO(1,3)$ and $a, b ...$ are extra dimensional indices for the tangent space group $SO(7)$.

Substituting (1.2.4), (1.2.5), (1.2.7), (1.2.8) and (1.2.11) into (1.1.15) we find that

$$\overset{\circ}{\tilde{D}}_\mu = D_\mu + m\gamma_\mu\gamma_5 \tag{1.2.14}$$

$$\overset{\circ}{\tilde{D}}_m = D_m - \frac{m}{2}\Gamma_m, \tag{1.2.15}$$

where $\overset{\circ}{\gamma}_\mu = \overset{\circ}{e}_\mu{}^\alpha\gamma_\alpha$ and $\overset{\circ}{\Gamma}_m = e_m{}^a\overset{\circ}{\Gamma}_a$.

Several comments are now in order. The constancy of m in the ansatz (1.2.7) is necessary in order to satisfy the field equations. However the ansatz (1.2.8) is not necessary and indeed we shall consider solutions with non-zero F_{mnpq} in Section 2. The maximally symmetric solution to (1.2.9) is, in fact, anti-de Sitter space since the cosmological constant $\Lambda = -12m^2$ is negative. There are infinitely many seven-dimensional Einstein spaces M_7 satisfying (1.2.10) and, for the moment, M_7 will be left arbitrary. The important point is that Einstein spaces of positive curvature and Euclidean signature are automatically compact [5] and hence spontaneous compactification to $d = 4$ has indeed been achieved.

2. Relation between d = 11 and d = 4 supergravity

2.1 The de Wit-Nicolai theory

In 1981 de Wit and Nicolai [6] constructed a four-dimensional supergravity theory by gauging the $SO(8)$ subgroup of E_7 in the global $E_7\times$ local $SU(8)$ supergravity of Cremmer and Julia [7]. In common with the Cremmer-Julia theory, this theory described the self-interaction of a single massless $N = 8$ supermultiplet of spin (2, 3/2, 1, 1/2, 0^+, 0^-) but with local $SO(8)\times$ local $SU(8)$ invariance. In addition to the gravitational constant, there was a new parameter; the $SO(8)$ gauge coupling constant e, and it was necessary to modify the Cremmer-Julia Lagrangian and transformation rules by other e dependent terms in order to preserve the $N = 8$ supersymmetry. In particular there were Yukawa-like interactions between the fermion and spin 0 fields, and a non-trivial effective potential for the scalars. As in the Cremmer-Julia theory, the $SU(8)$ gauge potentials were composite fields with no kinetic term of their own and were therefore expressible in terms of the spin 0 fields and their derivatives. The transformation of the various fields under $SO(8) \times SU(8)$ is summarized in Table I. We note, in particular that the fermions are *chiral* under $SU(8)$. The number of spin 0 *fields* is reduced to 133 by assigning the u's and v's to a "sechsundfunfsigbein" which is an element of E_7. The number of physical spin-0 states is $70 = 133 - 63$ since 63 fields may be gauged away by an $SU(8)$ rotation.

TABLE I

spin	2	3/2	1	1/2	0	0
field	$e_\mu^{\ a}$	$\psi_\mu^{\ i}$	$B_\mu^{\ IJ}$	χ^{ijk}	U^{ijIJ}	$V_{ij}^{\ IJ}$
SO(8)	1	1	28	1	28	28
SU(8)	1	8	1	56	28	$\overline{28}$

Since E_7 is no longer a symmetry, however, we cannot parametrize these 70 scalars by the coset $E_7/SU(8)$ as Cremmer-Julia did. They are no longer Goldstone bosons.

The existence of the effective potential naturally prompted the question of its extrema. As discussed by de Wit and Nicolai [6], the obvious symmetric extremum for which both the 35 scalars and 35pseudoscalars have zero VEV's yields an $SO(8)$ vacuum state with $N = 8$ supersymmetry. ($SU(8)$ is not, or course, a symmetry of the vacuum.) The physical states belong to the familiar massless $N = 8$ supermultiplet with $(1, 8_s, 28, 56_s, 35_v, 35_c)$ representations of $SO(8)$. The effective potential then yields a non-zero cosmological constant for this vacuum given by $4\pi G\Lambda = -3e^2$, and the Yukawa couplings yield a "mass term" for the gravitino. The relevant algebra is therefore the $OSp(4|8)$ de Sitter supersymmetry rather than the $N = 8$ Poincaré sypersymmetry of the Cremmer-Julia vacuum. Note, however, that the de Wit-Nicolai theory goes smoothly over to the Cremmer-Julia theory in the limit $e \to 0$. We shall return to this point shortly.

For full details of the de Wit-Nicolai theory, see [6]. By truncation, one may also obtain gauged $SO(N)$ supergravities for $N < 8$, in particular $N = 5$ and 6. The gauged $N \leq 4$ theories were already known to exist [8,9,10]. The existence of gauged extended supergravity for $N > 4$ was already strongly suggested, before the explicit construction of de Wit and Nicolai, by the discovery that such theories would necessarily have vanishing one-loop β function [11]. This was strongly reminiscent of the vanishing one-loop β function in $N > 2$ super Yang-Mills theories, and indeed both were subsequently explained by the device of spin-moment sum rules [12,13].

2.2 Conjectures on the seven-sphere

Shortly after the de Wit-Nicolai theory was constructed, Duff and Pope [13,14,16,17] observed that $d = 11$ supergravity admitted vacuum solutions of the form $AdS \times S^7$, and that since S^7 has isometry group $S0(8)$ and admits 8 Killing spinors, this gives rise via the Kaluza-Klein mechanism to an effective four-dimensional theory with local $S0(8)$ invariance and $N = 8$ supersymmetry. It followed automatically that this $d = 4$ theory described the interaction of one massless $N = 8$ supermultiplet of spins $(2, 3/2, 1, 1/2, 0^+, 0^-)$ in $S0(8)$ reps $(1, 8_s, 28, 56_s, 35_v, 35_c)$ with an infinite tower of massive $N = 8$ supermultiplets. It also followed that in this $N = 8$ vacuum Λ was proportional to $-Ge^2$ with a calculable, but at the time not yet calculated, coefficient.

All this led to the conjecture [13,14,16,17] that the dynamics of the massless $N = 8$ supermulitpiet of this Kaluza-Klein theory was nothing but the $N = 8$ theory of de Wit and Nicolai. Let us call this conjecture I. The arrival of the Englert solution [15] led to the observation [16] that the round S^7 and the round S^7 with "torsion" (*i.e.* non-vanishing A_{mnp}) were merely different phases of the same four-dimensional theory, and that "torsion" corresponded to a non-zero VEV for the 35_c pseudoscalars. This led to conjecture II: Just as the symmetric $N = 8$ extremum of the de Wit-Nicolai effective potential in $d = 4$ corresponds to the round S^7 without torsion, so non-symmetric extrema correspond to other solutions of the $d = 11$ theory with different geometry but the same S^7 topology, and that non-zero VEVs for scalars and pseudoscalars in $d = 4$ correspond to a deformation of and/or torsion on S^7 in $d = 11$ [16,17] (Of course, the deformation or torsion in question would have to correspond to the 35_v scalars and/or the 35_c pseudoscalars. This means, in particular, that the squashed S^7 of Awada, Duff and Pope [19] would not admit any such de Wit-Nicolai interpretation, since here it is the 300 of $SO(8)$ which acquires the non-zero VEV. This latter fact was not appreciated until later [21], however). Thus one had a $d = 11$ geometrical Kaluza-Klein interpretation of the $d = 4$ Brout-Englert-Higgs-Kibble effect. This relationship was further explored in [18,20,22].

While some held the truth of these conjectures to be self-evident, others raised objections. Their verification would require a complete non-linear analysis of the $d = 4$ equations of motion obtained from $d = 11$ and, to date, only a partial results are known mainly because of the complicated non-polynomial dependence on the scalars and pseudoscalars. Let us therefore list some of these objections and see how far they have been overcome to date. First of all, implicit in conjecture I is the assumption that it is in fact possible to truncate the full Kaluza-Klein theory to its massless sector in a consistent fashion. As it turns out, such fears are in general well founded. It is a remarkable property of S^7 that, with the exception of T^7, it provides the only examples of M_7 compactifications of $d = 11$ supergravity to admit such consistent trucations. This is the subject of Section 3.

A second objection to conjecture I concerned the cosmological constant. A necessary condition for agreement is that the spacetime cosmological constant Λ in the round S^7 vacuum must be related to the $SO(8)$ gauge coupling constant e by the same formula $4\pi G\Lambda = -3e^2$ as obtained in the $N = 8$ phase of the de Wit-Nicolai theory. Weinberg [23] has shown how the coupling constant may be calculated starting from pure Einstein gravity in higher dimensions. Applied to the round S^7 of radius m^{-1}, the formula gives $e^2 = 64\pi Gm^2$ which, combined with $\Lambda = -12m^2$ from (1.2.9) yields $16\pi G\Lambda = -3e^2$ which disagrees by a factor of 4 with the de Wit-Nicolai value: an apparent disaster for conjecture I. However, it turns out that the presence of the A_{MNP} field in $d = 11$ supergravity leads to a modification of Weinberg's pure gravity calculation [24]. As described in Sections 2.3 and 3.2 when this is taken into account it changes the relation between e^2 and m^2 to $e^2 = 16\pi Gm^2$ which yields precisely the de Wit-Nicolai relation.

A third objection to conjecture I, closely related to the other two, involved the massless ansatz and the supersymmetry transformation rules. One important consistency check, pointed out in [16], was whether the $d = 11$ supersymmetry tranformation rules consistently yielded the correct $d = 4$ transformation rules when the ansatz for the massless supermultiplet was substituted in. This massless ansatz was known in full at the linearized level [17,20] and is given in Section 2.3, so one could check the transformation rules to the same order of approximation. In fact, as we shall see in Section 2.3, everything works out well [25,26] despite earlier claims that there was a mismatch in the y dependence.

A surprising result, not anticipated at the time conjecture I was made, is the appearance of the 294_v massless 0^+ in the round S^7 spectrum not belonging to the massless $N = 8$ supermultiplet [27]. The inclusion of these fields in the massless ansatz would clearly violate supersymmetry and lead to inconsistencies. When discussing conjecture I it is always important therefore to distinguish the ansatz for the mass *supermultiplet* from the ansatz for the massless *sector*.

At the time conjecture II was made, only one non-symmetric extremum of the de Wit-Nicolai theory was known [28]. This was Warner's $SO(3) \times SO(3)$ extremum with $N = 0$. Since then Warner has discovered many more and has classified all those which have at least $SU(3)$ symmetry [29]. See Section 2.4.

Up until recently, these results posed the gravest threat to conjecture II. For although all comparisons of the $SO(7)_c$ extremum in $d = 4$ with the Englert solution in $d = 11$ proved positive [30,31], the other non-symmetric extrema defied all attempts to yield to a $d = 11$ origin. It was even proved that no Freund-Rubin type solution with or without "torsion" could have G_2 symmetry unless it was the round S^7 [32]. The resolution of this problem is the subject Section 2.4 of and turns out to have.far-reaching consequences for Kaluza-Klein.

2.3. Linearized massless ansatz and transformation rules

The ansatz for the massless supermultiplet at the linearized level can be obtained by retaining only the lowest modes in the Fourier expansion with the harmonics specialized to S^7. Making the field redefinitions

$$h'_{MN} = h_{MN} + \frac{1}{2} g_{MN} h^q_q \tag{2.3.1}$$

$$\Psi'_M = \Psi_M - \frac{1}{2} \hat{\Gamma}_M \hat{\Gamma}^q \Psi_q \tag{2.3.2}$$

the ansatz is [17,20];

$$h'_{\mu\nu}(x,y) = h_{\mu\nu}(x) \tag{2.3.3}$$

$$h'_{\mu n}(x,y) = \frac{1}{2} B^{IJ}_\mu(x) \eta^{IJ}_n \tag{2.3.4}$$

$$h'_{mn}(x,y) = S^{IJKL}(x) \eta^{IJKL}_{mn} \tag{2.3.5}$$

$$f_{\mu\nu\rho\sigma}(x,y) = \frac{3m}{2} \epsilon_{\mu\nu\rho\sigma} (h'^{\,\tau}_{\tau} - \frac{2}{9} h'^{\,t}_{t}) \tag{2.3.6}$$

$$f_{\mu\nu p q}(x,y) = \frac{1}{24m} \epsilon_{\mu\nu\rho\sigma} \nabla^\sigma \nabla_q h'^{\,t}_{t} \tag{2.3.7}$$

$$f_{\mu\nu p q}(x,y) = -\frac{1}{2m} \epsilon_{\mu\nu\rho\sigma} \nabla^\rho \nabla_{[p} h'_{q]}{}^{\sigma} \tag{2.3.8}$$

$$f_{\mu n p q}(x,y) = -\frac{1}{2} \partial_\mu P^{IJKL}(x) \eta^{IJKL}_{mnpq} \tag{2.3.9}$$

$$f_{mnpq}(x,y) = 2mP^{IJKL}(x)\eta^{IJKL}_{mnpq} \tag{2.3.10}$$

$$\Psi'_\mu(x,y) = \psi^I_\mu(x)\eta^I \tag{2.3.11}$$

$$\Psi'_m(x,y) = \chi^{IJK}(x)\eta^{IJK}_m \tag{2.3.12}$$

The y-dependent tensors η are defined in terms of "Killing spinors." A Killing spinor η on M^7 is defined to be a solution of the equation

$$\overset{\circ}{\tilde{D}}_m\eta = 0$$

where $\overset{\circ}{\tilde{D}}_m$ is the $d = 7$ supercovariant derivative of (1.2.15). On the round S^7 there exists the maximum number (8) of Killing spinors, denoted $\eta^I (I = 1...8)$ where I denotes the 8_s representation of $SO(8)$. From these we may construct

$$\eta^{IJ}_m = \bar{\eta}^I\Gamma_m\eta^J \quad , \quad \eta^{IJKL}_{mn} = \eta^{[IJ}_m\eta^{KL]}_n$$
$$\eta^{IJKL}_{mnp} = \eta^{IJ}_{[mn}\eta^{KL}_{p]} \quad , \quad \eta^{IJKL}_{mnpq} = \eta^{IJ}_{[mn}\eta^{KL}_{pq]} \tag{2.3.13}$$

where we also define

$$\eta^{IJ}_{mn} = \bar{\eta}^I\Gamma_{mn}\eta^J \quad , \quad \eta^{IJK}_m = \eta^{[I}\Gamma^{JK]}_m \tag{2.3.14}$$

The vectors η^{IJ}_m are in fact the 28 Killing vectors on the round S^7.

The expansion of the supersymmetry parameter, which we shall modify later, will for now be taken to be

$$\epsilon(x,y) = \epsilon^I(x)\eta^I \tag{2.3.15}$$

The idea now is to substitute these ansätze into the $d = 11$ supersymmetry transformation rules (1.1.23), (1.1.24) and (1.1.25) in order to verify to first order in fields, four dimensional supersymmetry rules are obtainable *i.e.* that the y dependence of the left and right hand sides of each of these equations matches. Here we follow the paper of Awada, Nilsson and Pope [25]. Starting with (1.1.23) we first convert it into transformation law for the linearized metric

$$\delta h_{MN} = -2i\bar{\epsilon}\hat{\Gamma}^{(0)}_{(M}\Psi_{N)} \tag{2.3.16}$$

where $\hat{\Gamma}^{(0)}_M$ is obtained from $\hat{\Gamma}_M$ using the background vielbein. The calculations here are straightforward, and indeed yield equations in which the y dependence matches.

There are three cases to consider, corresponding to $MN = \mu\nu$, μn, and mn, and so dropping the y dependence these give, after some algebra,

$$\begin{aligned}
\delta h_{\mu\nu} &= -2i\bar{\epsilon}^I \gamma_{(\mu}\psi^I_{\nu)} \\
\delta B^{IJ}_\mu &= -2i\bar{\epsilon}^{[I}\gamma_5\psi^{J]}_\mu - 2i\bar{\epsilon}^K\gamma_\mu\chi^{KIJ} \\
\delta S^{IJKL} &= -2i\bar{\epsilon}^{[I}\gamma_5\chi^{JKL]_+}
\end{aligned} \tag{2.3.17}$$

where in the final equation $[IJKL]_+$ denotes the self-dual projection of $[IJKL]$. This projection follows from the duality property of η^{IJKL}_{mn} defined in (2.3.13).

We now convert (1.1.25) into a transformation law for F_{MNPQ} and so after linearization we obtain

$$\delta f_{MNPQ} = 6D_{[M}(\bar{\epsilon}\hat{\Gamma}^{(0)}_{NP}\Psi_{Q]}). \tag{2.3.18}$$

On substituting the ansätze into (2.3.18), one again finds after some straightforward but tedious algebra that the y dependence matches. There are five cases to consider, corresponding to the various possible combinations of spacetime and internal indices. All except the cases of 3 or 4 internal indices reproduce the previously obtained transformation laws (2.3.17). The remaining cases produce the last bosonic transformation law

$$\delta P^{IJKL} = -2\bar{\epsilon}^{[I}\chi^{JKL]}_- \tag{2.3.19}$$

We now turn to the fermion transformation law (1.1.24). This is the equation which at first sight appears to give rise to a mismatch of the y dependence, and hence an inconsistent result, and so here we shall describe the calculation in greater detail. Linearizing (1.1.24), we obtain

$$\delta\Psi_M = \overset{\circ}{\tilde{D}}_M\epsilon - \frac{1}{4}(2e_A{}^N \overset{\circ}{D}_{[M}V_{N]B} - e_A{}^P e_B{}^Q e_M{}^D \overset{\circ}{D}_P V_{QD})\hat{\Gamma}^{AB}\epsilon - \frac{i}{144}(\hat{\Gamma}_M^{(0)NPQR} + 8\delta_M^N\hat{\Gamma}^{(0)PQR})f_{NPQR}\epsilon - \frac{i}{144}(\hat{\Gamma}_M^{(1)\,NPQR} + 8\delta_M^N\hat{\Gamma}^{(1)PQR})F^{(0)}_{NPQR}\epsilon \qquad (2.3.20)$$

where $V_M{}^A$ is the fluctuation around the background vielbein $e_M{}^A$ corresponding to the metric fluctuation h_{MN}. It is convenient to impose Lorentz gauge condition $V_m{}^\alpha = 0$. The matrices $\hat{\Gamma}_M^{(1)NPQR}$ and $\hat{\Gamma}^{(1)PQR}$ are the terms linear in $V_M{}^A$ which result from converting tangent space to world indices on the Dirac matrices $\hat{\Gamma}_A{}^{BCDE}$ and $\hat{\Gamma}^{CDE}$.

We now substitute the ansätze into (2.3.20), and consider first the case $M = m$. On general grounds we expect the variation $\delta\Psi_m$ to involve terms containing B_μ, $\nabla_{[\mu}B_{\nu]}$, $\partial_\mu S$, $\partial_\mu P$, S, and P. Detailed calculation shows that the term in B_μ vanishes, while after a Fierz transformations the y dependence of the term in $\nabla_{[\mu}B_{\nu]}$ is seen to match with the left hand side of (2.3.20). For the scalar and pseudoscalar terms, we require the Fierz transformation

$$\begin{aligned} &- 144(\delta^{I[J}\overset{\circ}{\eta}{}_m^{KLM]} + \frac{1}{9}\Gamma_m\delta^{I[J}\eta^{KLM]}) \\ &= \Gamma_m{}^{npq}\eta^I\eta^{JKLM}_{npq} + 6\Gamma^{np}\eta^I\eta^{JKLM}_{mnp} + 18\Gamma^n\eta^I\eta^{JKLM}_{mn} \\ &- 2\Gamma_m\eta^I\eta^{JKLM}_{pq}g^{pq} + \Gamma_m{}^p\eta^I\eta^{[JK}_{np}\eta^{LM]}_q g^{nq} \\ &- 2\eta^I\eta^{JK}_{mn}\eta^{LM}_p g^{np} \end{aligned} \qquad (2.3.21)$$

where we have dropped the ${}^{(0)}$ superscript on the $d = 7$ Γ matrices. It is crucial to note that the first two terms on the right hand side of (2.3.21) are anti-self-dual in $JKLM$, since they are associated with the 35_c of three-forms of the pseudoscalar ansatz, while the remaining terms are self-dual in $JKLM$, since they are associated with the 35_v of Killing tensors of the scalar ansatz. The terms in $\delta\Psi_m$ involving $\partial_\mu S$ and $\partial_\mu P$ are precisely the self-dual and anti-self-dual parts of (2.3.21), respectively, and hence the y dependence of these terms matches with $\delta\Psi_m$; they occur in the combination $\partial_\mu(S + i\gamma_5 P)$.

Repeating the procedure for the S and P terms in $\delta\Psi_m$, one runs into an apparent inconsistency. The relevant terms in the transformation rule are

$$\delta\chi^{IJK}|_{S,P}\,(\eta_m^{IJK}+\frac{1}{9}\Gamma_m\eta^{IJK})$$
$$=\frac{m}{72}S^{JKLM}\epsilon^I(-5\Gamma_m{}^p\eta^I\eta_{np}^{[JK}\eta_q^{LM]}g^{nq}-2\Gamma_m\eta^I\eta_{pq}^{JKLM}g^{pq}-18\Gamma^n\eta^I\eta_{mn}^{JKLM})$$
$$+\frac{im}{72}P^{JKLM}\gamma_5\epsilon^I(-8\Gamma_m^{npq}\eta^I\eta_{npq}^{JKLM}-12\Gamma^{pq}\eta^I\eta_{mpq}^{JKLM}) \qquad (2.3.22)$$

Using (2.3.21), one can now show that the y dependence on each side of (2.3.22) does not match. To remedy this, we note that the ansatz (2.3.15) for $\epsilon(x,y)$ implies that $\bar{D}_m\epsilon=0$, but that the ansatz can be modified by the addition of S- and P-dependent terms in such a manner that $\tilde{D}_m\epsilon$ now produces precisely the required extra terms in (2.3.22) in order that the y dependence coincides with that of the Fierz equation (2.3.21). The modified ansatz is

$$\epsilon(x,y)=[1+\frac{1}{72}S^{JKLM}(-2\eta_{mn}^{JKLM}g^{mn}+3\eta_{mn}^{[JK}\eta_p^{LM]}g^{np}\Gamma^m)$$
$$+\frac{i}{72}\gamma_5P^{JKLM}(-3\eta_{mnp}^{JKLM}\Gamma^{mnp})]\epsilon^I(x)\eta^I \qquad (2.3.23)$$

Note that this does not upset any of the previously derived linearized transformation rules, since these ϵ always appear in terms already linear in fields. Summarizing, the transformation rule for $\delta\Psi_m$ gives, in terms of $F_{\mu\nu}^{IJ}=2\nabla_{[\mu}B_{\nu]}^{IJ}$,

$$\delta\chi^{IJK}=\frac{3}{8}F_{\mu\nu}^{[IJ}\gamma^{\mu\nu}\epsilon^{K]}-2\gamma_5\partial\mu(S^{IJKL}+i\gamma_5P^{IJKL})\gamma^\mu\epsilon^L$$
$$-4m(S^{IJKL}+i\gamma_5P^{IJKL})\epsilon^L. \qquad (2.3.24)$$

Finally, substituting the ansätze for the massless fields, and the modified ansatz (2.3.23) for $\epsilon(x,y)$, into the transformation rule for $\delta\Psi_\mu$, one finds that the y dependence matches on each side of the equation, and

$$\delta\psi_\mu^I=\bar{D}_\mu\epsilon^I-2mB_\mu^{IJ}\epsilon^J-[\frac{(1-\gamma_5)}{2}F_{\mu\nu}^{+IJ}-\frac{(1+\gamma_5)}{2}F_{\mu\nu}^{-IJ}]\gamma^\nu\epsilon^J, \qquad (2.3.25)$$

where $F^{\pm}_{\mu\nu} = \frac{1}{2}(F_{\mu\nu} \pm {}^*F_{\mu\nu})$ and ${}^*F_{\mu\nu} = \frac{i}{2}\epsilon_{\mu\nu\rho\sigma}F^{\rho\sigma}$. The derivative $\tilde{D}_\mu$ is the anti-de Sitter covariant derivative $D_\mu + m\gamma_\mu\gamma_5$ including terms up to first order in fluctuations. The transformation rules must, of course, be $SO(8)$ gauge covariant, and defining an $SO(8)$ gauge covariant derivative $\mathcal{D}_\mu$ by

$$\mathcal{D}_\mu \epsilon^I = D_\mu \epsilon^I - eB^{IJ}_\mu \epsilon^J \tag{2.3.26}$$

then (2.3.25) may be written as

$$\delta\psi^I_\mu = \tilde{\mathcal{D}}_\mu \epsilon^I - [\frac{(1-\gamma_5)}{2}F^{+IJ}_{\mu\nu} - \frac{(1+\gamma_5)}{2}F^{-IJ}_{\mu\nu}]\epsilon^J \tag{2.3.27}$$

provided the $SO(8)$ gauge coupling constant e is chosen to be

$$e = 2m. \tag{2.3.28}$$

Here we have defined $\tilde{\mathcal{D}}_\mu$ in the same manner as $\tilde{D}_\mu$, *i.e.* $\tilde{\mathcal{D}}_\mu = \mathcal{D}_\mu + m\gamma_\mu\gamma_5$. (As we shall see in Section 3.2, the normalization chosen for the massless ansatz of this section is such that $4\pi G = 1$ where G is the $d = 4$ Newton's constant. This implies $e^2 = 16\pi G m^2$ *i.e.* $4\pi G\Lambda = -3e^2$ as required for consistency with conjecture I.)

It is important to check that the redefinition of $\epsilon(x, y)$ in (2.3.23) does not upset any of our previous results. In particular, the criterion for unbroken supersymmetry $< \delta\Psi_M >= 0$ remains unaltered. This is because$< S >=< P >= 0$ in the round S^7 background.

Thus we see that the truncation of the $d = 11$ theory to include just the massless supermultiplet is indeed consistent with the transfromation rules at this order of approximation. Moreover, to the same order of approximation, the $d = 4$ transformation rules coincide with those of the de Wit-Nicolai theory as may be seen after some trivial rescalings of the fields. In particular, note that the $SO(8)$ covariantization of the derivative in (2.3.27) and the appearance throughout of the combination of fields $(S + i\gamma_5 P)$. When one considers the very different $d = 11$ origins of S and P, this latter fact is truly remarkable and is suggestive of some deeper connection between g_{MN} and A_{MNP} in $d = 11$ which may explain the $SU(8)$ symmetry in $d = 4$.

One may wonder how the *linearized* ansatz can give information about couplings constants and covariant derivatives. The answer is that, although we worked to only linear order in *fields*, the transformation rules involve products of fields and supersymmetry parameters. Hence they yield information about cubic terms in the Lagrangian. Thus consistency of the above transformation rules is already probing the non-linear structure. It is to the non-linear theory that we now turn.

2.4 New S^7 solutions and Warner's extrema

Warner [29] has now completed the list of all the extrema of the de Wit-Nicolai potential in $d = 4$ which induce a breaking of $SO(8)$ to some smaller group containing $SU(3)$. It is not yet known how many other extrema there are, but the original example of $SO(3) \times SO(3)$ is known as mentioned in Section 2.1. The known extrema are shown in Table 2.

Let us now reexamine conjecture II in the light of these extrema. All comparisons which have been made to date regarding gauge symmetries, supersymmetries, cosmological constants and partial mass spectra are consistent with the identification of : E_0 with the Duff-Pope S^7 solution [13,14,16,17]; E_1 with the Englert S^7 solution [15]; and E_4 with the Pope-Warner S^7 solution [33].

TABLE 2

Known extrema of the scalar potential

Extremum	Gauge Symmetry	Supersymmetry	Non-zero VEVs
E_0	SO(8)	N=8	-
E_1	$SO(7)_c$	N=0	P
E_2	$SO(7)_s$	N=0	S
E_3	G_2	N=1	S,P
E_4	SU(4)	N=0	P
E_5	SU(3)xU(1)	N=2	S,P
E_6	SO(3)xSO(3)	N=0	S,P

In all cases, non-zero VEVs for the pseudoscalars P in $d = 4$ correspond to non-zero F_{mnpq} in $d = 11$. The fact that $< S >= 0$ does not necessarily imply the *round* S^7 geometry since the seven-sphere is "stretched" in the Pope-Warner solution. This reflects the fact that the pseudoscalar will enter at the non-linear level in the ansatz for $g_{mn}(x,y)$.

Much more mysterious from the point of view of conjecture II are those extrema with $< S > \neq 0$, namely E_2, E_3, E_5 and E_6. Indeed, it can be proved that with the usual product ansatz of Section 1.2, there is no Freund-Rubin type solution, with or without "torsion", capable of reproducing these extrema [32]. This led to claims that conjecture II was wrong. It was realized in [34] that the resolution to the problem was the "warp-factor" $f(y)$ [35,32,36] discussed in Section 1.2, which we have ignored up until now. In other words, one should generalize the ground state ansatz for the metric $\hat{g}_{MN}$ to be

$$< \hat{g}_{MN}(x,y) > = \begin{pmatrix} f(y)g_{\mu\nu}(x) & 0 \\ 0 & g_{mn}(y) \end{pmatrix} \tag{2.4.1}$$

One can now find an S^7 solution with $SO(7)_s$ symmetry and $N = 0$ which can be identified with the E_2 extremum [34]. In [37], it was shown that this de Wit-Nicolai solution can be written in the simple form

$$\begin{aligned} f(y) &= a\left(1 - \frac{4}{5}\sin^2\alpha\right)^{2/3} \\ g_{mn}dy^m dy^n &= bf^{-1/2}\sin^2\alpha d\Omega_6^2 + cfd\alpha^2 \end{aligned} \tag{2.4.2}$$

where $d\Omega_6^2$ is the metric on the round S^6, and α is the seventh coordinate with $0 \leq \alpha \leq \pi$ and a, b, and c are constants. The E_3 extremum can be explained in a similar way [38], as can presumably the E_5 and E_6. It is also interesting to note that all the known solutions with $f(y) \neq 1$ correspond to *inhomogeneous* deformations of S^7 *i.e.* they are not coset spaces. This is because any extremum with $< S > \neq 0$ corresponds to a deformation h_{mn} of S^7 induced by the 35_v Killing tensors [17]. As explained in [21,22], the trace h_m^m is non-constant for the 35_v (in contrast to the 1 and 300) and hence the deformed S^7 is inhomogeneous.

As for stability, E_0, E_3, and E_5 are automatically stable; E_1 and E_2 are known to be unstable [26]; the rest have not yet been analyzed.

3. *Consistency of the Kaluza – Klein Ansatz*

3.1 Generic Kaluza-Klein Theories

In attempting to establish the truth of conjecture I and II, we are forced to re-examine some of the basic assumptions of Kaluza-Klein theories in general. Implicit in much of the Kaluza-Klein literature is the assumption that it is in fact possible to truncate the full theory to its massless sector in a consistent fashion. This is the subject of this section and is based on the paper by Duff, Nilsson, Pope and Warner [39], and a forthcoming paper by Duff, Nilsson, and Pope [40].

Let us begin by recalling that in modern approaches to Kaluza-Klein theories, the extra (k) dimensions are treated as physical and are not to be regarded merely as a mathematical device. In this framework, therefore, it is essential that at every stage in the derivation of the effective four-dimensional field theory one maintains consistency with the higher-dimensional field equations. To derive this effective theory one selects the ground state of (spacetime) × (compact manifold M_k) and performs a generalized Fourier expansion of all the fields in terms of harmonics on M_k. Provided one retains all the modes in this expansion (*i.e.* all the massive states) then no such problems of inconsistency can arise. Moreover, it is well-known that if M_k has isometry group G then the theory includes massless Yang-Mills bosons with gauge group G. (Assuming, as we shall for the time being, that any other matter fields non-zero in the ground-state are singlets under G.)

In practice, however, one is often interested in extracting an effective "low energy" theory by discarding all but a finite number of states including the massless graviton, the massless gauge fields and other matter fields which are usually (but not necessarily) massless. For historical reasons this procedure is often called the "Kaluza-Klein ansatz". Despite the name, this should not be an ad hoc procedure but should correspond to retaining only the appropriate Fourier modes, for example, the zero eigenvalue modes for the mass operators in the case of the massless particles. It is generally believed that the correct ansatz for the metric tensor $\hat{g}_{MN}(x,y)$ is

$$\begin{aligned}\hat{g}_{\mu\nu}(x,y) &= g_{\mu\nu}(x) + A^{\alpha}_{\mu}(x)A^{\beta}_{\nu}K^{m\alpha}(y)K^{n\beta}(y)\mathring{g}_{mn}(y)\\ \hat{g}_{\mu n}(x,y) &= A^{\alpha}_{\mu}(x)K^{m\alpha}(y)g_{mn}(y)\\ \hat{g}_{mn}(x,y) &= \mathring{g}_{mn}(y)\end{aligned} \tag{3.1.1}$$

where the coordinates $x^{\mu}(\mu = 1...4)$ refer to spacetime and $y^m(m = 1...k)$ to the extra dimensions and $\mathring{g}_{mn}(y)$ is the metric on M_k. The quantities $K^{m\alpha}(y)$ are the Killing vecores corresponding to the isometries of this metric and α runs over the dimension of the isometry group G. The claim that this is the correct ansatz is based on the observation that substituting (3.1.1) into the higher dimesional action and integrating over y, one obtains the four dimensional Einstein-Yang-Mills action with metric $g_{\mu\nu}(x)$ and gauge potential $A_\mu^\alpha(x)$.

But as we have already emphasized the correct Kaluza-Klein ansatz must be consistent with the higher dimensional *field equations* and, as we shall now demonstrate, (3.1.1) does not in general satisfy this criterion. As has already been stressed elsewhere [16,17] one obvious source of inconsistency is the neglect of scalar fields in (3.1.1). For example, setting $\hat{g}_{55} = 1$ in the $d = 5$ pure gravity theory is inconsistent with the $\hat{R}_{55} = 0$ components of the Einstein equation which would force $F_{\mu\nu}F^{\mu\nu}$ to vanish. In this case, the remedy is simple: one includes the single massless scalar field ϕ via $\hat{g}_{55} = \phi(x)$, then after a suitable Weyl rescaling the $\hat{R}_{55}$ equation becomes

$$\Box(log\phi) \sim \phi F_{\mu\nu}F^{\mu\nu}.$$

In this example, the process of restoring consistency stops here; there is no need to include any higher massive modes in the Fourier expansion.

The purpose of this section is to point out a new, and much more serious, source of inconsistency arising from (3.1.1) . To illustrate it, let us consider the field equations of pure gravity with a positive cosmological constant Λ in $d = 4 + k$ dimensions

$$\hat{R}_{MN} = \Lambda \hat{g}_{MN} \tag{3.1.2}$$

This theory admits a classical ground state solution of de Sitter spacetime $\times$ compact manifold M_k. (The fact that spacetime is de Sitter rather than anti-de Sitter or Minkowski space need not concern us here since we are concerned only with the mathematical consistency of the ansatz.) Substituting (3.1.1) into (3.1.2) we find the $d = 4$ Einstein equation

$$R_{\mu\nu} - \frac{1}{2}g_{\mu\nu}R + \Lambda g_{\mu\nu} = \frac{1}{2}(F_{\mu\rho}{}^\alpha F_\nu{}^{\rho\beta} - \frac{1}{4}g_{\mu\nu}F_{\rho\sigma}{}^\alpha F^{\rho\sigma\beta})K_n{}^\alpha K^{n\beta} \tag{3.1.3}$$

where $F_{\mu\nu}{}^\alpha$ is the Yang-Mills field strength. The inconsistency is now apparent. The left hand side of (3.1.3) is independent of y while the right-hand side depends on y via the Killing vector combination $K_m{}^\alpha K^{m\beta}$. For example, when $M_k = S^k$ with its $SO(k+1)$ invariant metric,

$$K^\alpha_m(y)K^{m\beta}(y) = \delta^{\alpha\beta} + Y^{\alpha\beta}(y) \tag{3.1.4}$$

where $Y^{\alpha\beta}(y)$ is that harmonic of the scalar Laplacian with next to lowest non-vanishing eigenvalue $2\Lambda(k+1)/(k-1)$ belonging to the $k(k+3)/2$ dimensional representation of $SO(k+1)$.

How can consistency be restored? It is not difficult to see that including scalars in the ansatz (3.1.1) is not sufficient to cure the problem with the gauge field's stress tensor. One might try including the next-to-lightest massive graviton which involves the harmonic $Y^{\alpha\beta}(y)$ in an attempt to provide the correct y dependence on the left hand side of (3.1.3). However, it is well known that only with zero [41] or an infinite number of massive gravitons is the theory consistent. In other words, one would end up having to put back all the massive states.

If one insists on making a truncation to a finite number of states including gauge bosons, one may alternatively select a subgroup G' of G with Killing vectors $K^{\alpha'}_m$, where α' runs over the dimension of G', for which

$$K^{\alpha'}_m K^{m\beta'} = \delta^{\alpha'\beta'} \tag{3.1.5}$$

One obvious way to achieve this arises when M_k is itself the non-abelian group manifold G' which, with its bi-invariant metric, has isometry group $G = G' \times G'$. Then the left-invariant or right-invariant vector fields are Killing vectors which separately obey (3.1.5), but taken together the cross terms generate a y-dependent piece as in (3.1.4). Contrary to many claims in the literature, therefore, the Kaluza-Klein ansatz for group manifolds yields the gauge bosons only of G' and not $G' \times G'$. The full invariance of $G' \times G'$ is restored only by including the massive modes. For abelian group manifolds G', the isometry group is only G' and all Killing vectors automatically satisfy (3.1.5). So for compactification on the k-torus T^k, the ansatz is always consistent.

Another way to achieve (3.1.5) arises when M_k is a principle bundle with G' as the fibre. For example $M_k = S^{4n+3}$ (n = positive integer) is an $SU(2)$ bundle over HP^n. In this case there are two Einstein metrics: the round sphere with $G = SO(4n+4)$ and the squashed sphere with $G = Sp(n+1) \times SU(2)$. In either case, the $SU(2)$ Killing vectors on the fibre satisfy (3.1.5). Once again, the residual gauge group is much smaller than the isometry group.

Note that (3.1.5) implies the existence of everywhere non-vanishing vector fiels and so (3.1.5) can never be satisfied on spaces not admitting such fields *i.e.* on spaces with non-zero Euler number χ (odd dimensional spaces always have $\chi = 0$). For example S^{2n} has $\chi = 2$, and CP^n has $\chi = n + 1$. In these cases, the Kaluza-Klein ansatz would involve no gauge bosons at all. Since (3.1.5) is a sum of k terms, it follows that for a general space, the number of Killing vectors satisfying (3.1.5) is less than or equal to k, the dimension of the space, with equality for group manifolds.

Of course, having cured the inconsistency of the $\mu\nu$ components of the Einstein equation, one must also ensure the consistency of the remaining components. For $\hat{R}_{mn}$, this would involve at least the inclusion of a scalar singlet in the ansatz for $\hat{g}_{mn}$.

3.2 $d = 11$ supergravity and the seven -sphere

The situation we have described so far changes radically when we turn our attention to $d = 11$ supergravity. The reason for this difference is the presence of the three-index gauge field $\hat{A}_{MNP}$ in addition to the metric $\hat{g}_{MN}$. Recall that the bosonic field equations are

$$\hat{R}_{MN} - \frac{1}{2}\hat{g}_{MN}\hat{R} = \frac{1}{3}(\hat{F}_{MNPQ}\hat{F}_N{}^{PQR} - \frac{1}{8}\hat{g}_{MN}\hat{F}^2) \tag{3.2.1}$$

$$\nabla_M \hat{F}^{MPQR} = -\frac{1}{576}\epsilon^{M_1 \ldots M_8 PQR}\hat{F}_{M_1 \ldots M_4}\hat{F}_{M_5 \ldots M_8} \tag{3.2.2}$$

where $\hat{F}_{MNPQ} = 4\partial_{[M}\hat{A}_{NPQ]}$. It is this $\hat{A}_{MNP}$ field, rather than an explicit cosmological constant, which is responsible for the spontaneous compactificaiton to $d = 4$. In particular, let us consider the Freund-Rubin type compactifications of Section I with ground state $AdS \times M_7$ obtained from

$$< \hat{g}_{\mu\nu} > = \mathring{g}_{\mu\nu}(x) \qquad < \hat{g}_{mn} >= \mathring{g}_{mn}(y)$$
$$< \hat{F}_{\mu\nu\rho\sigma} > = 3m\epsilon_{\mu\nu\rho\sigma} \tag{3.2.3}$$

with all other components vanishing, where $\mathring{g}_{\mu\nu}$ is the metric on AdS and $\mathring{g}_{mn}(y)$ the Einstein metric on M_7 satisfying $\mathring{R}_{mn} = 6m^2\mathring{g}_{mn}$.

The critical observation is that the standard Kaluza-Klein ansatz for the gauge bosons must now be augmented by the additional ansatz [17,25]

$$\hat{F}_{\mu\nu pq} = -\frac{1}{4}m^{-1}\epsilon_{\mu\nu\rho\sigma}F^{\rho\sigma\alpha}\nabla_{[p}K^{\alpha}_{q]}. \tag{3.2.4}$$

(In the case of the round S^7 this can be seen from (2.3.4) and (2.3.8)). Since we are concerned primarily with possible inconsistencies in the gauge field sector, we shall postpone the inclusion of the scalar fields. Substituting (3.1.1) and (3.2.4) into (3.2.1) we find the $d = 4$ Einstein equation [24]

$$\begin{aligned} R_{\mu\nu} - \frac{1}{2}g_{\mu\nu}R - 12m^2 g_{\mu\nu} = \\ \frac{1}{2}(F_{\mu\nu}{}^{\alpha}F_{\nu}{}^{\rho\beta} - \frac{1}{4}g_{\mu\nu}F_{\rho\sigma}{}^{\alpha}F^{\rho\sigma\beta})(K_m{}^{\alpha}K^{m\beta} \\ + \frac{1}{2}m^{-2}\nabla_m K^{\alpha}_n \nabla^m K^{n\beta}) \end{aligned} \tag{3.2.5}$$

Comparing (3.2.5) with (3.1.3) we see that instead of the consistency condition (3.1.5) which permits at most 7 Killing vectors we now have

$$K^{\alpha'}_m K^{m\beta'} + \frac{1}{2}m^{-2}\nabla_m K^{\alpha'}_n \nabla^m K^{n\beta'} = \delta^{\alpha'\beta'} \tag{3.2.6}$$

which permits at most 28 since it is the sum of 28 terms. Under what circumstances is (3.2.6) satisfied? Let us begin by examining the maximally symmetric round S^7 solution. Since S^7 admits the maximum of 8 unbroken supersymmetries i.e. 8 Killing spinors $\eta^I (I = 1...8)$ satisfying $D_m\eta^I = \frac{m}{2}\Gamma_m\eta^I$, we may form all 28 Killings vectors of $SO(8)$ via [17]

$$K^{IJ}_m = \bar{\eta}^I\Gamma_m\eta^J \tag{3.2.7}$$

and for convenience, we choose the η's to be orthonomal $\bar{\eta}^I\eta^J = \delta^{IJ}$. Using the Fierz identity for $d = 7$ commuting Majorana spinors [17]

$$\psi\bar{\phi}\chi = \frac{1}{8}(\chi\bar{\phi}\psi - \Gamma_m\chi\bar{\phi}\Gamma^m\psi - \frac{1}{2}\Gamma_{mn}\chi\bar{\phi}\Gamma^{mn}\psi + \frac{1}{6}\Gamma_{mnp}\chi\bar{\phi}\Gamma^{mnp}\psi) \tag{3.2.8}$$

one may verify that condition (3.2.6) is indeed satisfied for all 28 Killing vectors of $SO(8)$ and that the ansatz (3.1.1) and (3.2.4) are therefore consistent. Of course one must also include the scalars and pseudoscalars as well in order to verify complete consistency, not only of the $\mu\nu$ components of (3.2.1) but also the other components of (3.2.1) and all components of (3.2.2). One must also check the fermion ansatz. On the round S^7 the total ansatz for the massless $N = 8$ supermultiplet of spins $(2, 3/2, 1, 1/2, 0^+, 0^-)$ is not yet known to all orders owing to the complicated non-polynomial dependence on the scalars and pseudoscalars. However, to the order to which it is known, everything is consistent. This seems to be guaranteed by the fact that the ansatz for all spins can be expressed in terms of the Killing spinors η^I [7].

Indeed, whenever M_7 admits $N \leq 8$ Killing spinors one may always find Killing vectors satisfying (3.2.6) namely those constructed as in (3.2.7). For such compactifications the total gauge group G is always of the form $G = SO(N) \times K$ for some group K [42]. The construction (3.2.7) singles out $G' = SO(N)$ as the subgroup surviving in the consistent ansatz. These are the gauge bosons of the N-extended gravity supermultiplet.

In fact the round S^7 seems to be the only M_7 which exploits the extra freedom permitted by (3.2.6) as compared with (3.1.5). To see this, we first rewrite (3.2.6) as

$$\frac{1}{4m^2}(\Box + 16m^2)K_m^{\alpha'}K^{m\beta'} = \delta^{\alpha'\beta'}. \tag{3.2.9}$$

Thus whenever M_7 admits Killing vectors $K_m^{\alpha'}$ satisfying (3.1.5) then $\frac{1}{2}K_m^{\alpha'}$ will also satisfy (3.2.9). In order to see whether there are any more Killing vectors satisfying (3.2.9) but not (3.1.5), let us observe that in the case of a general coset space the appropriately normalized Killing vectors satisfy

$$K_m^{\alpha'}K^{m\beta'} = \frac{1}{4}\delta^{\alpha'\beta'} + \phi^{\alpha'\beta}(y) \tag{3.2.10}$$

where $\phi^{\alpha'\beta'}(y)$ is tracefree. The extra freedom of (3.2.9) over (3.1.5) is that $\phi^{\alpha'\beta'}(y)$ can now be non-zero, provided that is satisfies

$$-\Box\phi^{\alpha'\beta'} = 16m^2\phi^{\alpha'\beta'} \tag{3.2.11}$$

Moreover, a case by case analysis of all homogeneous M_7's reveals that, with the exception of the round S^7, none of them permits eigenfunctions of the Laplacian in the right representation and with the right eigenvalues to satisfy (3.2.11). Interestingly enough, condition (3.2.11) implies the existence of massless scalars in the $O^{+(1)}$ tower of [43]. In the case of the round S^7, these are just the 35_v massless scalars of the $N = 8$ supermultiplet. None of the other M_7's has massless scalars of this kind. The surviving gauge groups G' compatible with the consistent ansatz are listed in Table 3 for some typical M_7 space.

TABLE 3

Gauge groups surviving in the consistent massless ansatz

M_7	G	G'
round S^7	SO(8)	SO(8)
squashed s^7	SO(5)xSU(2)	SU(2)
M(m,n)	SU(3)xSU(2)xU(1)	U(1)
N(k,l)	SU(3)xU(1)	U(1)
N(1,1)	SU(3)xSU(2)	SU(2)
Q(p,q,r)	$[SU(2)]^3$xU(1)	U(1)
T^7	$[U(1)]^7$	$[U(1)]^7$

We have seen that all gauge bosons appearing in the consistent ansatz for pure gravity in $d = 11$ will also appear in the consistent ansatz for supergravity but with a normalization differing by a factor of 2. We remark incidentally that this is responsible for the modification of Weinberg's [23] formula relating the gauge coupling constant to the radius of the extra dimensions. One finds [24]

$$e^2(supergravity) = \frac{1}{4}e^2(pure\ gravity). \tag{3.2.12}$$

Note that by integrating (3.1.3) and (3.2.5) over y, one may verify that Wienberg's formula and (3.2.12) remain valid for all the gauge bosons of G and not merely those of G'. Note also that the ansatz (3.1.1) together with the normalization of the Killing vector chosen in (3.2.6) establishes the convention $4\pi G = 1$ in the Einstein equation (3.2.5). This result was used in Section 2.3 in deriving the relation $e^2 = 16\pi G m^2$ on the round S^7, a result which could also be obtained from (3.2.12) without reference to the transformation rules. To obtain the coupling constant directly, one can calculate the Yang-Mills structure constants from the algebra of the Killing vectors having once established their normalization.

3.3 Significance of consistency

We conclude this section with some comments on the significance of Kaluza-Klein consistency.

(i) The consistency condition (3.2.6) was derived under the assumption of Freund-Rubin compactifications given by (3.2.3), but we might also consider more general AdS solutions with $< \hat{g}_{\mu\nu} >= f(y)g_{\mu\nu}(x)$ and/or $< \hat{F}_{mnpq} >\neq 0$. We have little to say about the consistency of the Kaluza-Klein ansatz for general solutions of this kind but we can make some plausible conjectures in the case of those which are topologically S^7. These conjectures are related to conjectures I and II of Section 2. It seems highly plausible that just as the round S^7 ansatz for the massless fields of the de Wit-Nicolai theory was already completely consistent without any reduction of the vacuum symmetry G, so the ansatze for the other S^7 vacua corresponding to Table 3 will also be completely consistent provided we retain the same de Wit-Nicolai fields, some of which will now be massive. Indeed, it seems that S^7 and $d = 11$ supergravity go hand-in-hand.

(ii) For a generic Kaluza-Klein theory, it would seem that our consistency condition is equivalent to the requirement that the entire Kaluza-Klein ansatz (and not merely the ground state metric) is left invariant under a transitively acting subgroup of the isometry group. This is a version of what is sometimes called "K-invariance" [44] of the ansatz. For example, this would yield only the gauge bosons of G' in the case of a group manifold G' and only the gauge bosons of $SU(2)$ in the case of S^{4n+3}. The reason for this equivalence is that the consistency condition amounts to requiring that the Lagrangian obtained by substituting the ansatz into the higher dimensional Lagrangian be independent of y. Although this is guaranteed by "K-invariance", it is clearly too strong a condition as the example of S^7 and $d = 11$ supergravity illustrates. [The smallest subgroup of $SO(8)$ which acts transitively on S^7 is $SO(5)$ and so "K-invariance" would yield only the gauge bosons of $SU(2)$ This gives rise to new consistent truncations of $d = 11$ supergravity with $N \leq 3$, not contained in the $N = 8$ truncation [45].]

(iii) In summary we have seen that the standard Kaluza-Klein ansatz for massless gauge bosons is in general inconsistent with the higher dimensional field equations. Consistency can be restored either by putting back the infinite tower of massive states or else by reducing the gauge group G to some subgroup G'. Exeptions to this rule are provided by certain S^7 compactifications of $d = 11$ supergravity. An interesting question, to which we do not know the answer, is whether there are other Kaluza-Klein theories with consistent truncations. We emphasize that we are not disputing the Yang-Mills content of the complete Kaluza-Klein theory but only the ability to truncate consistently while retaining the full gauge symmetry.

So far we have been concerned with mathematical consistency. Does inconsistency matter from a physical point of view? Here it is important to distinguish between Kaluza-Klein theories which predict a Minkowski spacetime ground state and those, of the kind discussed in this paper, which predict AdS with a Planck-sized cosmological constant. In the former case the answer is probably no if we are interested in energies small compared with the compactification scale, since the inconsistency will only be relevant for operators of dimension > 4. (We are grateful to E. Witten and S. Weinberg for pointing this out.) In the case of a cosmological constant $\Lambda \sim e^2/G$ (typical of $d = 11$ supergravity and all those Kaluza-Klein theories for which Λ is not fine-tuned to zero), then this is no longer the case. For example dimension six operators like $GR\, F_{\mu\nu}F^{\mu\nu}$ may be converted to dimension four operators like $e^2\, F_{\mu\nu}F^{\mu\nu}$ on using the field equations $R = e^2/G + \ldots$ (This observation was also relevant when computing the $\beta(e)$ function in gauged supergravity and its relation to the renormalization of $G\Lambda$ [11].) In this case, of course, it is no longer clear that the massless sector is in any case a good approximation to the full theory.

In our opinion, until such time as the cosmological constant problem in Kaluza-Klein theories has been solved (as opposed to being merely fine-tuned away), the physical significance of the Kaluza-Klein ansatz remains obscure.

In summary, it seems that conjectures I and II are almost certainly true but we are as far away from understanding their *physical* significance as we were when they were made. For example, the squashed S^7 of [19] is a perfectly good solution of the $d = 11$ theory but has nothing to do with the de Wit-Nicolai theory. This S^7 solution has $< S > \neq 0$ and $< P > = 0$ and it is interesting to compare it with the $< S > \neq 0, < P > = 0$ S^7 solutions in Table 2. Warner [38] has pointed out that these S^7 geometries can all be embedded in flat $\mathbf{R}^8$ in contrast to the squashed S^7 [19,21].

Recent Results

The problem of consistent truncations in Kaluza-Klein theories in general, and $d = 11$ supergravity in particular, is treated in a recent paper by Duff and Pope [45]. Here it is stressed that if the truncation to the $N = 8$ massless supermultiplet is consistent in the sense described in the paper, then conjecture II is an automatic consequence of conjecture I. It is also made clear that the problem of inconsistency is not one of non-linearity: an inconsistent ansatz goes wrong already at the linear level! Further progress in demonstrating the truth of these conjectures *to all orders* has recently been made by Nilsson [46] and by de Wit, Nicolai and Warner [47]. The need for a final proof is amply demonstrated by a recent paper of Boucher [48] which claims to prove that there is no consistent $N = 8$ truncation on S^7! In our opinion, Boucher has failed to realize that although the K-invariance of Section 3.3 is a sufficient condition for consistency, it is not necessary [45].

The final proof should also clear up on or two other naggging points. Does the $N = 8$ theory obtained from $d = 11$ on S^7 exhibit the $SU(8)$ structure? (The Cremmer-Julia theory obtained from T^7 does not. Further field redefinitions are necessary [7]). Let us bear in mind that in this formulation the de Wit-Nicolai theory has anomalies [49] even though the $d = 11$ theory is anomaly-free. What role, if any, does the $d = 11$ Chern-Simons term [3] play in all of this?

ACKNOWLEDGEMENTS

I am grateful to my collaborators M. Awada, B. Nilsson, C. Pope and N. Warner. Conversations with C. Pope and N. Manton are especially acknowledged. This work was supported in part by the National Science Foundation under Grant No. PHY77-27084, supplemented by funds from the National Aeronautics and Space Administration. I am grateful for the hospitality of the Institute for Theoretical Physics, Santa Barbara.

REFERENCES

1. E. Cremmer, B. Julia and J. Scherk, "Supergravity Theory in 11 Dimensions," Phys. Lett. **76**B, 409 (1978).

2. E. Cremmer and S. Ferrara, "Formulation of $d = 11$ Supegravity in Superspace," Phys. Lett. **91**B, 61 (1980).

3. L. Alvarez-Gaumé and E. Witten "Gravitational Anomalies," Nucl. Phys. B**234**, 269 (1984); L. Alvarez-Gaumé, S. Della Pietra, G. Moore, HUTP-84/A028.

4. P.G.O. Freund and M.A. Rubin, "Dynamics of Dimensional Reduction," Phys. Lett. B**97**, 233 (1980).

5. This is Myers' theorem.

6. B. de Wit and H. Nicolai, "$N = 8$ Supergravity," Nucl. Phys. B**208**, 323 (1982).

7. E. Cremmer and B. Julia, "The $SO(8)$ Supergravity," Nucl. Phys. B**159**, 141 (1979).

8. D.Z. Freedman and A. Das, "Gauge Internal Symmetry in Extended Supergravity," Phys. Lett. **74**B, 333 (1977).

9. E.S. Fradkin and M. Vasiliev, unpublished.

10. A. Das, M. Fischler and M. Rocek, "Super-Higgs Effect in a new Class of Scalar Models and a Model of Super QED," Phys. Rev. D**16**, 3427 (1977).

11. S.M. Christensen, M.J. Duff, G.W. Gibbons and M. Rocek, "Vanishing One-Loop β Function in Gauged $N > 4$ Supergravity," Phys. Rev. Lett. **45**, 161 (1980).

12. T. Curtwright, "Charge Renormalization and High Spin Fields," Phys. Lett. **102**B, 17 (1981).

13. M.J. Duff, "Ultraviolet Divergences in Extended Supergravity" in : Supergravity, 81, eds. S. Ferrara and J.G. Taylor, 257 (C. U. P., 1982).

14. M.J. Duff and D.J. Toms, "Kaluza-Klein Kounterterms," in "Unification of the Fundamental Interactions II," eds. J. Ellis and S. Ferrara (Plenum 1982).

15. F. Englert, "Spontaneous Compactification of 11-Dimensional Supergravity," Phys. Lett. **119**B, 339 (1982).

16. M.J. Duff, "Supergravity, the Seven-Sphere and Spontaneous Symmetry Breaking," Proceedings of the Marcel Grossman Meeting on General Relativity, Shanghai August 1982, ed. Hu Ning (Science Press and North Holland, 1983) and Nucl. Phys. B**219**, 389 (1983).

17. M.J. Duff and C.N. Pope, "Kaluza-Klein Supergravity and the Seven Sphere," in Supersymmetry and Supergravity '82 eds. S. Ferrara, J.G. Taylor and P. van Nieuwenhuizen (World Scientific Publishing, 1983).

18. R. D'Auria, P. Fré and P. van Nieuwenhuizen, "Symmetry Breaking in $d = 11$ Supergravity on the Parallelized Seven-Sphere," Phys. Lett. **122**B, 225 (1983).

19. M.A. Awada, M.J. Duff, and C.N. Pope, "$N = 8$ Supergravity Breaks Down to $N = 1$," Phys. Rev. Lett. **50**, 294 (1983).

20. B. Biran, F. Englert, B. de Wit, and H. Nicolai, "Gauged $N = 8$ Supergravity and its Breaking from Spontaneous Compactification," Phys. Lett. **124**B, 45 (1983).

21. M.J. Duff, B.E.W. Nilsson and C.N. Pope, "Spontaneous Supersymmetry Breaking by the Squashed Seven-Sphere," Phys. Rev. Lett. **50**, 2043 (1983); Erratum **51**, 846 (1983).

22. M.J. Duff, B.E.W. Nilsson and C.N. Pope, "Superunification from Eleven Dimensions," Nucl. Phys. B**233**, 433 (1984).

23. S. Weinberg, "Charges from Extra Dimensions," Phys. Lett. **124**B, 265 (1983).

24. M.J. Duff, C.N. Pope and N.P. Warner, "Cosmological and Coupling Constants in Kaluza-Klein Supergravity," Phys. Lett. **130**B, 254 (1983).

25. M.A. Awada, B.E.W. Nilsson and C.N. Pope, "The Supersymmetry Transformation Rules for Kaluza-Klein Supergravity on the Seven-Sphere," Phys. Rev. (rapid communication) D**29**, 335 (1984).

26. B. de Wit and H. Nicolai, "On the Relation Between $d = 4$ and $d = 11$ Supergravity," Nucl. Phys. B**243**, 91 (1984).

27. B. Biran, A. Casher, F. Englert, M. Rooman and P. Spindel, "The fluctuat ing Seven Sphere in Eleven Dimensional Supergravity," Phys. Lett. **134**B, 17 (1984).

28. N.P. Warner, "Some New Extrema of the Scalar Potential of Gauged $N =$ Supergravity," Phys. Lett. **128**B, 169 (1983).

29. N.P. Warner, "Some Properties of the Scalar Potential in Gauged Super gravity Theories," Nucl. Phy. B**231**, 250 (1984).

30. B. de Wit and H. Nicolai, "The Parallelizing S^7 Torsion in Gauged $N =$ Supergravity," Nucl. Phys. B**231**, 506 (1984).

31. B. Biran and Ph. Spindel, "Instability of the Parallelized Seven-Sphere An Eleven-Dimensional Approach," Bruxelles-Mons preprint ULB-TH 84/00 (March 1984).

32. M. Gunaydin and N.P. Warner, Nucl. Phys. B (to appear).

33. C.N. Pope and N.P. Warner, "An $SU(4)$ Invariant Compactification c $d = 11$ Supergravity on a Stretched Seven-Sphere," Phys. Lett. B, to appear.

34. B. de Wit, and H. Nicolai, "A New $SO(7)$ Invariant Solution of $d = 1$ Supergravity," TH-3956-CERN (1984).

35. V.A. Rubakov and M.E. Shaposhnikov, "Extra Space-Time Dimensions Toward a Solution to the Cosmological Constant Problem," Phys. Lett. **125**B 139 (1983).

36. P. van Nieuwenhuizen, "An Introduction to Simple Supergravity and th Kaluza-Klein Program" in Relativity, Groups and Topology II, eds. B.S. d Witt and R. Stora, (North Holland, 1984).

37. P. van Nieuwenhuizen and N.P. Warner, ITP-SB-84-75 (unpublished).

38. N.P. Warner (unpublished).

39. M.J. Duff, B.E.W. Nilsson, C.N. Pope and N.P. Warner, "On the Consis tency of the Kaluza-Klein Ansatz," Phys. Lett. B, to appear.

40. M.J. Duff, B.E.W. Nilsson, and C.N. Pope, "Kaluza-Klein Supergravity" Phys. Reports (to appear).

41. D. Boulware and S. Deser, Ann. Phys. (N.Y.) **81**, 193 (1975).

42. L. Castellani, R. D'Auria and P. Fré, "$SU(3) \otimes SU(2) \otimes U(1)$ From $D = 11$ Supergravity," Nucl Phys. B**239**, 610 (1984).

43. M.J. Duff, B.E.W. Nilsson and C.N. Pope, "The Criterion for Vacuum Stability in Kaluza-Klein Supergravity," Phys. Lett. **139**B, 154 (1984).

44. P. Forgacs and N.S. Manton, "Space-Time Symmetries in Gauge Theories," Commun. Math Phys. **72**, 15 (1980).

45. M.J. Duff and C.N. Pope, "Consistent Truncations in Kaluza-Klein Theories," ITP preprint NSF-ITP-84-166.

46. B.E.W. Nilsson, preprint, Gothenburg 84-52.

47. B. de Wit, H. Nicolai and N. Warner (to appear).

48. W. Boucher, D.A.M.T.P. preprint.

49. S. Ferrara, P. di Vecchia, and L. Girardello, CERN preprint TH 4026/84.

VACUUM STABILITY IN SUPERGRAVITY

Daniel Z. Freedman
Department of Mathematics
Massachusetts Institute of Technology
Cambridge, MA 02139, U.S.A.

ABSTRACT

A discussion of the vacuum stability problem in supergravity theories is given. The definition of energy in a gravitational field theory, and the small fluctuation stability argument for anti de Sitter space are outlined. The notion of Killing spinors and the related proof of global stability are discussed briefly. References to the recent literature are given.

1. Introduction

There are several branches of the subject of supergravity, all of which are now being studied quite actively. These are

i) N = 1 phenomenological models in which the supergravity multiplet of spins 2 and 3/2 is coupled to lower spin matter multiplets. These models contain predictions for new particles which are the superpartners of known particles.

ii) Higher N extended supergravity models in which the graviton is unified with lower spin particles.

iii) Higher dimensional Kaluza-Klein models.

One problem which is common to all branches of the subject is the question of vacuum stability. This problem must be approached quite differently in a field theory which includes gravity, than in a flat-space theory. A reasonable understanding of the vacuum stability problem has been attained in the last few years with results that are quite counter intuitive in the case of the anti de Sitter backgrounds which are common in theories of type ii) and iii) above. In this report I will outline recent results and give references to the recent literature.

2. Motivation

The motivation for my work on vacuum stability came from the family of gauged extended supergravity theories with $4 \leq N \leq 8$. Scalar fields ϕ^i are present in these theories, unified with the graviton, but there is a potential $V(\phi^i)$ which is unbounded below. In a flat space field theory this would mean that there is no stable vacuum state, a theoretical disaster. Gravity changes the situation, and one can show that all of these theories have ground states which are stable against fluctuations which obey certain boundary conditions which we believe are the natural boundary conditions for the problem.

The multiplicity table of extended supergravity theories is

N =	1	2	3	4	5	6	7	8
spin								
2	1	1	1	1	1	1	1	1
3/2	1	2	3	4	5	6	7+1	8
1		1	3	6	10	15+1	21+7	28
1/2			1	4	10+1	20+6	35+21	56
0				1+1	5+5	15+15	35+35	35+35

The entries in each column denote the dimensions of SO(N) irreps of the particles of spin $S \leq 2$, all of which are unified in one irrep of the supergroup OSp(N,4). This unification of bosons and fermions and of the spin 2 graviton with spin 1 particles whose exchanges might mediate conventional gauge interactions is very striking.

The field theories involving these particles contain two coupling constants, Newton's constant κ (with $\kappa^2 = 4\pi G$) and an SO(N) gauge constant e. The theories are complicated for higher N and the only complete Lagrangian we will show is that of N = 1 cosmological supergravity which emerges as the common "truncation" of all these theories when all fields except the graviton and one gravitino are discarded. The action

$$S = \int d^4x\{\frac{-(\det V)R}{4\kappa^2} - \frac{1}{2}\varepsilon^{\lambda\mu\nu\rho}\bar{\psi}_\lambda\gamma^5\gamma_\mu\hat{D}_\nu\psi_\rho + \frac{3e^2\det V}{2\kappa^4}\}$$

$$R = V_a{}^\mu V_b{}^\nu R_{\mu\nu ab} \tag{1}$$

$$R_{\mu\nu ab} = \partial_\mu\omega_{\nu ab} - \partial_\nu\omega_{\mu ab} + \omega_{\mu a}{}^c\omega_{\nu cb} - \omega_{\nu a}{}^c\omega_{\mu cb}$$

$$\hat{D}_\nu\psi_p = (\partial_\nu + \frac{1}{2}\omega_{\nu ab}\sigma^{ab} + \frac{ie}{2\kappa}\gamma_\nu)\psi_\rho$$

is invariant under the transformation rules

$$\delta V_{a\mu}(\alpha) = -i\kappa\bar{\varepsilon}(x)\gamma_a\psi_\mu(x) \tag{2}$$

$$\delta\psi_\mu(x) = \frac{1}{\kappa}\hat{D}_\mu\varepsilon(x) = \frac{1}{\kappa}(\partial_\mu + \frac{1}{2}\omega_{\mu ab}\sigma^{ab} + \frac{ie}{2\kappa}\gamma_\mu)\varepsilon \tag{3}$$

Here one should notice the large cosmological constant $\Lambda \sim e^2/\kappa^2$ which is characteristic of these theories. Notice also the modified covariant derivative with a γ_μ term. Its significance will become clear below.

Next we will consider the gravity and scalar part of the gauged SO(4) theory (with simplex scalar field $z(x) = \kappa(A(x) + iB(x))$. The action is

$$S = \frac{1}{2\kappa^2}\int d^4x\sqrt{-g}\{-\frac{1}{2}R + g^{\mu\nu}\frac{\partial_\mu\bar{z}\partial_\nu z}{(1-\bar{z}z)^2} + \frac{e^2}{K^2}(3 + \frac{2\bar{z}z}{1-\bar{z}z})\} \tag{4}$$

Here one sees that the scalar dynamics is that of a nonlinear σ-model on the non-compact space of constant -ve curvature which is known by several names -- H_2 or $SU(1,1)/U(1)$ or the Lobachevsky plane. In the parameterization used, the field values are restricted by the inequality $|z| < 1$. One also sees a scalar potential with coefficient $\sim e^2$ which arises in order that the gauging of SO(4) (by conventional covariantization of derivatives etc. elsewhere in the Lagrangian) is compatible with local supersymmetry. Notice that the potential is unbounded below, and that the only stationary point is the global maximum at $z = 0$, where one finds a cosmological term of the same value as in the $N = 1$ action.

The situation just described is typical of the gauged supergravity theories for $N \geq 5$. The scalar dynamics is always that of a nonlinear σ-model on a non-compact coset space. The potentials $V(\phi^i)$ are complicated functions of the scalar fields ϕ^i. They are always unbounded below. For every N there is a critical point with full SO(N) invariance which is a maximum and has the same cosmological constant as above. All other critical points $\bar{\phi}^i$ which are known are either maxima or saddle points. With one exception, all have $V(\bar{\phi}^i) < 0$.

Thus we are faced with the task of demonstrating that there are stable vacuum states in field theories where the energy density can be arbitrarily negative and in which the Lagrangian masses of scalar fluctuations can be negative.

3. Backgrounds and Their Stability

Let us begin the discussion in a standard flat space scalar field with action

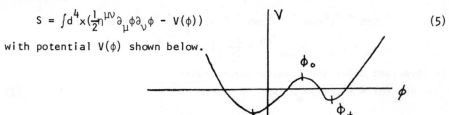

$$S = \int d^4x\left(\tfrac{1}{2}\eta^{\mu\nu}\partial_\mu\phi\partial_\nu\phi - V(\phi)\right) \tag{5}$$

with potential $V(\phi)$ shown below.

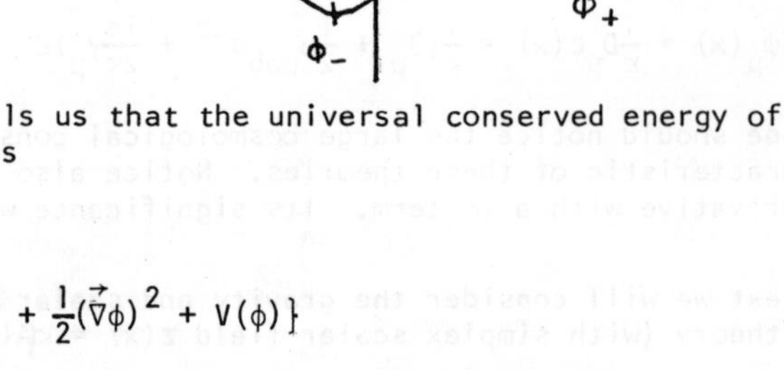

Noether's theorem tells us that the universal conserved energy of a field configuration is

$$E = \int d^3x T^0{}_0$$

$$= \int d^3x\{\tfrac{1}{2}(\partial_t\phi)^2 + \tfrac{1}{2}(\vec{\nabla}\phi)^2 + V(\phi)\} \tag{6}$$

In the semi-classical approach the ground state of the system has scalar vacuum expectation value $\phi(x)$ which is the classical solution of the field equations

$$\Box\phi + \frac{\partial V}{\partial\phi} = 0 \tag{7}$$

of lowest energy. Clearly this is obtained by taking $\phi(x) = \phi_-$, i.e., a translation invariant vacuum at the global minimum of the potential. Fluctuations about the background, i.e. $h(x) = \phi(x) - \phi_-$ are then quantized perturbatively, and the background is stable because such fluctuations have positive energy.

If we considered the critical point $\phi(x) = \phi_+$, then it is known that the corresponding quantum state is a metastable false vacuum which decays by tunneling to a bubble with field $\phi(x) \approx \phi_-$ at its center. This bubble then expands to fill all space because it is energetically favorable to do so. If one chose the local maximum $\phi(x) = \phi_0$, then there would be a classical instability; small fluctuations have negative energy and grow exponentially with time. The quantum theory contains tachyons. Finally, if $V(\phi)$ is bounded below, no stable ground state exists.

We now consider the same scalar field theory coupled to gravity in units with $\kappa = 1$. The action is now

$$S = \frac{1}{2}\int d^4x\sqrt{-g}\{-\frac{1}{2}R + g^{\mu\nu}\partial_u d\rho\partial_v\phi - 2V(\phi)\} \tag{8}$$

where $g_{\mu\nu}(x)$ is now a dynamical field.

The previous procedure must be modified because with gravity present there is no "universal" conserved energy functional which allows us to compare the energies of two field configurations. The modified procedure is the following.

A. Find a background solution of the classical field equations. The background metric determines the macroscopic geometry of space-time. Fluctuations about the background are quantized as "particles."

B. If the background is static (i.e., time-translation invariant) one can define a conserved energy functional which measures the energy of the fluctuations.

C. If E > 0 for fluctuations obeying suitable boundary conditions, then the background is stable.

Let us look into these steps more closely.

<u>A. Background Solutions</u>. The field equations of the coupled gravity-scalar systems are

$$R^{\mu\nu} - \frac{1}{2}g^{\mu\nu}R = 2T^{\mu\nu} = 2\partial^\mu\phi\partial^\nu\phi - g^{\mu\nu}[(\partial\phi)^2 - 2V(\phi)] \tag{9}$$

$$\frac{1}{\sqrt{-g}}\partial_\mu(\sqrt{-g}\, g^{\mu\nu}\partial_\nu\phi) + \frac{\partial V}{\partial\phi} = 0 \tag{10}$$

Guided by symmetry considerations we assume space-time homogeneity -- i.e., no preferred points in space-time. This implies that the background scalar configuration is constant, i.e., $\phi(x) = \bar{\phi}$, and the field equation (10) requires that

$$\left.\frac{\partial V}{\partial \phi}\right|_{\bar{\phi}} = 0 \tag{11}$$

So $\bar{\phi}$ must be a critical part of the potential which we will allow to be either a minimum, maximum or saddle.

Substituting back in the Einstein equation (9) we find

$$R^{\mu\nu} - \frac{1}{2}g^{\mu\nu}R - g^{\mu\nu}2V(\bar{\phi}) = 0 \tag{12}$$

where the naive "vacuum energy" determines the cosmological constant $\Lambda = -2V(\bar{\phi})$. We consider the maximally symmetric solution of (12), namely,

$$\bar{g}_{\mu\nu} = \begin{cases} \text{flat Minkowski space if } V(\bar{\phi}) = 0 \\ \text{de Sitter space (isometry group } O(4,1)) \text{ if } V(\bar{\phi}) > 0 \\ \text{anti de Sitter space (isometry group } O(3,2)) \text{ if } V(\bar{\phi}) < 0 \end{cases}$$

We shall restrict attention to the last case here, which is relevant to the critical points of gauged extended supergravity.

Anti de Sitter space is the 4-dimensional hyperboloid embedded in R^5 with Cartesian coordinates y^A ($A = 0, 1, 2, 3, 4$) and defined by the equation

$$\eta_{AB}y^Ay^B = \frac{1}{a^2} \tag{13}$$

with $O(3,2)$ invariant metric $\eta_{AB} = (+---+)$ and scale a^2 related to the cosmological constant by $\Lambda = -2V(\bar{\phi}) = 3a^2$.

There is a global choice of intrinsic coordinates in which the line element induced from the embedding takes the form:

$$ds^2 = \bar{g}_{\mu\nu}dx^\mu dx^\nu$$

$$= \frac{1}{a^2\cos^2\rho}\left[dt^2 - d\rho^2 - \sin^2\rho(d\theta^2 + \sin^2\theta d\phi^2)\right] \tag{14}$$

The time coordinate t is an angular coordinate in the $y^0 - y^4$ plane. Here we extend the range to $-\infty < t < \infty$ and thus pass to the covering space of the hyperboloid in order to circumvent an obvious problem with causality. The radial coordinate ρ has finite range $0 \leq \rho < \pi/2$, and $\rho = \pi/2$ is spatial infinity. The polar angles θ,ϕ are standard. Note that the metric is static, and that the t = const sections are conformal to the upper hemisphere of S_3. Thus spatial infinity can be viewed as the equator of S_3 (i.e., an S_2), and boundary conditions will be imposed here.

It is clear from the definition that the isometry group of $(AdS)_4$ is the group $SO(3,2)$ with 10 generators

$$K_{AB} = y_A\frac{\partial}{\partial y^B} - y_B\frac{\partial}{\partial y^A} \tag{15}$$

Each generator can be expressed in intrinsic coordinates as a Killing vector $K_{AB} = K_{AB}{}^{\mu}\partial/\partial x_{\mu}$ which satisfies

$$D_{\mu}K_{\nu} + D_{\nu}K_{\mu} = 0 \ . \tag{16}$$

The time translation generator is $K_{04} = \partial/\partial t$ or $K_{04}{}^{\mu} = (1,0,0,0)$. Clearly it is globally time-like, $(K_{04})^2 > 0$.

B. What is Energy? For every infinitesimal generator of the isometry group of a spacetime one can use the Killing vector to construct a formally conserved "charge" carried by the fluctuations.[1] To do this one defines fluctuations of the metric and matter fields by

$$g_{\mu\nu}(x) = \bar{g}_{\mu\nu}(x) + h_{\mu\nu}(x)$$
$$\phi(x) = \bar{\phi} + h(x) \tag{17}$$

One then separates the full Einstein equation into terms linear in $h_{\mu\nu}$ and nonlinear terms in $h_{\mu\nu}(x)$ and $h(x)$. It then takes the form

$$R_L^{\mu\nu} - \frac{1}{2}\bar{g}^{\mu\nu}R_L - \Lambda h^{\mu\nu} = 2\Theta^{\mu\nu}$$
$$= 2(t^{\mu\nu} + T^{\mu\nu}) \tag{18}$$

where

i) the left side involves the linearized Ricci tensor formed with background covariant derivatives $\bar{D}$ acting on h,

ii) $t^{\mu\nu}$ contains terms of order $(h_{\mu\nu})^2$ and higher. It is the analog of the gravitational energy-momentum pseudotensor of relativity texts.

iii) $T^{\mu\nu}$ is the stress tensor of matter fluctuations $h(x)$ but also depends on $h_{\mu\nu}$.

iv) all indices are raised and lowered by $\bar{g}_{\mu\nu}$.

By expanding the standard Bianchi identity in powers of $h_{\mu\nu}$ one sees that the left side of (18) satisfies

$$\bar{D}_{\mu}(R_L^{\mu\nu} - \frac{1}{2}\bar{g}^{\mu\nu}R_L - \Lambda h^{\mu\nu}) = 0 \tag{19}$$

identically, and therefore $\bar{D}_{\mu}\Theta^{\mu\nu} = 0$ must be satisfied as a consequence of the field equations of the fluctuation.

If we contract $\Theta^{\mu\nu}(x)$ with a Killing vector K_{AB} we see, using (15), that $\partial_{\mu}(\sqrt{-\bar{g}}g^{\mu\nu}K_{AB\nu}) = 0$, so that the Killing charge

$$M_{AB} = \int_t d^3x\sqrt{-\bar{g}}\Theta^{0\nu}\xi_{AB\nu} \tag{20}$$

is formally conserved. It will actually be conserved in time if the fluctuations $h_{\mu\nu}(x)$ and $h(x)$ vanish sufficiently fast at spatial infinity.

One can show using canonical commutation rules[2] that the ten M_{AB} generate the group SO(3,2). We shall be particularly interested in the energy operator

$$E = M_{04} = \int_t d^3x\sqrt{-\bar{g}}\Theta^0{}_0 \tag{21}$$

whose positivity defines the stability criterion of the background.

C. Stability. A direct proof that the Killing energy is positive is very difficult both because the energy density $\Theta^0{}_0$ is not positive (neither for the purely gravitational part $t^0{}_0$ nor for $T^0{}_0$ in gauged supergravity), and also because there are non-linear constraints among the fluctuation variables on the initial time surface. For this reason the first approach to the stability question in gauged supergravity was to establish positivity of energy for small fluctuations,[3] i.e., quadratic order in $h_{\mu\nu}{}^2$ and h^2. This gives some insight why E can be positive even at a local maximum of the scalar potential $V(\phi)$. The second step was to extend the Witten positive energy argument of ordinary general relativity[5] to the situation of gauged supergravity. This argument established the global stability of a class of $(AdS)_4$ backgrounds which correspond to critical points of the potentials which are supersymmetric. Finally, the Witten argument was developed further[6] enabling the proof of global stability of at least one background where the supersymmetry of the Lagrangian is spontaneously broken.

Let us discuss the small fluctuation stability argument briefly. In this approximation the contribution of scalar fluctuations to the energy is

$$\begin{aligned} E(T^{\mu\nu}) &= \int_t d^3x\sqrt{-\bar{g}}T^0{}_0 \\ &= \frac{1}{2}\int d^3x\sqrt{-\bar{g}}\{\bar{g}^{00}\partial_0 h\partial_0 h - \bar{g}^{ij}\partial_i h\partial_j h + V''(\bar{\phi})h^2\} \end{aligned} \tag{22}$$

where $V''(\bar{\phi})$ is the Lagrangian $(\text{mass})^2$, and the explicit $(AdS)_4$ metric coefficients of (14) are inserted. Note that $E(T^{\mu\nu})$ is meaningful because the scalar contribution to the energy is independently conserved in the small fluctuations approximation. The total scalar energy consists of a positive-definite kinetic energy and a potential energy which can be negative definite if the critical point is a maximum. Since we are mainly interested in this case let us write $V''(\bar{\phi}) = -\alpha a^2$ where $\alpha > 0$ is a dimensionless parameter characterizing the negative curvature of the potential at its maximum.

Now the main requirement is that the integrated total scalar energy must be positive for stability, and this can be established for fluctuations which fall-off sufficiently fast at spatial infinity. First one notices that the energy integral converges for fluctuations with asymptotic behavior $h(x) \xrightarrow[\rho\to\pi/2]{} (\cos\rho)^\mu$ if $\mu > 3/2$. There is then a fairly simple scaling argument involving the substitution $h(x) = (\cos\rho)^\mu h'(x)$, partial integration, and then optimization of the parameter μ to prove the desired positivity property which holds provided that $\alpha < 9/4$. The physical reason for positive total energy involves the uncertainty principle. Since the fluctuations must vanish at infinity, the field derivatives cannot be arbitrarily small, and the scaling argument merely shows that under the required asymptotic conditions the positive kinetic energy dominates over the negative potential

To check that the boundary conditions required in the scaling argument are natural, one considers the free quantum scalar field theory in an $(AdS)_4$ background with action

$$S = \frac{1}{2}\int d^4x\sqrt{-\bar{g}}(\bar{g}^{\mu\nu}\partial_\mu h\partial_\nu h + \alpha a^2 h^2) \quad (23)$$

One then studies solutions of the wave equations of this action, and imposes boundary conditions[7] so that the Cauchy problem has a unique solution, and the standard formally conserved quantities of a scalar field theory, the scalar product

$$(h_1,h_2) = i\int d^3x\sqrt{-\bar{g}}g^{0\nu}(h_1\partial_\nu h_2) \quad (24)$$

of a pair of solutions of the equation, and the ten SO(3,2) generators

$$M_{AB}[h] = \int d^3x\sqrt{-\bar{g}}T^{0\nu}K_{AB\nu} \quad (25)$$

are actually conserved in time.

Standard boundary analysis then allows one to find a complete set of positive frequency modes[3]

$$\phi_{\omega\ell m}(x) = e^{-i\omega t}Y_\ell^{\,m}(\theta,\phi)R_{\omega\ell}(\rho) \quad (26)$$

such that the expansion

$$h(x) = \sum_{\omega\ell m}\{a_{\omega\ell m}\phi_{\omega\ell m}(x) + a^*_{\omega\ell m}\phi^*_{\omega\ell m}(x)\} \quad (27)$$

with $a_{\omega\ell m} = (\phi_{\omega\ell m},h)$ gives the solution of the Cauchy problems. This program can be carried through provided that the previous condition $\alpha < 9/4$ is satisfied, and one can show that the mode functions are a complete mathematical basis for a unitary infinite dimensional positive energy irrep of SO(3,2). This is exactly how we would expect the "particle states" in an $(AdS)_4$ field theory to behave.

Actually, for certain values of the mass parameter α, one finds two complete sets of mode functions corresponding roughly to Dirichlet or Neumann boundary conditions on the solution of the field equations. One set corresponds to the boundary conditions found in the scaling argument outlined above. The second set has asymptotic behavior $h(x) \sim (\cos\rho)^\mu$ with $\mu < 3/2$. Nevertheless, one can show that such modes imply that the total energy constructed from an "improved" stress tensor is conserved and positive. We refer interested readers to the original literature[3] concerning the complication of the improved energy. This is not a completely settled matter, and there is room for further research.

There can be little doubt concerning the main result of this work. It has been shown that an $(AdS)_4$ background corresponding to a local maximum of a scalar potential $V(\phi)$ is stable against small fluctuations, provided that the downward curvature of the potential is not too large compared with the curvature of the space-time background. The specific stability condition $\alpha < 9/4$ has been applied to determine the stability of many of the critical points of gauged extended supergravity in 4-dimensions and to 4-dimensional space-time backgrounds coming from Kaluza-Klein solutions of 11-dimensional supergravity.[8] The argument has also been extended to scalar fields in $(AdS)_d$ backgrounds in d dimensions, and the stability criterion is then $\alpha < (d-1)^2/4$. The boundary conditions which were required to prove stability are that the SO(3,2) generators actually converge and that there is no leakage of SO(3,2) fluxes through spatial infinity. These boundary conditions appear natural because they lead to a complete orthonormal set of mode functions for the free-field problem.

4. Supersymmetric backgrounds and Killing spinors

The small fluctuations stability argument gives some physical understanding of why an $(AdS)_4$ background can be stable even at a maximum of the scalar potential. However, it is still necessary to establish global stabilityfor fluctuations which are arbitrarily large in the interior but vanish at spatial infinity. This leads naturally to the question of whether a background solution of a supergravity theory preserves part of the supersymmetry of the theory.

To examine this question in the simplest context let us consider the N = 1 cosmological supergravity theory of (1-3). One natural background solution of the theory is that in which the gravitino field $\psi_\mu(x)$ vanishes and the background verbein $\bar{V}_{a\mu}(x)$ is a frame on $(AdS)_4$ with scale $a^2 = e^2/\kappa^2$. This solution is invariant under the subset of local sypersymmetry transformations (2-3) generated by spinors $\varepsilon(x)$ which satisfy

$$\bar{D}_\mu\varepsilon(x) = (\partial_\mu + \frac{1}{2}\bar{\omega}_{\mu ab}\sigma^{ab} + i\frac{a}{2}\bar{\gamma}_\mu)\varepsilon(x) = 0 \tag{28}$$

where $\bar{\omega}_{\mu ab}$ is the spin correction of the chosen frame and $\bar{\gamma}_\mu = \gamma^a\bar{V}_{a\mu}(x)$. The integrability condition of this equation is

$$[\bar{D}_\mu, \bar{D}_\nu]\varepsilon = (\frac{1}{2}\bar{R}_{\mu\nu ab}\sigma^{ab} - \frac{1}{4}a^2[\bar{\gamma}_\mu,\bar{\gamma}_\nu])\varepsilon(x) = 0 \tag{29}$$

and it is satisfied because $(AdS)_4$ is a maximally symmetric space-time for which the curvature tensor satisfies

$$\bar{R}_{\mu\nu ab} = a^2(\bar{V}_{a\mu}\bar{V}_{b\nu} - \bar{V}_{a\nu}\bar{V}_{b\mu}) \tag{30}$$

In a particular frame[3] one can show that the four linearly independent solutions of (29) are given by

$$\begin{aligned} \varepsilon(x) &= S(x)_{\alpha\beta}\xi_\beta \\ S(x) &= (\cos\rho)^{-1/2}[\cos\frac{\rho}{2} + i\hat{\gamma}\cdot\hat{x}\sin\frac{\rho}{2}]e^{-\frac{i\gamma^0 t}{2}} \end{aligned} \tag{31}$$

where ξ_β is a 4-component constant spinor.

A spinor $\varepsilon(x)$ which satisfies (29) is called a Killing spinor. Earlier we discussed the fact that given a Killing vector K_{AB} of a background, one can construct the corresponding conserved generator

M_{AB}. Analogously one can use[1] the Killing spinor to construct a conserved supercharge in the schematic form

$$Q_\alpha = \int_t d^3x\sqrt{-g}\bar{S}(x)_{\alpha\beta}J^0_\beta \tag{32}$$

where $J^\mu_\beta(x)$ is the supercurrent of the theory which plays the same role as $\theta^{\mu\nu}$ of (18). This supercharge annihilates the vacuum state just as the analogous bosonic charges M_{AB} do. Therefore in a very precise sense, each independent Killing spinor determines a residual supersymmetry of the background.

Given two Killing spinors $\varepsilon_1(x)$ and $\varepsilon_2(x)$, one can form the bilinear $K_\mu = \bar{\varepsilon}_1(x)\gamma_\mu\varepsilon_2(x)$. It is easy to show, using (28), that $D_\mu K_\nu + D_\nu K_\mu = 0$. Hence K is a Killing vector which must be a linear combination [3] of the ten Killing vectors K_{AB} of $(AdS)_4$. Using this fact and the canonical formalism, one can show that the supercharges have anti-commutator

$$\{Q_\alpha,\bar{Q}_\beta\} = (\gamma^a M_{a4} + i\sigma^{ab}M_{ab})_{\alpha\beta} \tag{33}$$

which is one of the structure relations of the superalgebra OSp(1,4).

In gauged supergravity for higher N and for Kaluza-Klein supergravity, Killing spinors may again be defined as the subset of gauge parameters for which a background configuration remains invariant under the supersymmetry transformation rules of the theory. One must study these transformation rules, evaluated in the background, to see whether Killing spinors exist or not. When they do exist the Killing spinors can be used to construct supercharges, and the vacuum state is then invariant under a superalgebra such as OSp(N,4). If Killing spinors do not exist then the supersymmetry of the Lagrangian is spontaneously broken in the background, and one has a super-Higgs effect by which the gravitino fields become massive. The gauged N = 5 supergravity theory is a nice example of these ideas, because there are two critical points of the scalar potential.[3,5] One of them is invariant under OSp(5,4). For the other, supersymmetry is broken.

When the background is supersymmetric the energy operator is related to the supercharges as in (33), and one may trace with the matrix $\gamma^0{}_{\alpha\beta}$ to obtain

$$E = M_{04} = \frac{1}{4}\sum_\alpha(Q_\alpha Q_\alpha{}^+ + Q_\alpha{}^+Q_\alpha) \tag{34}$$

Thus the energy operator is positive, which implies that a supersymmetric background is globally stable. The formal statement of positivity, (34), may be converted to a precise analytic argument[5] based on Witten's proof of positive energy in ordinary relativity. This argument requires certain boundary conditions on the fluctuations[11] as in our discussion of linearized stability.

5. Summary

There are now available three distinct stability arguments which can be applied to ascertain the stability or instability of the many known backgrounds of 4-dimensional supergravity theories and Kaluza-Klein theories.

i) The small fluctuations stability argument[3] is based on the criterion $\alpha < 9/4$ for the mass eigenvalues of scalar field fluctuations. If this criterion is satisfied, one must go on to investigate global stability. If it is not satisfied, the background may be rejected because it is unstable.

ii) If a background has Killing spinors, then the generalized Witten argument[5] may be applied to establish global stability.

iii) If supersymmetry is spontaneously broken in a particular background, then the technique of Boucher[6] may be applied to study the stability question.

In two exotic versions of gauged N = 4 supergravity[12] there are globally stable backgrounds, but they are not of the $(AdS)_4$ type discussed here. Rather, they are dimensionally reduced backgrounds such as $(AdS)_2 \times S_2$ with non vanishing electric or magnetic fields. There are now several papers on these electrovacs[13] or magnetovacs.[14] The first named paper contains a pedagogical discussion of the generalized Witten positive energy argument. Some exotic gaugings of N = 8 supergravity based on non-compact groups such as SO(7,1) have recently been obtained.[15] The background configurations of these theories and their stability are only partially known.

6. References

1] L.F. Abbott and S. Deser, "Stability of Gravity with a Cosmological Constant," Nucl. Phys. B195 (1982) 76.

2] M. Henneaux and C. Teitelboim, "Hamiltonian Treatment of Asymptotically Anti-de Sitter Spaces," Phys. Lett. 142B (1984) 355.

3] P. Breitenlohner and D.Z. Freedman, "Positive Energy in Anti-de Sitter Backgrounds and Gauged Extended Supergravity," Phys. Lett. 115B (1982)197; "Stability in Gauged Extended Supergravity," Annals of Physics 144 (1982) 249.

4] E. Witten, "A New Proof of the Positive Energy Theorem," Comm. Math. Phys. 80 (1981) 381; J.M. Nester, "A New Gravitational Energy Expression with a Simple Positivity Proof," Phys. Lett. 83A (1981) 241.

5] G.W. Gibbons, C.M. Hull and N.P. Warner, "The Stability of Gauged Supergravity," Nucl. Phys. B218 (1983) 173.

6] W. Boucher, "Positive Energy Without Supersymmetry," Nucl. Phys. B242 (1984) 282.

7] S.J. Avis, C.J. Isham, and D. Storey, "Quantum Field Theory in Anti de Sitter Space-Time," Phys. Rev. D18 (1978) 3565.

8] For example, see B. de Wit and H. Nicolai, "The Parallelizing S^7 Torsion in Gauged N = 8 Supergravity," Nucl. Phys. B231 (1984) 504; M.J. Duff, B.E.W. Nilsson and C.N. Pope, "The Criterion for Vacuum

Stability in Kaluza-Klein Supergravity," Phys. Lett. 139B (1984) 154; B. Biran and Ph. Spindel, "Instability of the Parallelized Seven-Sphere: An Eleven-Dimensional Approach," Phys. Lett. 141B (1984) 181; K. Ito and O. Yasuda, "Instability of a Class of Freund-Rubin Type Solutions in d = 11 Supergravity," University of Tokyo preprint (1984); D.N. Page and C.N. Pope, "Instabilities in Englert-Type Supergravity Solutions," Imperial College preprint (1984).

9] L. Mezincescu and P.K. Townsend, "Stability at a Local Maximum in the Higher Dimensional Anti-de Sitter Space and Applications to Supergravity," Annals of Physics, Vol. 159, No. 1 (1985).

10] B. de Wit and H. Nicolai, "Extended Supergravity with Local SO(5) Invariance," Nucl. Phys. B188 (1981) 98.

11] S.W. Hawking, "The Boundary Conditions for Gauged Supergravity," Phys. Lett. 126B (1983) 175.

12] D.Z. Freedman and J.H. Schwarz, "N = 4 Supergravity Theory with Local SU(2) x SU(2) Invariance," Nucl. Phys. B137 (1978) 333; S.J. Gates Jr. and B. Zwiebach, "Gauged N = 4 Supergravity Theory with a New Scalar Potential," Phys. Lett. 123B (1983) 200; B. Zwiebach, "The Inequivalent Gauged SO(4) Supergravities," Nucl. Phys. B238 (1984) 367.

13] D.Z. Freedman and G.W. Gibbons, "Electrovac Ground State in Gauged SU(2) x SU(2) Supergravity," Nucl. Phys. B233 (1984) 24.

14] D.Z. Freedman and B. Zwiebach, "Magnetovacs in Gauged N = 4 Supergravity," Nucl. Phys. B237 (1984) 573.

15] C.M. Hull, "Non-compact Gaugings of N = 8 Supergravity," Phys. Lett. 142B (1984) 39.

SUPERSYMMETRY AND COSMOLOGY*

Stuart Raby

Los Alamos National Laboratory

Theoretical Divsision

T-8, MS B285

Los Alamos, New Mexico 87545

Outline

I. REVIEW OF MINIMAL LOW ENERGY SUPERGRAVITY (MLES) SPECTRUM.

II. DEFINE R-PARITY AND DISCUSS CANDIDATES FOR LIGHTEST SUPERSYMMETRIC PARTNER (LSP).

III. REVIEW STANDARD "BIG BANG" COSMOLOGY AND RELATIVISTIC BOLTZMANN EQUATION.

IV. COSMOLOGICAL BOUNDS FOR THE MASS OF THE LSP.

V. COSMOLOGICAL BOUNDS FOR THE GRAVITINO.

I. Spectrum of MLES

In Table I we list the states in a MLES model.[1] On the left-hand side are all the ordinary particles; three families of quarks and leptons, higgs bosons and gauge bosons of $SU_3 \times SU_2 \times U_1$. In addition, we note that supersymmetry requires two sets of Higgs doublets in order to a) give mass to both up and down quarks, and b) to have no weak hypercharge anomally. We also have the graviton and the scalar partners of the goldstino. The right-hand side contains all the super-partners. Experimentally, we know that the slepton mass, $m_{\tilde{\ell}^\pm} \gtrsim 20$ GeV,[2] and the gluino has a mass, $m_{\tilde{g}} \geq 5$ GeV.[3] The fact is that the right-hand side has remained quite invisible; although recently some authors have argued that the jet events at CERN may indicate squarks or gluinos with mass of order 40 GeV.[4]

II. R-parity and the LSP

Fayet and Farrar[5] have defined a discrete symmetry, called

*Two lectures given at the XV GIFT INTERNATIONAL SEMINAR ON THEORETICAL PHYSICS, JUNE 4-8, 1984.

R-parity, which distinguishes the ordinary particles from their superpartners. Consider a chiral superfield, $\phi(x,\theta)$. R-parity is defined by the transformation

$$\phi'(x,\theta) = \eta\phi(x,-\theta) \quad , \tag{II.1}$$

which for the components ϕ, ψ, F gives

$$\begin{aligned} \phi'(x) &= \eta\phi(x) \quad , \\ \psi'(x) &= -\eta\psi(x) \quad , \\ F'(x) &= \eta F(x) \quad . \end{aligned} \tag{II.2}$$

Thus the R-charge of a particle with spin s is

$$(-1)^{2s}\eta \quad . \tag{II.3}$$

For gauge fields $V(x,\theta,\bar{\theta})$, we have

$$V'(x,\theta,\bar{\theta}) = V(x,-\theta,-\bar{\theta}) \quad , \tag{II.4}$$

which for the components χ, A_μ, D gives

$$\begin{aligned} \chi'(x) &= -\chi(x) \quad , \\ A_\mu'(x) &= A_\mu(x) \quad , \\ D'(x) &= D(x) \quad . \end{aligned} \tag{II.5}$$

The Lagrangian for the MLES model is given (schematically) by the expression

$$\begin{aligned} \mathcal{L}_{MLES} &\sim \int d^4\theta\phi_i^*(x,\theta)e^{-gV(x,\theta,\bar{\theta})}\phi_i(x,\theta)(1 + m_G^2\theta^2\bar{\theta}^2) \\ &+ \{ \int d^2\theta[mH_1\bar{H}_2 + gH_1Q\bar{U} + \bar{g}\bar{H}_2Q\bar{D} + \bar{g}'\bar{H}_2L\bar{E}](1 + m_G\theta^2) + \text{h.c.}\} \\ &+ \{ \int d^2\theta W^\alpha W_\alpha(1 + m_G\theta^2) + \text{h.c.}\} \quad , \end{aligned} \tag{II.6}$$

where i labels the chiral superfield $Q, \bar{U}$, etc. and $W_\alpha \equiv \bar{D}^2[e^{gV}D_\alpha e^{-gV}]$ is the gauge field strength. $\mathcal{L}_{MLES}$ is invariant under the R-parity transformation with $\eta = -1$ for the fields $Q,\bar{U},\bar{D},L,\bar{E}$; $\eta = +1$ otherwise. Note that, in Table I, all fields on the left-hand side (right-hand side) are R even (odd). Thus, the MLES model has an exact R-parity. It was noted by Farrar and Weinberg[6] that the generator of the R-parity defined above is equivalent to the operator

$$R \equiv (-1)^{3(B-L)}(-1)^F \quad , \tag{II.7}$$

where B is baryon number, L is lepton number, and F is fermion number. This observation makes it clear that if (B-L) and F are conserved, then R-parity is an exact symmetry.

In the following text, we shall assume that R-parity is conserved. We shall not consider the possibility of explicitly breaking R-parity with operators of the form

$$\int d^2\theta H_1 L \quad , \tag{II.8}$$

or spontaneously broken R-parity via a non-vanishing $\tilde{\nu}$ expectation value. If R-parity is conserved, then the Lightest Supersymmetric Partner (LSP) is absolutely stable. We are then obliged to ascertain the present expected abundance of the LSP (using standard early universe assumptions) and see if it is consistent with observations. In particular, we shall obtain limits on the mass of the LSP by requiring that its energy density ρ_{LSP} is less than the cosmological critical energy density ρ_c. Before we do this, however, let us discuss the possible candidates for the LSP.

What are the likely candidates for the LSP? We shall only consider electrically neutral, color singlet states. The reason is that a) since they have less interactions, they are probably lighter (for example, if gluinos and photinos are degenerate at $M_{p\ell}$, then one loop renormalization group corrections give the ratio $m_{\tilde{g}}:m_{\tilde{\gamma}} = \alpha_3:\alpha$ at the weak scale); and b) if they are charged, then they have already been ruled out. Wolfram[7] has calculated the expected abundance of stable heavy leptons (n_L) or hadrons (n_H). He finds

$$n_L \sim 10^{-5} n_b \left(\frac{m_L}{1 \text{ GeV}}\right),$$

$$n_H \sim 10^{-10} n_b \left(\frac{m_H}{1 \text{ GeV}}\right), \tag{II.9}$$

where n_b is the observed baryon number density and m_L (m_H) is the mass of the heavy lepton (hadron). A search has been made for such stable particles[8] with the result that for a mass $m < 350$ GeV, the observed density is less than $10^{-21} n_b$. Hence, a charged LSP is not consistent with observation.

We shall henceforth consider the following LSP candidates:

1. photino-Higgsino,
2. sneutrino (scalar neutrino),
3. gravitino. (II.10)

III. Review standard "Big Bang" cosmology

We shall use units such that $\hbar = c = k = 1$. Newton's constant G_N defines

$$M_{p\ell} \equiv G_N^{-1/2} = 1.9 \times 10^{19} \text{ GeV, or sometimes } M \equiv \frac{M_{p\ell}}{\sqrt{8\pi}} = 2.4 \times 10^{18} \text{ GeV.}$$

We assume space-time is described, on large scales, by the Friedmann-Robertson-Walker metric

$$ds^2 = -dt^2 + a^2(t)\left[\frac{dr^2}{1-kr^2} + r^2 d\Omega^2\right], \tag{III.1}$$

where $d\Omega^2 \equiv d\theta^2 + \sin^2\theta \, d\phi^2$, and the parameter k determines the spatial topology of the universe: $k = +1$, closed; $k = 0$, open and flat; $k = -1$, open and curved. $\underline{a}$ is the cosmological scale parameter. Recall that the FRW metric is uniquely determined by requiring space to be both homogeneous and isotropic.

The dynamical equations of cosmology are given by:

Einstein's equation

$$G_{\mu\nu} = \frac{1}{M^2} T_{\mu\nu} \quad , \quad T^{\mu\nu}{}_{;\nu} = 0 \quad , \tag{III.2}$$

which implies $\left(\dot{a} \equiv \frac{da}{dt} \quad , \quad \dot{\rho} \equiv \frac{d\rho}{dt}\right)$

$$H^2(t) \equiv \left(\frac{\dot{a}}{a}\right)^2 = \frac{\rho}{3M^2} - \frac{k}{a^2} \quad , \tag{III.3a}$$

$$\dot{\rho} = -3(\rho + p)H \quad , \tag{III.3b}$$

where

$$T_{\mu\nu} \equiv (\rho + p)u_\mu u_\nu + pg_{\mu\nu} \quad ,$$

$$u^\mu \equiv \frac{dx^\mu}{d\tau} = (1,\vec{0}) \quad . \tag{III.4}$$

ρ,p are the energy, pressure densities, respectively, for a perfect fluid, u^μ is the fluid velocity field 4-vector, and $H(t)$ is the Hubble expansion rate. The geodesic equations of motion for a particle (or fluid element) is given by

$$\frac{du^\gamma}{d\tau} + \Gamma^\gamma_{\mu\nu} u^\mu u^\nu = 0 \quad . \tag{III.5}$$

In a FRW universe, we find

$$\frac{du^\gamma}{d\tau} = 0 \quad \text{for } u^\gamma = (1,\vec{0}) \quad . \tag{III.6}$$

Thus, the FRW coordinate position r,θ,ϕ in Eq. (III.1) describes a coordinate frame which is co-moving with the fluid elements. The distance between two such elements is then increasing with the scale factor $a(t)$.

The dynamical equations (III.3) must be supplemented with an equation of state, $p = p(\rho)$. We shall consider three forms:

1) radiation;

$$p_r = \frac{1}{3}\rho_r \quad , \tag{III.7}$$

(using III.3b implies)

$$\rho_r \sim \frac{1}{a^4} \quad .$$

2) non-interacting conserved massive particles, i.e., matter;

$$p_m = 0 \quad , \tag{III.8}$$

(using III.3b implies)

$$\rho_m \sim \frac{1}{a^3} \quad .$$

3) cosmological constant;

$$p_\Lambda = -\rho_\Lambda \quad . \tag{III.9}$$

(using III.3b implies)

$$\rho_\Lambda \equiv \text{constant} \quad .$$

The cosmological constant Λ is typically defined by

$$\Lambda \equiv \frac{\rho_\Lambda}{3M^2} \quad . \tag{III.10}$$

We now have a complete system of equations (III.3,7-9).

Let us now discuss some observed properties of the present universe which shall be useful later.

1. The present value of the Hubble expansion parameter is

$$H_0 = h100km/s/Mpc \quad , \tag{III.11}$$

with $1/2 \leq h \leq 1$. (1pc = 3.26 light years, 1Mpc $\simeq 3 \times 10^{24}$cm). H_0 determines the observed red shift, $Z \equiv \frac{\lambda_{observed} - \lambda_{emitted}}{\lambda_{emitted}} \left(= \frac{a(t_0)}{a(t_e)} - 1 \right)$ of distant stars using the relation

$$d = H_0^{-1}\left[Z + \frac{1}{2}(1 - q_0)Z^2 + 0(Z^3)\right] \quad , \tag{III.12}$$

where d is the present distance to the star and q_0 is the present value of the decelleration parameter $\left[q_0 \equiv - \frac{\ddot{a}a}{\dot{a}^2}\bigg|_{today} \leq 1 \right]$.

2. We define a critical enery density

$$\rho_c(t) \equiv 3H^2(t)M^2 \quad . \tag{III.13}$$

Using (III.3a) we see that for

$$\rho \begin{Bmatrix} > \\ = \\ < \end{Bmatrix} \rho_c \quad , \; k = \begin{Bmatrix} 1 \\ 0 \\ -1 \end{Bmatrix} \quad . \tag{III.14}$$

Hence, ρ_c is the boundary between an open or closed universe. Using H_0 (III.11) we find today

$$\begin{aligned} \rho_c^0 &\simeq 2 \times 10^{-29} g/cm^3 h^2 \\ &\simeq (3 \times 10^{-3} eV)^4 h^2 \quad . \end{aligned} \tag{III.15}$$

3. The observed energy density in 2.7°k black body radiation is

$$\rho_r^0 = 4.5 \times 10^{-34} g/cm^3 \quad . \tag{III.16}$$

Cosmologists typically define the ratio, for matter of type i,

$$\Omega_i \equiv \frac{\rho_i}{\rho_c} \quad . \tag{III.17}$$

Thus

$$\Omega_r^0 \equiv 2.25 \times 10^{-5} h^{-2} \quad . \tag{III.18}$$

For luminous matter ("baryons") a typical value is

$$\Omega_b^0 \simeq .02 \quad . \tag{III.19}$$

However, the dominant form of energy in the universe is not luminous. This "dark matter" contributes an amount

$$\Omega_{dm}^0 \simeq .2 \quad . \tag{III.20}$$

Note that the energy in the universe is apparently matter dominated today. (III.19, 20) are subject to considerable debate (for a review see ref. 9). Note that for $p_0 << \rho_0$, the decelleration parameter satisfies

$$q_0 \equiv \frac{1}{2} \frac{\rho_{total}^0}{\rho_c^0} \equiv \frac{1}{2} \Omega_{total}^0 \leq 1 \quad , \tag{III.21}$$

which implies

$$\Omega_{total}^0 \leq 2 \quad . \tag{III.22}$$

This is the fundamental cosmological constraint which shall be imposed on model building. As a first example, if we require

$$\rho_\Lambda^0 < \rho_c^0 \quad , \tag{III.23}$$

we find

$$\Lambda < 10^{-120} M^2 \ . \tag{III.24}$$

Hence the cosmological constant is one of the most finely-tuned parameters in all physics.

4. The ratio of the number of baryons to photons is

$$\left.\frac{n_b}{n_\gamma}\right|_0 \simeq 10^{-9} \ , \tag{III.25}$$

where $n_\gamma^0 \simeq 400/\text{cm}^3$, $n_b^0 \simeq \rho_b^0/1$ GeV .

As a result, the thermodynamic variables, pressure, and entropy are radiation dominated.

5. Since ρ is matter dominated, we have, using (III.3,8)

$$\left(\frac{\dot{a}}{a}\right)^2 \sim \frac{\Omega_m^0 \rho_c^0}{3M^2}\left(\frac{a_0}{a}\right)^3 \ , \tag{III.26}$$

where a_0 is the present value of a(t), and we neglected k. The solution is

$$a(t) \sim t^{2/3} \ . \tag{III.27}$$

Thus, $H_0 = 2/(3t_0)$, and the age of the universe t_0 is determined by the Hubble constant

$$t_0 \sim H_0^{-1} \sim 10^{10} \text{years} \ . \tag{III.28}$$

This calculation is valid for $\rho_m > \rho_r$ and $a_0/a > \Omega_0^{-1} - 1$. The second constraint has probably been satisfied until perhaps very recently.

Let us now determine how long the universe has been matter dominated. We define the time, t_{eq}, and the scale factor, a_{eq}, such that ρ_m equals ρ_r. We find

$$\frac{a_0}{a_{eq}} \sim 10^4 \quad , \quad \text{and } t_{eq} \sim 10^4 \text{ years} \quad . \tag{III.29}$$

Hence, the universe has been matter dominated throughout most of its history.

<u>Relevant statistical mechanics</u>[10]

For a particle in thermal equilibrium, the phase space density $f(\vec{p})$ is given by

$$f(\vec{p}) \equiv \frac{dN}{d^3\vec{p}d^3\vec{x}} = \frac{N_h}{(2\pi)^3} \frac{1}{\exp\left(\frac{E-\mu}{T}\right) + \theta} \quad , \tag{III.30}$$

where $\theta = \left\{ \begin{matrix} +1 & \text{fermions} \\ -1 & \text{bosons} \end{matrix} \right\}$, N_h is the number of helicity states,

$E \equiv \sqrt{\vec{p}^2 + m^2}$, T is the temperature and μ is the chemical potential.

The number density n and energy density ρ are then given by

$$\begin{aligned} n &= \int d^3\vec{p} f(\vec{p}) \quad , \\ \rho &= \int d^3\vec{p} E(\vec{p}) f(\vec{p}) \equiv n\langle E\rangle \quad , \end{aligned} \tag{III.31}$$

where brackets, $\langle\rangle$, denotes thermal average.

For $T \gg m$, we obtain the number and energy density appropriate for "radiation".

$$\begin{aligned} n_r &= \frac{g'\zeta(3)T^3}{\pi^2} \quad , \\ T \gg m & \\ \rho_r &= g \frac{\pi^2}{30} T^4 \quad . \end{aligned} \tag{III.32}$$

$g' = N_{hB} + 3/4\, N_{hF}$, $g = N_{hB} + 7/8\, N_{hF}$ are weighted sums of boson (N_{hB})

and fermion (N_{hF}) helicities and $\zeta(3) = \sum_{i=1}^{3} i^{-3} \sim 1.2$. Note that the thermal averaged photon energy is

$$\langle E\gamma\rangle \simeq 2.6T \quad . \qquad \text{(III.33)}$$

For $T \ll m$, we obtain densities appropriate to non-relativistic "matter".

$$n_m = N_h\left(\frac{mT}{2\pi}\right)^{3/2} e^{-m/T} \ ,$$

$T \ll m$
$\mu = 0$ (III.34)

$$\rho_m = mn_m + \frac{3}{2} n_m T \ \ , \ p_m = n_m T \ \ ,$$

where p_m is the pressure density. As noted previously, $p_m/p_r|_0 \sim 10^{-9}$. Comparing Eqs. (III.7) and (III.32) we see that

$$T \sim \frac{1}{a} \ , \qquad \text{(III.35)}$$

for radiation in thermal equilibrium. For matter, on the other hand, $T \sim 1/a$ only if $\mu \neq 0$, or it is out of thermal equilibrium (see III.8 and III.34). In thermal equilibrium, n_m is Boltzmann suppressed ($e^{-m/T}$) and thus decreases much faster than $1/a$. It is not always possible, however, for the reaction rates, which govern the approach to equilibrium, to keep up with the expansion and the system goes out of equilibrium. Let us, therefore, briefly discuss the approach to thermal equilibrium.

Consider two species of particles, x (which is out of equilibrium), and r (in thermal equilibrium). We want to know how x approaches equilibrium.[10]

The relativistic version of Boltzmann's equation is given by

$$\frac{dn_x}{dt} + 3\frac{\dot{a}}{a}n_x = \Lambda^{x}_{r_1r_2}[f_{r_1}f_{r_2}(1-\theta_x f_x)|\mathcal{M}(r_1r_2\to x)|^2$$

$$- f_x(1-\theta_r f_{r_1})(1-\theta_r f_{r_2})|\mathcal{M}(x\to r_1r_2)|^2]$$

$$+ 2\Lambda^{34}_{12}[f_{r_1}f_{r_2}(1-\theta_x f_{x_3})(1-\theta_x f_{x_4})|\mathcal{M}(r_1r_2\to x_3x_4)|^2$$

$$- f_{x_1}f_{x_2}(1-\theta_r f_{r_3})(1-\theta_r f_{r_4})|\mathcal{M}(x_1x_2\to r_3r_4)|^2] \ , \tag{III.36}$$

where the phase space factor Λ is defined by

$$\Lambda^{b_1b_2}_{a_1a_2} \equiv \int \frac{d^4p_{a_1}}{(2\pi)^3}\frac{d^4p_{a_2}}{(2\pi)^3}\frac{d^4p_{b_1}}{(2\pi)^3}\frac{d^4p_{b_2}}{(2\pi)^3}\,\delta^4(p^2_{a_i}-m^2_{a_i})$$

$$\times\ \delta^4(p^2_{b_i}-m_{b_i})(2\pi)^4\delta^4(\Sigma_i\, p_{a_i} - \Sigma_i\, p_{b_i}) \ ,$$

$\theta_{r,x}$ is defined in Eq. (III.30) and $\mathcal{M}$ is the scattering amplitude for the indicated process. The four terms in Eq. (III.36) represent the four processes in fig. 1, respectively. If we assume that the interactions are both CPT and CP invariant, use the fact that f_r is an equilibrium distribution and assume that the gas is non-degenerate, we obtain

$$\frac{dn_x}{dt} + 3\frac{\dot{a}}{a}n_x = (n^{eq}_x - n_x)\langle\Gamma_{x\to r_1r_2}\rangle$$

$$+ 2[(n^{eq}_x)^2 - (n_x)^2]\langle v\sigma_{xx\to rr}\rangle \ . \tag{III.37}$$

n^{eq}_x is the thermal equilibrium number density. Note that equilibrium is achieved <u>if</u> $H \equiv \dot{a}/a \ll \langle\Gamma\rangle$ or $\langle\sigma vn\rangle$, i.e., the expansion rate is less than a typical reaction rate. Otherwise we find $n_x \sim 1/a^3$, as in Eq. (III.8).

We now have the necessary tools to review the standard picture of the early universe. In Tables II and III we have listed the major events. We shall assume that all particles are in thermal equilibrium at the Planck temperature. As the universe expands it goes through several phase transitions listed in Table III. There may be an inflationary phase. Typically during this transition the universe cools adiabatically and then reheats to approximately the initial transition temperature. We reset our clocks at that time and then continue as before. We shall not discuss inflation more than to say that in such a transition a huge amount of entropy, of order 10^{84}, is produced.

By the time the universe gets down to a temperature of order 100 MeV, particles with mass greater than this temperature are severely Boltzmann suppressed or are out of equilibrium. At about 100 MeV, muons begin to annihilate[11] (see Table II). Neutrinos will soon decouple, leaving an asymptotic ratio of neutrons to neutrons plus protons of about 1/6. This will be the initial ratio for the subsequent process of helium synthesis, (at the temperature, $T_{He} \sim .1$ MeV), where all the remaining neutrons will be incorporated into helium nuclei. At a temperature $T \sim 1/2$ MeV, electrons annihilate and heat up the photons but not the neutrinos, since they have already decoupled. Using entropy conservation we find that neutrinos are thus colder than photons,

$$T_\nu = \left(\frac{4}{11}\right)^{1/3} T_\gamma \quad . \qquad \text{(III.38)}$$

Throughout the epoch from $T_{p\ell}$ to $T_{eq} \sim \Omega_m h^2$ eV, the universe is radiation dominated (i.e., the first 3×10^4 years). Thereafter, the energy density is matter dominated. At about $T \sim 1/3$ eV, free protons and electrons combine and form neutral Hydrogen. This is the so-called recombination temperature. Once this occurs, photons decouple. These photons continue to red-shift for the next $\sim 10^{10}$ years. They are presently observed as the 2.7°k, microwave background radiation.

IV. Cosmological bounds on particle masses

We are now ready to obtain the bounds, on particle masses, coming from cosmology. Before we discuss the bounds on the possible candidates for the LSP (see II.10), we shall first warm up by discussing the standard cosmological bounds on the "neutrino" mass.

Neutrinos

"Neutrinos" are stable, weakly interacting, neutral fermions. We want to calculate the present energy density in neutrinos, ρ_ν^0. We shall require $\rho_\nu^0 < \rho_c^0 (\Omega_\nu^0 < 1)$. At early times, neutrinos are in thermal equilibrium with the thermal bath. If they are light enough so as to decouple when they are still relativistic, their annihilation rate is given by

$$\langle\sigma v n\rangle \sim G_F^2 T^5 \quad . \tag{IV.1}$$

Assuming the universe is radiation dominated at the decoupling temperature, the Hubble expansion rate is given by

$$H \sim \frac{T^2}{M} \quad . \tag{IV.2}$$

The decoupling temperature, T_d, is then obtained by requiring (see III.37)

$$\langle\sigma v n\rangle = H \quad . \tag{IV.3}$$

We obtain

$$T_d = 1 \text{ MeV} \quad . \tag{IV.4}$$

At $T \sim 1$ MeV, the universe is radiation dominated, hence (IV.2) is valid. Moreover, if we require $m_\nu < 1$ MeV then the approximation (IV.1) is also valid. When neutrinos decouple we have

$$n_\nu = \frac{3}{4} n_\gamma \quad . \tag{IV.5}$$

After electron annihilation at T ~ 1/3 MeV, we have

$$n_\nu = \frac{3}{4} \cdot \frac{4}{11} n_\gamma \quad . \qquad \text{(IV.6)}$$

This relation remains valid to the present, so that today

$$\rho_\nu^0 = (\Sigma m_\nu) n_\nu^0 = (\Sigma m_\nu) \frac{3}{11} n_\gamma^0 \cdot \qquad \text{(IV.7)}$$

(Σm_ν represents the sum over all neutrino species.) Using $\rho_\gamma^0 = (2.6T_0) n_\gamma^0$, Eq. (III.33), we obtain

$$\rho_\nu^0 = (\Sigma m_\nu) \frac{3}{11} \frac{\rho_\gamma^0}{(2.6T_0)} \leq \rho_c^0 \quad ,$$

or

$$(\Sigma m_\nu) \leq 100 \text{ eVh}^2 \quad . \qquad \text{(IV.8)}$$

[12]

Before we continue with heavy neutrinos, we note that the discussion of light gravitinos is very similar to that of light neutrinos. The only difference is that G_F is replaced by Newton's constant, G_N in (IV.1). As a result the decoupling temperature for light gravitinos is of order 100 GeV. There are many more states in thermal equilibrium at 100 GeV($g \gtrsim 100$) and thus the effective gravitino temperature is much lower than that of light neutrinos. Pagels and Primack[13] obtain

$$m_{\tilde{G}} \leq 1 \text{ KeV} \quad , \qquad \text{(IV.9)}$$

or

$$\Lambda_{ss} \leq 10^6 \text{ GeV}$$

(since $m_{\tilde{G}} \simeq \Lambda_{ss}^2/M$). This is the origin of our lower limit for the supersymmetry breaking scale in Table III.

We now consider neutrinos with mass greater than 1 MeV. (We use the symbol N_0 to denote heavy neutrinos). From the previous analysis, we conclude neutrinos (N_0) are non-relativistic when they decouple. We expect a Boltzmann suppression for heavy neutrinos of order $\exp(-m_{N_0}/T_d$ and thus for some value of $m_{n_0} > T_d$ we should find a rea-reasonable neutrino energy density. We must however reevaluate the decoupling temperature, using the correct annihilation rate for a non-relativistic process.

Following Lee and Weinberg,[14] we use

$$\langle\sigma v\rangle \sim G_F^2\, m_{N_0}^2\, \frac{N_A}{2\pi} \quad , \tag{IV.10}$$

for processes of the type indicated in fig. 2. (N_A is the number of final states in the annihilation of N_0.) The rate equation for N_0 is given by (see Eq. III.37):

$$\frac{dn}{dt} + 3\,\frac{\dot{a}}{a}\,n = \left(n_{eq}^2 - n^2\right)\langle\sigma v\rangle \quad , \tag{IV.11}$$

where n is the number of N_0 per unit volume. Using the relations

$$aT = \text{constant} \quad , \tag{III.35}$$

$$\frac{\dot{a}}{a} = -\,\frac{\dot{T}}{T} = \left(\frac{\rho}{3M^2}\right)^{1/2} \quad , \tag{III.3a}$$

with

$$\rho = g\,\frac{\pi^2}{30}\,T^4 \quad , \tag{III.32}$$

and defining the new parameters

$$x \equiv \frac{T}{m_{N_0}} \quad , \quad f \equiv \frac{n}{T^3} \quad ,$$

$$f_0 \equiv \frac{n_{eq}}{T^3} = \begin{cases} g'\,\frac{\zeta(3)}{\pi^2} & x \gg 1\ , \quad \text{(III.32)} \\ 2(2\pi x)^{-3/2}\, e^{-1/x} & x \lesssim 1\ , \quad \text{(III.34)} \end{cases} \tag{IV.12}$$

we obtain

$$\frac{df}{dx} = \left(\frac{3M^2 m^2_{N_0}}{g\pi^2/30}\right)^{1/2} \langle\sigma v\rangle (f^2 - f_0^2)\ . \tag{IV.13}$$

In fig. 3 we illustrate the solution. For $x \gg 1$, $f \sim f_0 \sim$ constant; $df/dx \sim 0$. However as T decreases and $x \lesssim 1$, f_0 becomes Boltzmann suppressed and df/dx increases. At some temperature, T_f, the effective density f can no longer keep up with f_0. At this point N_0 decouples or <u>freezes out</u> of thermal equilibrium. This occurs approximately at the temperature, T_f, defined by the relation

$$\frac{df_0}{dx} = c f_0^2\ , \tag{IV.14}$$

with $c \equiv \left(\frac{3M^2 m^2_{N_0}}{g\pi^2/30}\right)^{1/2} \langle\sigma v\rangle$.

We find

$$\frac{T_f}{m_{N_0}} \equiv x_f \sim \frac{1}{\ell n c} \sim \frac{1}{20}\ . \tag{IV.15}$$

For $x \ll x_f$, f_0 can be neglected in (IV.13) and we obtain the analytic result

$$f(0) \sim \left[c x_f + \frac{1}{f(x_f)}\right]^{-1} \sim \frac{20}{c}\ . \tag{IV.16}$$

Let us now evaluate the energy density, ρ_{N_0}. We have

$$\rho^0_{N_0} = m_{N_0} n(T_0) \simeq m_{N_0} f(0) T_0^3 < \rho^0_c \quad , \tag{IV.17}$$

or using (IV.16),

$$\left(\frac{\pi^2}{15} T_0\right)^{-1} \frac{\rho^0_\gamma}{\rho^0_c} < \left[m_{N_0} f(0)\right]^{-1} \sim \frac{1}{20} \left(\frac{3M^2}{g\pi^2/30}\right)^{1/2} \langle\sigma v\rangle \quad . \tag{IV.18}$$

Since $\langle\sigma v\rangle$ is proportional to $m^2_{N_0}$ (IV.10), we obtain for $N_A \sim 14$, $g \sim 4.5$

$$m_{N_0} \geq 2 \text{ GeV} \quad . \tag{IV.19}$$

<u>Photinos-Higgsinos</u>

The previous calculations are a paradigm for the following discussion. The only changes will be in the value of the annihilation rates, which greatly depends on the particle physics details.

In a MLES Lagrangian, the four neutral fermions $\tilde{W}_3$, $\tilde{B}$, $\tilde{H}^0$, $\tilde{\bar{H}}^0$ (see Table 1) have a mass matrix of the form

$$\begin{array}{c} \\ \tilde{W}_3 \\ \tilde{B} \\ \tilde{H}^0 \\ \tilde{\bar{H}}^0 \end{array} \begin{array}{c} \begin{array}{cccc} \tilde{W}_3 & \tilde{B} & \tilde{H}^0 & \tilde{\bar{H}}^0 \end{array} \\ \left(\begin{array}{cc|cc} M_2 & 0 & \frac{-g_2 v_1}{\sqrt{2}} & \frac{g_2 v_2}{\sqrt{2}} \\ 0 & \frac{5}{3}\frac{\alpha_1}{\alpha_2} M_2 & \frac{g_1 v_1}{\sqrt{2}} & \frac{-g_1 v_2}{\sqrt{2}} \\ \hline \frac{-g_2 v_1}{\sqrt{2}} & \frac{g_1 v_1}{\sqrt{2}} & 0 & \varepsilon \\ \frac{g_2 v_2}{\sqrt{2}} & \frac{-g_1 v_2}{\sqrt{2}} & \varepsilon & 0 \end{array}\right) \end{array} \tag{IV.20}$$

ε, M_2, M_1 are defined by the Lagrangian terms

$$\mathcal{L} \supset \varepsilon \tilde{H}_1 \tilde{\bar{H}}_2 - M_2 \tilde{W}_a \tilde{W}_a - M_1 \tilde{B}\,\tilde{B} \quad , \qquad \text{(IV.21}$$

where, as a result of one-loop renormalization group analysis,

$$M_1 = \frac{5}{3}\,\frac{\alpha_1}{\alpha_2}\, M_2 \text{ at } M_W \qquad \text{(IV.22}$$

if $M_1 = M_2$ at $M_{p\ell}$. $v_1 = \langle H_0 \rangle$ and $v_2 = \langle \bar{H}_0 \rangle$.

Diagonalizing (IV.20) we find the four eigenstates

$$\tilde{Z}_i = \alpha_i \tilde{W}_3 + \beta_i \tilde{B} + \gamma_i \tilde{H}_0 + \delta_i \tilde{\bar{H}}_0$$

$$i = 1, \ldots, 4 \text{ where } \alpha_i, \beta_i, \gamma_i, \delta_i \qquad \text{(IV.23}$$

are the mixing angles.

We shall consider the lightest $\tilde{Z}_i$ (with $i = i^*$):

1)a <u>Higgsino</u>: if it is predominantly a higgsino ($\gamma_{i^*}, \delta_{i^*} \gg \alpha_{i^*}, \beta_{i^*}$) and hence couples with Yukawa couplings or,

2)a <u>Photino</u>: if it is predominantly a gaugino ($\alpha_{i^*}, \beta_{i^*} \gg \gamma_{i^*}, \delta_{i^*}$) and hence couples with gauge couplings.

We note that in the limit $\varepsilon, M_2 \to 0$, there are two light states

$$1. \quad \tilde{S}_0 \equiv \frac{v_2 \tilde{H}_0 + v_1 \tilde{\bar{H}}_0}{v} \quad , \quad v \equiv \sqrt{v_1^2 + v_2^2} \quad , \qquad \text{(IV.24)}$$

with mass

$$m_{\tilde{S}_0} \sim \frac{2 v_1 v_2}{v}\, \varepsilon \quad ,$$

and

2. $$\tilde{\gamma} \equiv \frac{g_1 \tilde{W}_3 + g_2 \tilde{B}}{\sqrt{g_1^2 + g_2^2}} \qquad \text{(IV.25)}$$

with mass

$$m_{\tilde{\gamma}} \sim \frac{8}{3} \frac{g_1^2}{g_1^2 + g_2^2} M_2 \ .$$

<u>Higgsino</u>

Higgsino annihilation proceeds via the graphs of fig. 4. The annihilation rate for a relativistic higgsino is of the form

$$\langle \sigma v \rangle \sim G_F^2 T^2 \ . \qquad \text{(IV.26)}$$

If $m_{\tilde{S}_0} < 1$ MeV then, like light neutrinos, Higgsinos are relativistic when they decouple. We thus obtain

$$m_{\tilde{S}_0} < 100 \text{ eVh}^2 \ . \qquad \text{(IV.27)}$$

For $m_{\tilde{S}_0} > 1$ MeV, Higgsinos are non-relativistic at decoupling. However, the results are not identical to those of Lee-Weinberg, since higgsinos (and photinos) are Majorana fermions. As a result there is a p-wave suppression in their annihilation rate.[15] We find

$$\langle \sigma v \rangle \sim G_F^2 \left[\langle v_{rel}^2 \rangle + m_f^2 \right] \ , \qquad \text{(IV.28)}$$

unlike the rate for N_0 annihilation Eq. (IV.10). m_f is the mass of the fermion in the final state, and v_{rel} is the relative velocity of the two higgsinos. (Note, IV.28 is valid for $v_1 \neq v_2$. For $v_1 = v_2$, Z_0 exchange is suppressed and the dominant contribution comes from the Yukawa terms, which are typically smaller.)

Before we continue let us briefly illustrate the reason for the p-wave suppression. Consider the graphs of fig. 4. The vertices conserve chirality (neglecting small chiral breaking corrections of order $m_f/m_{\tilde{G}}$ coming from $\tilde{f}\tilde{f}$ mixing). If $m_f = 0$, then helicity is also conserved for the final state. It is clear from fig. 5 that the final state must have helicity 1 and thus total angular momentum, $J_{final} \geq 1$. We then conclude $J_{initial} \geq 1$. Now since $\tilde{S}_0$ is a Majorana fermion, the initial state contains two identical fermions, and must thus be odd under interchange. The symmetry of the wave function of the initial state is given by

$$(-1)^{\ell+s+1} , \tag{IV.29}$$

where ℓ is the orbital angular momentum and s is the spin. If we assume $\ell = 0$ (s-wave), then $J_{initial} \geq 1$ implies $s = 1$, which is even. Thus we conclude that the initial state must have $\ell \geq 1$ and hence at least a p-wave suppression. If $m_f \neq 0$, there may then be an s-wave component proportional to m_f^2.

Using the exact cross-sections, Ellis et al.,[16] find

$$\rho^0_{\tilde{S}_0} < \rho^0_c \tag{IV.30}$$

if

$$m_{\tilde{S}_0} \geq m_b \sim 4.5 \text{ GeV} \tag{IV.31}$$

for $v_1 \neq v_2$, and

$$m_{\tilde{S}_0} \geq m_t \sim 30 \text{ GeV} \tag{IV.32}$$

for $v_1 = v_2$ (see note following IV.28).

Photinos

Photino annihilation proceeds via the graph of fig. 6. The annihilation rate is of the form

$$\langle\sigma v\rangle \sim G_F^2\left(\frac{M_W}{m_{\tilde{f}}}\right)^4 \left[v_{rel}^2 + m_f^2\right] \tag{IV.33}$$

The limits in this case depend on both m_f and $m_{\tilde{f}}$. Ellis et al.,[16] find for

$$\rho^0_{\tilde{\gamma}} < \rho^0_c$$

$$m_{\tilde{\gamma}} \geq \frac{1}{2} \quad ; \quad 1.8 \quad ; \quad 5 \text{ GeV} \quad , \tag{IV.34}$$

for $m_{\tilde{f}} \sim 20$; 40 ; 100 GeV, respectively.

As $m_{\tilde{f}}$ increases, m_f must also increase in order for $<\sigma v>$ to remain large enough (see IV.18), and thus $m_{\tilde{\gamma}}$ correspondingly increases. They also find for

$$\rho^0_{\tilde{\gamma}} < .1 \, \rho^0_c$$

$$m_{\tilde{\gamma}} \geq 1.8 \quad ; \quad 3 \quad ; \quad 15 \text{ GeV} \quad , \tag{IV.35}$$

for $m_{\tilde{f}} \sim 20$; 40; 100 GeV, respectively. We have discussed only higgsinos or photinos while the lightest $\tilde{Z}_i$ may not be identifiable as such. Ellis et al., find however, that in a more general treatment, the results are dominated by these two special limits.

Sneutrinos[17]

Sneutrinos annihilate via the processes of fig. 7. In fact, fig. 7a gives the dominant contribution:

$$<\sigma v> \sim G_F \left(\frac{M_W}{m_{\tilde{Z}}} \right)^2 \quad . \tag{IV.36}$$

This is because, this is the only process with no p-wave suppression. The annihilation rate (IV.36) is typically large (superweak), except in the limit $M_2, \varepsilon \to 0$. In this limit, the eigenstates $\tilde{Z}_i$ (Eq. III.23) are $\tilde{S}_0$, $\tilde{\gamma}$, and a dirac fermion. $\tilde{S}_0$ and $\tilde{\gamma}$ do not couple to neutrinos, as in fig. 7a. Moreover, the process of fig. 7a is proportional to the Majorana mass of $\tilde{Z}_i$ and thus the dirac fermion also doesn't contribute. We find, for M_2 and ε sufficiently large, no cosmological bound on $m_{\tilde{\nu}}$.

We have, however, assumed throughout that the sneutrino is the LSP. Thus $m_{\tilde{\nu}} < m_{\tilde{S}_0}, m_{\tilde{\gamma}}$. We find that if we require

$$\rho^0_{\tilde{\nu}} \sim \rho^0_c \quad , \tag{IV.37}$$

then

$$m_{\tilde{\nu}} < m_{\tilde{S}_0} \lesssim 2 \text{ GeV} \quad ,$$

or for

$$\rho^0_{\tilde{\nu}} \sim .1\, \rho^0_c \quad , \tag{IV.38}$$

then

$$m_{\tilde{\nu}} < m_{\tilde{S}_0} \lesssim 10 \text{ GeV} \quad .$$

Note, when we increase ε, we increase $\langle\sigma v\rangle$ and thus decrease $\rho^0_{\tilde{\nu}}$. Increasing ε, however, also increases $m_{\tilde{S}_0}$ (Eq. IV.24) and thus allows for a heavier sneutrino. Note also, that the result is less sensitive to varying M_2.

In summary, we have discussed the possibility that the LSP is either a higgsino, photino, or sneutrino. The cosmological bounds on higgsino and photino mass are collected in Eqs. (IV.27, 31, 32, 34, and 35). For sneutrinos we obtain no mass limits. However, if they are a significant contribution to the dark matter in our universe, then $m_{\tilde{\nu}}$ must satisfy the constraints of Eqs. (IV.37 and 38).

V. <u>Cosmological bounds for the gravitino</u>

In a MLES model the gravitino is typically not the LSP. For example, the sneutrino mass is given by the expression

$$m^2_{\tilde{\nu}} = m^2_{\tilde{G}} - \frac{1}{2} M^2_Z \left(\frac{v^2_1 - v^2_2}{v^2} \right) , \tag{V.1}$$

where, in general, $v_1 \geq v_2$. Thus in these theories the gravitino can decay via processes illustrated in fig. 8. The decay rate is of order[18]

$$\Gamma_{\tilde{G}} \simeq \frac{m_{\tilde{G}}^3}{M_{p\ell}^2} \quad . \tag{V.2}$$

Gravitinos can be produced singly, in scattering processes with lighter states, as in fig. 9. The production rate is, typically, of order[19]

$$\Gamma_{prod} \simeq \frac{\alpha N T^3}{M_{p\ell}^2} \equiv \langle \sigma v n_{\tilde{f}} \rangle \quad , \tag{V.3}$$

where N is the number of squarks and sleptons in the thermal bath.

Gravitinos decouple just below $M_{p\ell}$. If we require

$$\Gamma_{prod} = H \sim \frac{g^{1/2} T^2}{M_{p\ell}} \quad , \tag{V.4}$$

we find the decoupling temperature

$$T_d \sim \frac{g^{1/2}}{\alpha N} M_{p\ell} \quad . \tag{V.5}$$

If, for the moment, we neglect $\Gamma_{\tilde{G}}$, then

$$n_G a^3 = \text{constant} \tag{V.6}$$

for $T < T_d$. This can lead to problems, as we shall now show.

We <u>assume</u> that the initial gravitino number density is given by

$$n_{\tilde{G}}\Big|_{T_{p\ell}} = n_\gamma\Big|_{T_{p\ell}} \quad . \tag{V.7}$$

If there is no subsequent entropy production (which would increase n_γ

without affecting $n_{\tilde{G}}$) and we take $\Gamma_{\tilde{G}} \equiv 0$, then today we obtain

$$n^0_{\tilde{G}} = n^0_\gamma \quad . \tag{V.8}$$

Hence

$$\Omega_{\tilde{G}} \equiv \frac{\rho^0_{\tilde{G}}}{\rho^0_c} = \frac{m_{\tilde{G}} n^0_{\tilde{G}}}{\rho^0_c} = \left(\frac{m_{\tilde{G}}}{2.6 T_0}\right)\Omega^0_\gamma \quad ,$$

$$\Omega_{\tilde{G}} = 4 \times 10^9 h^{-2}\left(\frac{m_{\tilde{G}}}{10^2 \text{ GeV}}\right). \tag{V.9}$$

This is much too large. Of course, we have neglected $\Gamma_{\tilde{G}}$; perhaps $\Gamma_{\tilde{G}} \neq 0$ can solve the problem.[18] Otherwise we must have $n^0_{\tilde{G}}/n^0_\gamma \lesssim .25 \times 10^{-9} h^2 \; (m_{\tilde{G}}/10^2 \text{GeV})^{-1}$.

We define the parameters Y_G, Y'_G by the expressions

$$Y_G = \frac{n_{\tilde{G}}}{n_r} \quad , \tag{V.10}$$

or

$$n_{\tilde{G}} = Y_G \, \frac{g'\zeta(3)}{\pi^2} \, T^3 \equiv Y'_G T^3 \quad .$$

We shall obtain the following two possible solutions: $\Gamma_{\tilde{G}} \neq 0$ and

1) $Y'_G \lesssim 10^{-9}$, at the decay temperature T_D, or

2) $10^{-9} < Y'_G \lesssim 10^{-4}$ at T_D and $m_{\tilde{G}} \gtrsim 10^4$ GeV. (V.11)

Scenario 1:

Consider the possibility that

$$T_D > T_{eq} \quad , \tag{V.12}$$

i.e., the universe is radiation dominated when gravitinos decay. The decay time t_D is given in terms of T_D by the expression

$$\Gamma_{\tilde{G}}^{-1} = t_D \sim H_D^{-1} \quad , \tag{V.13}$$

where

$$H_D^2 = \frac{g_D\ (\pi^2/30) T_D^4}{3M^2} ,$$

and

$$\Gamma_{\tilde{G}} = \frac{N}{2\pi} \frac{m_{\tilde{G}}^3}{M_{p\ell}^2} .$$

We obtain

$$T_D = \left(\frac{N}{g_D^{1/2}}\right)^{1/2} 10^{-7}\text{GeV} \left(\frac{m_{\tilde{G}}}{10^2\ \text{GeV}}\right)^{3/2} . \tag{V.14}$$

Now using (V.12), i.e.,

$$\rho_{\tilde{G}}\Big|_{T_D} < \rho_r\Big|_{T_D} \quad , \tag{V.15}$$

or

$$m_{\tilde{G}} Y'_G < g_D \frac{\pi^2}{30} T_D \quad ,$$

we obtain

$$Y'_G < 10^{-9}\left(\frac{m_{\tilde{G}}}{10^2\ \text{GeV}}\right)^{1/2} . \tag{V.16}$$

Scenario 1 is acceptable.

1) The universe is radiation dominated at Helium synthesis, since $T_{H_e}4 \sim 10^{-4}$ GeV > T_D (V.14). Thus Helium synthesis is unchanged.
2) $\rho_{\tilde{G}}/\rho_r|_{T_D} < 1$ implies that when gravitinos decay, there is neglibible reheating of the radiation.
3) Gravitinos eventually decay into the LSP($\tilde{\gamma}$, $\tilde{S}_0$, or $\tilde{\nu}$) with

$$n_{\tilde{G}}\Big|_{T_D} = n_{LSP}\Big|_{T_D} \tag{V.17}$$

as a result of R-parity. However, since $n_G\Big|_{T_D} \leq 10^{-9} n_\gamma\Big|_{T_D}$ (V.16), there is not a problem with too many of the LSP. We have

$$\Omega^0_{LSP} \leq 10^{-9}\left(\frac{M_{LSP}}{2.6T_0}\right)\Omega^0_\gamma \quad , \tag{V.18}$$

which is safe [compare with (V.9)]

Scenario 2:

Consider:

$$T_D < T_{eq} \quad , \tag{V.19}$$

i.e., the universe is matter dominated when gravitinos decay. This implies

$$Y'_G > 10^{-9}\left(\frac{m_{\tilde{G}}}{10^2 \text{ GeV}}\right)^{1/2} \quad . \tag{V.20}$$

Since $\rho_{\tilde{G}}\Big|_{T_D} > \rho_r\Big|_{T_D}$, the energy which is released when the gravitinos decay will reheat the radiation. We define T'_D to be the reheat temperature for the radiation; $T'_D > T_D$. There are now three relevant possibilities:

$$\begin{aligned} &1)\quad T_D > T_{H_e^4} \sim 10^{-4} \text{ GeV}, \\ &2)\quad T'_D > T_{H_e^4} > T_D \text{ or}, \\ &3)\quad T_{H_e^4} > T'_D > T_D \; . \end{aligned} \tag{V.21}$$

Case 3 can be ruled out immediately. Since, in this case, the universe reheats after Helium synthesis, the ratio of baryons to photons at the time of Helium synthesis is larger than can be inferred from present observations. When a larger value, $\rho_b/\rho_r\Big|_{T_{H_e^4}}$, is put into the standard calculations, one obtains more Helium and less Deuterium as output. This is unacceptable since the standard result already agrees with the lower limit for the ratio of abundances, $D/H_e^4 \gtrsim 10^{-5}$.

<u>Consider case 1</u>; $T_D > T_{H_e^4}$.

We have

$$H_D^2 = \frac{m_G Y'_G T_D^3}{3M^2} = \Gamma_{\tilde{G}}^2 = \left(\frac{N}{2\pi}\frac{m_{\tilde{G}}^3}{M_{p\ell}^2}\right)^2 ,$$

or

$$T_D = \left(\frac{3N^2}{(16\pi^2)^2}\right)^{1/3} \frac{m_{\tilde{G}}^{5/3}}{M^{2/3}} Y_G'^{-1/3} . \qquad \text{(V.22)}$$

Thus

$$Y'_G < 2N^2 \times 10^{-19} \frac{m_{\tilde{G}}}{10^2\ \text{GeV}}^5 . \qquad \text{(V.23)}$$

But recall $Y'_G > 10^{-9}\left(\frac{m_{\tilde{G}}}{10^2 \text{GeV}}\right)^{1/2}$, Eq. (V.20). These two equations (V.20) and (V.23) are inconsistent unless

$$m_{\tilde{G}} \geq 10^4\ \text{GeV} . \qquad \text{(V.24)}$$

<u>Consider case 2</u>; $T'_D > T_{H_e^4} > T_D$.

Using energy conservation, we obtain an expression for T'_D:

$$\rho_{\tilde{G}}\Big|_{T_D} = g(T'_D)\frac{\pi^2}{15}(T'_D)^4 \quad . \tag{V.25}$$

Using

$$\rho_{\tilde{G}}\Big|_{T_D} = m_{\tilde{G}}\, Y'_G T_D^3 \quad , \tag{V.26}$$

and Eq. (V.22) we obtain

$$m_{\tilde{G}} = (16\pi^2)^{1/3}\left(\frac{\pi^2}{45N^2}\right)^{1/6} g(T'_D)^{1/6} M^{1/3} (T'_D)^{2/3} \quad . \tag{V.27}$$

Using $T'_D > T_{H_e}4$, we find

$$m_{\tilde{G}} > \left(\frac{g(T'_D)}{N^2}\right)^{1/6} \times 1.2 \times 10^4 \text{ GeV} \tag{V.28}$$

for $T_{H_e}4 \sim .1$ MeV, or

$$m_{\tilde{G}} > \left(\frac{g(T'_D)}{N^2}\right)^{1/6} \times 1.9 \times 10^4 \text{ GeV} \tag{V.29}$$

for $T_{H_e}4 \sim .2$ MeV.

In either case (1 or 2) we thus find

$$m_{\tilde{G}} \geq 10^4 \text{ GeV} \quad . \tag{V.30}$$

We now show that there is an upper limit on Y'_g.[19] In the standard scenario for baryogenesis, in grand unified theories, the ratio $n_b/n_\gamma \sim 10^{-9}$ is generated via the baryon violating decays of heavy leptoquark gauge bosons or color triplet Higgs bosons. In either case the value of n_b/n_γ which can be obtained is typically no greater than about 10^{-6}. The production of entropy, after the epoch of baryogenesis, will cause the ratio n_b/n_γ to decrease. We must therefore limit the amount of entropy production at low temperatures, i.e., T less than $\sim 10^{12}$ GeV.

We define $\Delta = s_f/s_i$, where s_i is the entropy before gravitino decay and s_f is the entropy after. Then

$$\Delta = \left(\frac{T'_D}{T_D}\right)^3 = \left(\frac{M}{m_{\tilde{G}}}\right)^{1/2} Y'_G \left(\frac{(16\pi^2)}{3N^2}\right)^{1/4} \left(\frac{15}{g\pi^2}\right)^{3/4} . \tag{V.31}$$

If we require $\Delta \leq 10^3$, we obtain an upper limit on Y'_G, i.e.,

$$Y'_G \leq 1.6 \times 10^{-4}(N^2g^3)^{1/4}\left(\frac{m_{\tilde{G}}}{10^4 \text{ GeV}}\right)^{1/2}\left(\frac{\Delta}{10^3}\right) . \tag{V.32}$$

Finally since $Y'_G > 10^{-9}$, we must also worry about the decay products of the gravitino.

In conclusion, scenario 1, i.e., $Y'_G \lesssim 10^{-9}$ at T_D is apparently the easiest one to live with. It is not difficult imagining mechanisms which might produce nine orders of magnitude of entropy, prior to baryogenesis, which would give $Y'_G \leq 10^{-9}$.

To complete our discussion of gravitinos, we should mention the limit $Y'_G \leq 10^{-9}$ places on inflation scenarios.[19] In order to solve the cosmological horizon and flatness problems, we need a mechanism to create $\sim 10^{84}$ orders of magnitude of entropy, prior to baryogenesis.[20] In standard scenarios the universe inflates exponentially and adiabatically, and then reheats to, approximately, the temperature just prior to inflation. During inflation the gravition number density vanishes as $1/a^3$. We now show, however, that the reheat temperature, T_R, can not be too high; otherwise, one runs into the problem of recreating the gravitinos.[19]

Consider the process of fig. 9 and the production rate (V.3). The rate equation for producing gravitinos is

$$\frac{dn_{\tilde{G}}}{dt} + 3\,\frac{\dot{a}}{a}\,n_{\tilde{G}} = \Gamma_{prod} n_r ,$$

$$\frac{dn_r}{dt} + 4\,\frac{\dot{a}}{a}\,n_r = 0 \quad . \tag{V.33}$$

(valid for $Y_G = n_{\tilde{G}}/n_r \ll 1$), or for Y_g,

$$\dot{Y}_G = HY_G + \Gamma_{prod} \quad . \tag{V.34}$$

The steady state solution, evaluated at T_r is

$$Y_G = -\,\frac{\Gamma_{prod}}{H_R} \sim -\alpha\sqrt{N}\,\frac{T_R}{M_{p\ell}} \tag{V.35}$$

where $H_R \sim \sqrt{N}\,T_R^2/M_{p\ell}$. Finally at T_D, we have

$$Y_G\Big|_{T_D} = \frac{2}{N}\;Y_G\Big|_{T_R} \sim \frac{2\alpha}{\sqrt{N}}\,\frac{T_R}{M_{p\ell}} \,. \tag{V.36}$$

The factor 2/N takes into account the annihilation of N-2 degrees of freedom between T_R and T_D. If we now require $Y_G \lesssim 10^{-9}$, we find

$$T_R \lesssim 10^{-12}\text{GeV}\left(\frac{Y_G}{10^{-9}}\right) . \tag{V.37}$$

This is the upper limit on the reheat temperature in an inflationary universe scenario.

We have discussed cosmological mass bounds for supersymmetric partners. The candidate lightest supersymmetric partners were higgsinos, photinos, and sneutrinos. Gravitinos were also discussed since they are also an important contribution to the energy density of the universe. We have demanded that $\rho_{LSP} \leq \rho_c$ today. If the LSP does in fact satisfy, $\rho^0_{LSP} \sim \rho^0_c$, then it is also an excellent candidate for the so-called dark matter which is apparently the dominant form of energy. Any one of the LSP candidates, with mass of order a few GeV or greater, would fall into the category of "cold" dark matter.[21] One particular property of "cold" dark matter is that it clusters on all scales as seems to be the case observationally (see Table IV). It may also be able to explain the large scale voids which have been observed.

References

1. L. Alvarez - Gaumé, J. Polchinski, and M. B. Wise, Nucl. Phys. B221, 495 (1983); J. Ellis, J. Hagelin, D. V. Nanopoulos, and K. Tamvakis, Phys. Lett. 125B, 275 (1983); L. E. Ibáñez and C. López, CERN preprint, Ref. Th. 3650 (1983), and L. E. Ibáñez, Nucl. Phys. B218, 514 (1983).
2. CELLO Collaboration, H. Behrend, et al., Phys. Lett. 114B, 287 (1982); JADE Collaboration, W. Bartel et al., Phys. Lett. 114B, 211 (1982); MARK J Collaboration, D. P. Barber et al., Phys. Rev. Lett. 45, 1904 (1981); Tasso Collaboration, R. Brandelik et al., Phys. Lett. 117B, 365 (1982); MARK II Collaboration, C. A. Blocker et al., Phys. Rev. Lett. 49, 517 (1982).
3. R. C. Ball et al., in Proceedings of the 1983 International Europhysics Conference on High Energy Physics, Brighton, p. 318, published by Rutherford Appleton Laboratory.
4. J. Ellis and H. Kowalski, CERN preprint, Ref. Th. 3843 (1984) and DESY preprint, DESY 84-045 (1984).
5. G. R. Farrar and P. Fayet, Phys. Lett. 76B, 575 (1978); 79B, 442 (1978).

6. G. R. Farrar and S. Weinberg, Phys. Rev. D27, 2732 (1983).
7. S. Wolfram, Phys. Lett. 82B, 65 (1979).
8. P. F. Smith and J. R. J. Bennett, Nucl. Phys. B149, 525 (1979).
9. G. R. Blumenthal, S. M. Faber, J. R. Primack, and M. J. Rees, SLAC preprint, SLAC-PUB-3307 (1984).
10. E. W. Kolb and S. Wolfram, Nucl. Phys. B172, 224 (1980).
11. S. Weinberg, Gravitation and Cosmology, John Wiley and Sons, Inc., 1972.
12. R. Cowsik and J. McClelland, Phys. Rev. Lett. 29 669 (1972) and Ap. J. 180, 7 (1973).
13. H. Pagels and J. Primack, Phys. Rev. Lett. 48, 223 (1982).
14. B. W. Lee and S. Weinberg, Phys. Rev. Lett. 39, 165 (1977).
15. H. Goldberg, Phys. Rev. Lett. 50, 1419 (1983).
16. J. Ellis, J. S. Hagelin, D. V. Nanopovlos, K. Olive, and M. Srednicki, Nucl. Phys. B (1983).
17. J. S. Hagelin, G. L. Kane, and S. Raby, Nucl. Phys. B241, 638 (1984); L. E. Ibáñez, FTUAM preprint 83-28 (1984).
18. S. Weinberg, Phys. Rev. Lett., 48, 1303 (1982).
19. L. M Krauss, Nucl. Phys. B227, 556 (1983); S. Weinberg (unpublished); J. Ellis et al. (ref. 16); J. Ellis, A. D. Linde, and D. V. Nanopoulos, Phys. Lett. 118B, 59 (1982); M. Yu. Khlopov and A. D. Linde, Phys. Lett 138B, 265 (1984).
20. A. Guth, Phys. Rev. D23, 347 (1981); A. Linde, Phys. Lett. 108B, 389 (1982); A. Albrecht and P. J. Steinhardt, Phys. Rev. Lett. 48, 1220 (1982).
21. G. R. Blumenthal, et al., (ref. 9); J. R. Primack and G. R. Blumenthal, UCSC-TH-162-83 (1983) and UCSC-TH-164-83 (1983).

Table I

$q = \begin{pmatrix} u \\ d \end{pmatrix}$	$\ell = \begin{pmatrix} \nu \\ e \end{pmatrix}$	$\tilde{q} = \begin{pmatrix} \tilde{u} \\ \tilde{d} \end{pmatrix}$	$\tilde{\ell} = \begin{pmatrix} \tilde{\nu} \\ \tilde{e} \end{pmatrix}$
$\bar{u}$	$\bar{e}$	$\tilde{\bar{u}}$	$\tilde{\bar{e}}$
$\bar{d}$		$\tilde{\bar{d}}$	
quarks and leptons		squarks and sleptons	
scalar partner of goldstino		G	
-graviton		goldstino	
		-gravitino	
$H_1 = \begin{pmatrix} H^+ \\ H^0 \end{pmatrix}$	$\bar{H}_2 = \begin{pmatrix} \bar{H}^- \\ \bar{H}^0 \end{pmatrix}$	$\tilde{H}_1 = \begin{pmatrix} \tilde{H}^+ \\ \tilde{H}^0 \end{pmatrix}$	$\tilde{\bar{H}}_2 = \begin{pmatrix} \tilde{\bar{H}}^- \\ \tilde{\bar{H}}^0 \end{pmatrix}$
Higgs bosons		Higgsinos	
g, γ, Z_0, $W^{\pm}$		$\tilde{g}$, $\tilde{W}_3$, $\tilde{B}$, $\tilde{W}^{\pm}$	
gauge bosons		guaginos	

Table II

t(s)	T(GeV)	comments
10^{-44}	10^{19}	"initial conditions" assume all particles in thermal equilibrium.
10^{-4}	10^{-1}	$\mu^+\mu^-$ annihilate: $e^{-m_\mu/T}$
10^{-2}	10^{-2}	ν_μ decouple: $<\sigma v n> < H$
		$\frac{n_n}{n_n + n_p} : \frac{1}{2} \to \frac{1}{6} : e^{-(m_n - m_p)/T}$
	2×10^{-3}	ν_e decouple : $<\sigma v n> < H$
4	$1/2 \times 10^{-3}$	e^+e^- annihilate : $e^{-m_e/T}$
		$(4(7/8) + 2)\, T_\nu^3 = 2T_\gamma^3$ (entropy conservation)
		$\to T_\nu = (4/11)^{1/3} T_\gamma$
100	10^{-4}	Helium synthesis
		$p + n \leftrightarrow d + \gamma$.
		$d + d \leftrightarrow H_e^3 + n \leftrightarrow H^3 + p$
		$H^3 + d \leftrightarrow H_e^4 + n$
		all free neutrons depleted $\to H_e^4$
		$\Rightarrow$ thermal bath includes
		p, H_e^4, d, γ, ν, e
		ratio $d/H_e^4 \sim 10^{-5}$

Table II (cont'd)

t(s)	T(GeV)	comments
$10^{12} = 3 \times 10^4 y$	$10^{-9}\Omega_m h^2$	$T = T_{eq}$, $\rho_m = \rho_r \simeq 1.68\ \rho_\gamma$
	$1/3 \times 10^{-9}$	$T = T_{recombination}$, $p + e^- \rightarrow H$ photons decouple

Table III

t(s)	T(GeV)	comments
10^{-38}	10^{16}	GUT scale
10^{-30}	10^{12}	inflation
		reset clocks
10^{-28}	10^{11}	SUSY breaking scale ↑
10^{-18}	10^{6}	SUSY breaking scale ↓
10^{-10}	10^{2}	weak breaking $g \gtrsim 100$
10^{-5}	1/3	quark-gluon plasma ↑ QCD phase transition ↓ Hadrons

Table IV

Dark matter [21]

Evidence	scale $M[M_\odot = 2 \times 10^{33} g]$	M/M_{lum}
large clusters	10^{15}	$8.4^{+7.0}_{-1.0}$
small spiral dominated groups	2×10^{13}	14.2^{+36}_{-6}
whole milky way	10^{12}	14
dwarf spheroidal galaxies	10^{6-8}	12

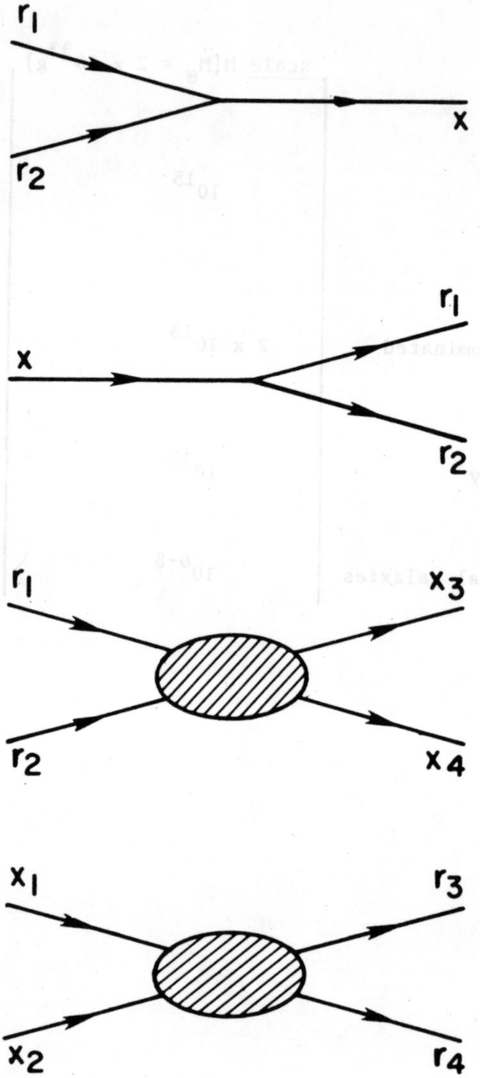

Fig. 1

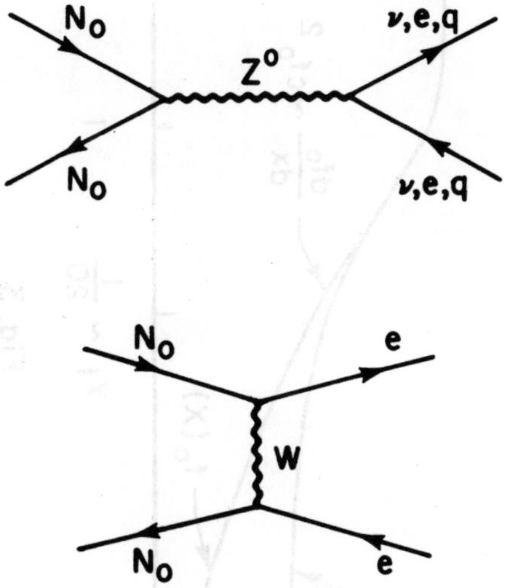

Fig. 2

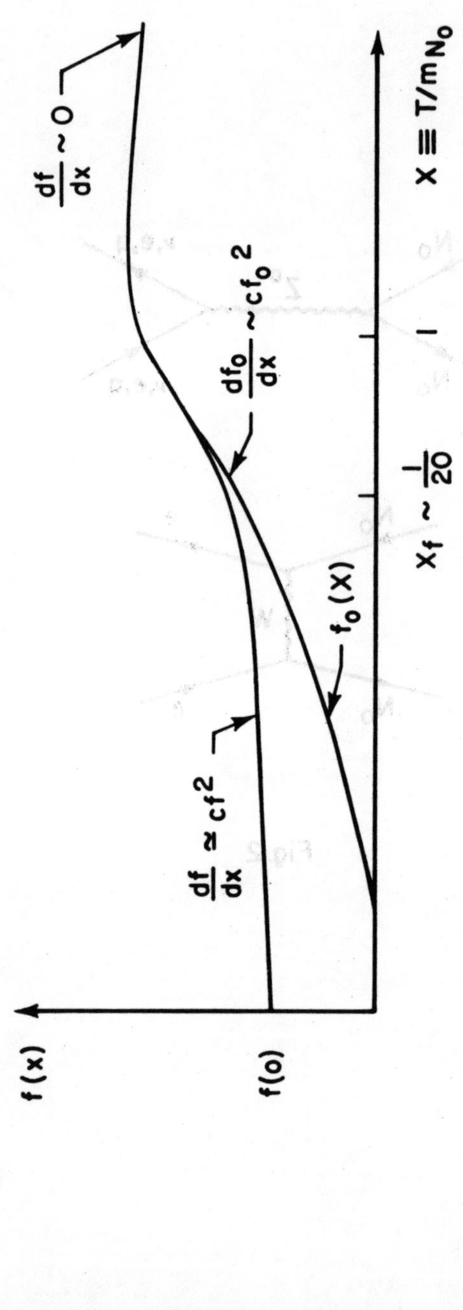

f(x)
f(o)
$\frac{df}{dx} \simeq cf^2$
$f_o(X)$
$\frac{df_o}{dx} \sim cf_o^2$
$\frac{df}{dx} \sim 0$
$X_f \sim \frac{1}{20}$
1
$X \equiv T/m_{N_o}$

Fig. 3

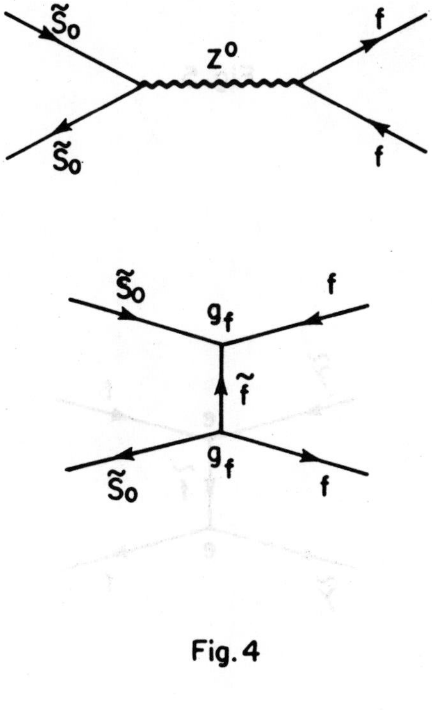

Fig. 4

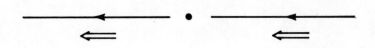

Fig. 5

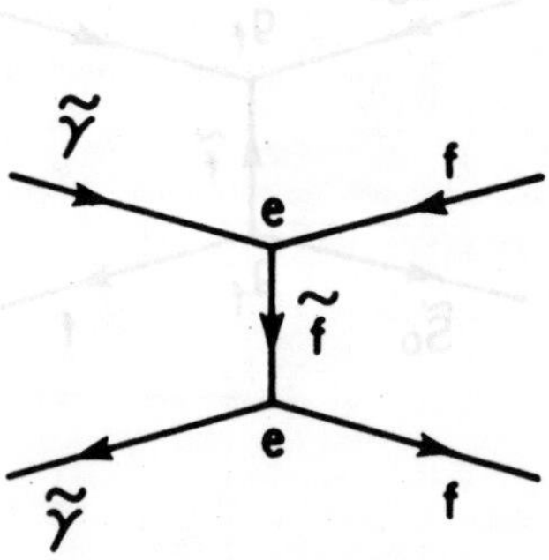

Fig. 6

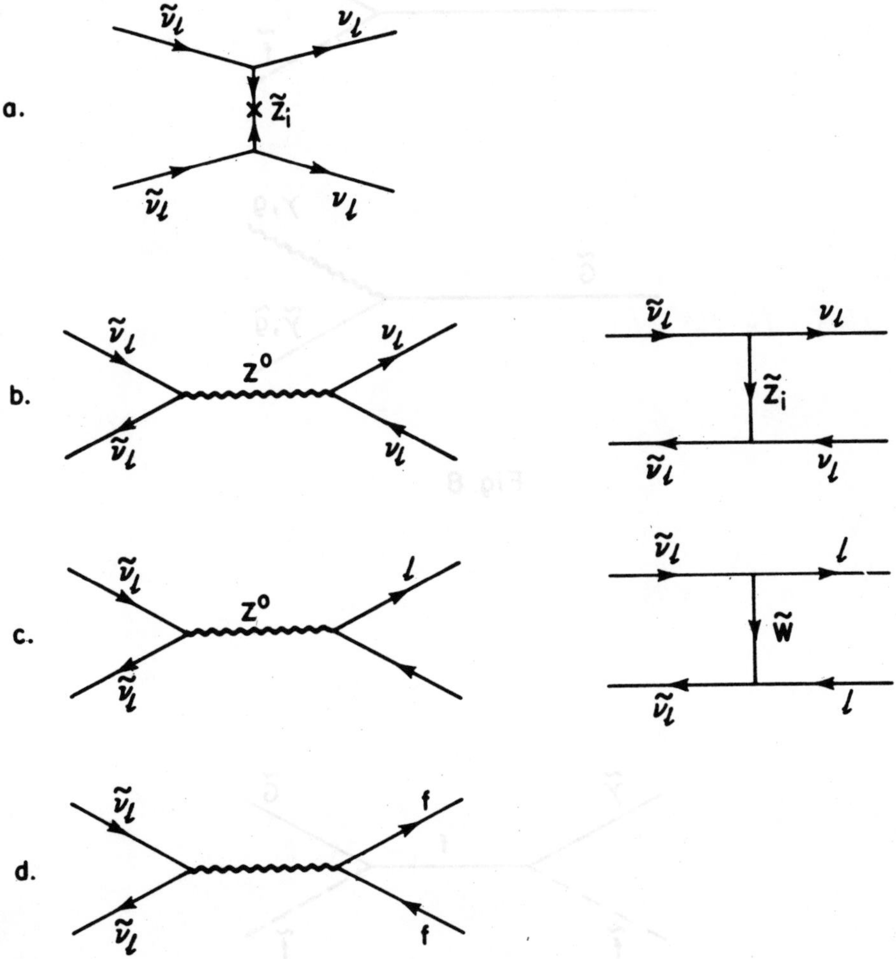

a.
$\tilde{\nu}_l$
ν_l
$\tilde{Z}_i$
$\tilde{\nu}_l$
ν_l
b.
$\tilde{\nu}_l$
Z^0
ν_l
$\tilde{\nu}_l$
ν_l
$\tilde{\nu}_l$
ν_l
$\tilde{Z}_i$
$\tilde{\nu}_l$
ν_l
c.
$\tilde{\nu}_l$
Z^0
l
$\tilde{\nu}_l$
$\tilde{\nu}_l$
l
$\tilde{W}$
$\tilde{\nu}_l$
l
d.
$\tilde{\nu}_l$
f
$\tilde{\nu}_l$
f

Fig. 7

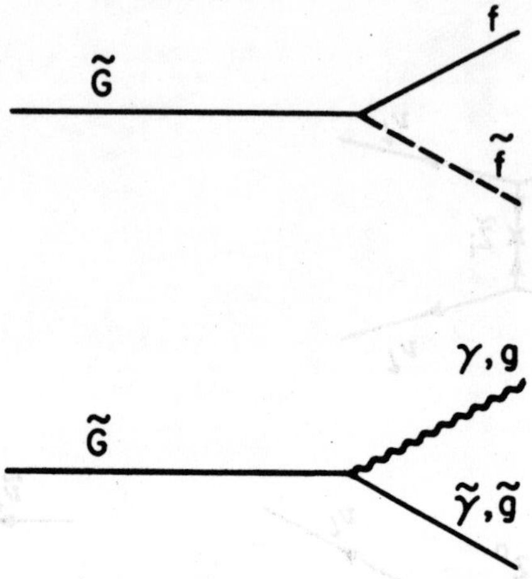

Fig. 8

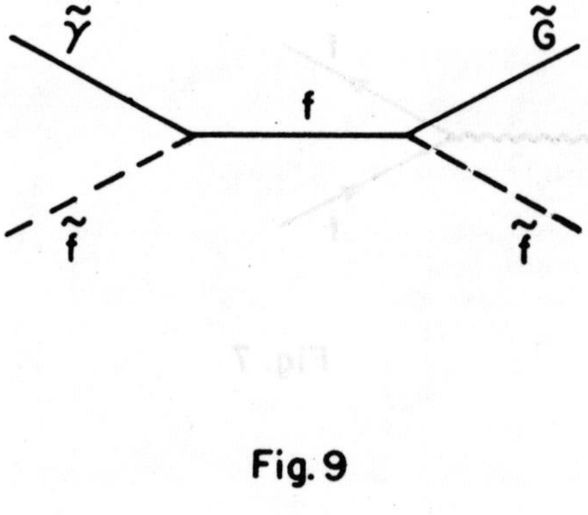

Fig. 9

TOWARDS A THEORY OF FERMION MASSES: A TWO-LOOP FINITE N = 1 GLOBAL SUSY GUT

D. R. Timothy Jones
Department of Physics
University of Colorado
Boulder, CO 80309
Stuart Raby*
Los Alamos National Laboratory
Theoretical Division
T-8, MS B285
Los Alamos, New Mexico 87545

One of the most outstanding problems of high energy physics is the family problem. Why are there (at least) three families of quarks and leptons with identical quantum numbers under $SU_3 \times SU_2 \times U_1$, and with mass ranging from .5 MeV to greater than ~30 GeV? Many attempts have been made to understand this puzzle, including global and local family groups or compositeness ideas. In the following, we discuss the constraints on fermion masses resulting from the requirement of finiteness in an N = 1 supersymmetric theory. Although SUSY may provide an explanation for the hierarchy $M_{p\ell}/M_W$, it has to date said little or nothing new about the family problem. This is a step in that direction.

This talk is organized as follows:

I. Conditions for finiteness.
II. Constructing a two-loop finite SU_5 GUT.
III. Fermion masses.
IV. Top quark mass.
V. A generalized two-loop finite SU_5 GUT.

---

* Third lecture given at the XV GIFT INTERNATIONAL SEMINAR ON THEORETICAL PHYSICS, June 4-8, 1984: presented by S. Raby.

I. Finiteness conditions

N = 4 global SUSY theories have been shown to be finite to all orders in perturbation theory.[1] N = 2 global SUSY theories are known, in general, to have one-loop divergences. However, for particular choices of representations they can be made one-loop finite. In these cases they are also finite to all orders in perturbation theory.[2]

What is known about N = 1 global SUSY theories? We consider an N = 1 SUSY Yang-Mills theory with gauge group G and N matter multiplets ϕ_i, i = 1, ..., N, (chiral superfields) which transform in a representation R (possibly reducible) of G. We also define a superspace potential

$$W = \frac{1}{6} C^{ijk}\phi_i\phi_j\phi_k \quad , \tag{I.1}$$

where C^{ijk} are Yukawa couplings and W is, of course, invariant under G. The conditions for one-loop finiteness are, as follows:[3]

$$\beta_g^{(1)} = \frac{g^3}{16\pi^2}[T(R) - 3C_2(G)] \equiv 0 \quad , \tag{I.2}$$

$$\gamma^{i\,(1)}_{\;j} = \frac{1}{2}[F^i_{\;j} - 4g^2C_2(R_\beta)(E_\beta)^i_{\;j}] \equiv 0 \quad . \tag{I.3}$$

$\gamma^i_{\;j}$ are the anomalous dimensions (using a SUSY regularization prescription), g is the gauge coupling, $F^i_{\;j} \equiv C^{imn}C^*_{jmn}$, $T(R)\delta_{ab} \equiv T_r(T_aT_b)$, and $\mathbb{1}C_2(R) \equiv \sum_a T_aT_a$, where T_a is the generator in the representation R of G, and $\mathbb{1}C_2(G) = \sum_a T_aT_a$ with T_a in the adjoint representation. $(E_\beta)^i_{\;j}$ is a projection onto the irreducible representation R_β. Recently it has been shown that the conditions for two-loop finiteness are:[3] $\beta_g^{(2)} = \gamma^{i\,(2)}_{\;j} = 0$ if $\beta_g^{(1)} = \gamma^{i\,(1)}_{\;j} = 0$, i.e., if the theory is one-loop finite it is then necessarily two-loop finite. It is not known, however, whether these theories are finite beyond two-loops.

II. Constructing a two-loop finite SU_5 GUT[4]

The gauge group $G = SU_5$ with $C_2(SU_5) = 5$. We consider the representations listed in Table I. In order to have a realistic theory we must include the following representations:

1) need 24 or 75 to break SU_5 to $SU_3 \times SU_2 \times U_1$;
2) need at least three families of quarks and leptons, i.e., $3(10 + \bar{5})$;
3) need Higgs fields in $\bar{5}$, 5 or $\overline{45}$, 45 to break $SU_2 \times U_1$ to U_1 and give quarks and leptons mass.

Consider the first condition for finiteness Eq. (I.2); we must satisfy $\sum_\beta T(R_\beta) = 3C_2(G) \equiv 15$. With one 24 and $3(10 + \bar{5})$ we have $\sum_\beta T(R_\beta) = 11$. We are then allowed $4(5 + \bar{5})$ of Higgs. It is clear from Table I that 45, 50, or 75 dimensional representations are much too large to obtain a finite theory. This is the minimal model. It will become clear later that a four family model cannot be made realistic.

In order to verify the second finiteness condition Eq. (I.3), we need to construct the superspace potential. We consider

$$W = q\mathrm{Tr}(24^3) + p\bar{K}24k + g_{ijk}10_i 10_j H_k + \bar{g}_{ijk}10_i \bar{5}_j \bar{H}_k$$

$$+ M\mathrm{Tr}(24^2) + M'\bar{K}K + \bar{K}(m_i H_i) + (\bar{m}_i \bar{H}_i)K + \bar{H}_i M_{ij} H_j \quad , \qquad \text{(II.1)}$$

where $i,j = 1,2,3$ and $\bar{K},\bar{H}_i$ (K,H_i) are the four $\bar{5}(5)$ of Higgs. The parameters $q,p,g_{ijk},\bar{g}_{ijk}$ must satisfy the finiteness condition (I.3). M, M', M_{ij} are mass parameters of order M_{GUT} (to be determined) and $m_i,\bar{m}_i$ will be an intermediate scale. This is <u>not</u> the most general superspace potential. It defines a particular model. In section V we shall discuss the most general superspace potential and the corrections on results obtained in sections II-IV. Now we shall discuss the mechanism for:

1) GUT breaking,

2) Higgs doublet, triplet splitting,

3) SUSY and $SU_2 \times U_1$ breaking.

1) GUT breaking: The terms in W, $qTr(24^3) + MTr(24^2)$ are necessary to affect the breaking of $SU_5 \rightarrow SU_3 \times SU_2 \times U_1$. In fact there is a SUSY minimum of the scalar potential for which

$$<24> = M_G \begin{pmatrix} 2 & & & & \\ & 2 & & & \\ & & 2 & & \\ & & & -3 & \\ & & & & -3 \end{pmatrix} , \tag{II.2}$$

where M_G is a function of M and q. There are also other SUSY minima which break $SU_5 \rightarrow SU_4 \times U_1$, or even preserve SU_5. Once SUSY is broken one of these minima will become lower than the others. For now we shall assume that the $SU_3 \times SU_2 \times U_1$ minimum is a global minimum.

2) Higgs doublet, triplet splitting: Consider the following terms in W,

$$\bar{K}[p24 + M']K + (\bar{m}_i\bar{H}_i)K + \bar{K}(m_iH_i) + \bar{H}_iH_jM_{ij} \quad . \tag{II.3}$$

In the SU_3 triplet sector we have the 4 x 4 mass matrix:

$$\begin{array}{c} \\ \bar{K} \\ \bar{H}_i \end{array} \begin{array}{c} \begin{array}{cc} K & H_j \end{array} \\ \left(\begin{array}{c|c} 0 & m_j \\ \hline \bar{m}_i & M_{ij} \end{array}\right) \end{array} \tag{II.4}$$

where we have fine-tuned $M' = -2pM_G$. We choose M_{ij} to have two non-zero eigenvalues of order M_G and one zero eigenvalue. As a result we obtain two massive triplets at M_G and two triplets with mass of order m. We define

$$H_i = \alpha_iH+\cdots \quad ,$$

$$\bar{H}_i = \beta_iH+\cdots \quad , \tag{II.5}$$

such that H and $\bar{H}$ are the zero eigenstates of M_{ij}. Then the off-diagonal mass terms are

$$(\bar{m}_i\beta_i)\bar{H}K + (m_i\alpha_i)\bar{K}H \quad . \tag{II.6}$$

What is the value of the parameters $m_i, \bar{m}_i$? Consider the SU_2 doublet mass matrix:

$$\begin{array}{c} \\ \bar{K} \\ \bar{H}_i \end{array} \begin{array}{c} \begin{array}{cc} K & H_j \end{array} \\ \left(\begin{array}{c|c} \bar{M} & m_j \\ \hline \bar{m}_i & M_{ij} \end{array} \right) \end{array} \tag{II.7}$$

where $\bar{M} \equiv -3pM_G + M' = 5/2\ M'$. We have three eigenvalues of order M_G and one eigenvalue of order m^2/M_G (if $m << M_G$). In particular, in the 2 x 2 sector (including $K, \bar{K}$ and the massless eigenstates of M_{ij}) we have

$$\begin{array}{c} \\ \bar{K} \\ H \end{array} \begin{array}{c} \begin{array}{cc} K & H \end{array} \\ \left(\begin{array}{cc} \bar{M} & m_i\alpha_i \\ \bar{m}_i\beta_i & 0 \end{array} \right) \end{array} \tag{II.8}$$

with the approximate diagonalized mass terms

$$\bar{M}\bar{K}K + \frac{(m\cdot\alpha)(\bar{m}\cdot\beta)}{\bar{M}}\bar{H}H \quad . \tag{II.9}$$

We then choose

$$\frac{(m\cdot\alpha)(\bar{m}\cdot\beta)}{\bar{M}} \sim M_W \tag{II.10}$$

in order to have one pair of light Higgs doublets. Thus $m \equiv (m\cdot\alpha) \sim (\bar{m}\cdot\beta)$ is of order 10^{10} GeV.

With m of order 10^{10} GeV, the color triplet Higgs scalars $(H,\bar{H})$ will contribute to nucleon decay with the dominant decay modes

$$p \to k^0\mu^+ \quad , \quad k^+\bar{\nu} \quad , \quad n \to k^0\bar{\nu} \quad . \tag{II.11}$$

The color triplet Higgs fermions will <u>not</u> contribute to proton decay at the tree level. This is because the mass terms Eq. (II.6) are off-diagonal and only H and $\bar{H}$ couple to quarks and leptons at the tree level. Once SUSY is broken, radiative corrections will induce the necessary dimension five operators contributing to proton decay. These are, however, expected to be suppressed.

3) <u>SUSY and $SU_2 \times U_1$ breaking</u>: We intend to break SUSY via "soft" SUSY breaking terms of the form[5]

$$M_W^{-2}\delta\mathcal{L} \sim \frac{1}{3}\sum_i \phi_i^*\phi_i + \chi\chi - (\frac{1}{6}C^{ijk}\phi_i\phi_j\phi_k + h.c.) \ , \tag{II.12}$$

which are known to preserve the one-loop finiteness of the theory. Once SUSY is broken, the weak breaking can proceed in the standard way by utilizing the renormalization group running of the parameters from M_G to M_W. This part of our program is now in progress.[6]

III. <u>Fermion masses</u>

We must now satisfy the second finiteness condition (I.3). Consider first the states $\bar{K}$,K,24 and the parameters p,q. The finiteness condition for K,$\bar{K}$ (see fig. 1) gives

$$p^2 = g^2 \tag{III.1}$$

For 24 we obtain (see fig. 2)

$$q^2 = \frac{5}{21}g^2 \quad . \tag{III.2}$$

Consider now the Higgs $H_i,\bar{H}_i$ and the three families of quarks and leptons. The relevant graphs are given in fig. 3a-d. We obtain the finiteness conditions for;

$$\bar{5}_i : \quad 4\bar{g}^*_{\ell im}\bar{g}_{\ell jm} = 2\delta_{ij}\frac{12}{5}g^2 \tag{III.3}$$

$$\bar{H}_i : \quad 4\bar{g}^*_{\ell mi}\bar{g}_{\ell mj} = 2\delta_{ij}\frac{12}{5}g^2 \tag{III.4}$$

$$10_i : \quad 2\bar{g}^*_{i\ell m}\bar{g}_{j\ell m} + 3g^*_{i\ell m}g_{j\ell m} = 2\delta_{ij}\frac{18}{5}g^2 \tag{III.5}$$

$$H_i : \quad 3g^*_{\ell mi}g_{\ell mj} = 2\delta_{ij}\frac{12}{5}g^2 \; . \tag{III.6}$$

Note that g_{ijk} and $\bar{g}_{ijk}$ <u>cannot</u> be set to zero!! The finiteness conditions have incorporated the Yukawa couplings into the general scheme. They are no longer completely independent or arbitrary parameters.

There are many ways of satisfying the conditions (III.3-6). We don't know if we have exhausted all possibilities. Consider, as an example, the 3 x 3 matrices (where the elements indicate the non-zero terms in g_{ijk} and $\bar{g}_{ijk}$, respectively):

<u>up</u>

$$\begin{array}{c} \\ 10_1 \\ 10_2 \\ 10_3 \end{array}\begin{array}{c} \begin{array}{ccc} 10_1 & 10_2 & 10_3 \end{array} \\ \begin{pmatrix} H_1 & 0 & 0 \\ 0 & H_2 & H_3 \\ 0 & H_3 & H_2 \end{pmatrix} \end{array} \tag{III.7}$$

<u>down</u>

$$\begin{array}{c} \bar{5}_1 \\ \bar{5}_2 \\ \bar{5}_3 \end{array}\begin{pmatrix} \bar{H}_1 & \bar{H}_2 & 0 \\ \bar{H}_2 & \bar{H}_1 & \bar{H}_3 \\ 0 & \bar{H}_3 & \bar{H}_2 \end{pmatrix} \tag{III.8}$$

The above anzatz automatically satisfies the partial condition that the left-hand side of III.3-6 are proportional to δ_{ij}. In order to satisfy this partial condition, it is sufficient to use the following rule:

1) only one Higgs per element in the 3 x 3 matrix, and
2) any Higgs may appear only once in any row or column.

With the above rule, the set of non-linear conditions III.3-6 become a set of 12 linear equations for the non-vanishing couplings (in an obvious notation) of III.7,8, respectively.

up

$$\begin{pmatrix} g_{111} & 0 & 0 \\ 0 & g_{222} & g_{233} \\ 0 & g_{233} & g_{332} \end{pmatrix} \qquad \text{(III.9)}$$

down

$$\begin{pmatrix} \bar{g}_{111} & \bar{g}_{212} & 0 \\ \bar{g}_{122} & \bar{g}_{221} & \bar{g}_{323} \\ 0 & \bar{g}_{233} & \bar{g}_{332} \end{pmatrix} \qquad \text{(III.10)}$$

The solution, for the up and down mass matrices, is

up

$$\sqrt{\frac{8}{5}}\, g \begin{pmatrix} \alpha_1 & 0 & 0 \\ 0 & \cos\theta\, \alpha_2 & \frac{\alpha_3}{\sqrt{2}} \\ 0 & \frac{\alpha_3}{\sqrt{2}} & \sin\theta\, \alpha_2 \end{pmatrix} \qquad \text{(III.11)}$$

down

$$\sqrt{\frac{6}{5}}\, g \begin{pmatrix} \left(\frac{1+\cos^2\Theta}{2}\right)^{1/2} \beta_1 & \frac{\sin\Theta}{\sqrt{2}}\beta_2 & 0 \\ \frac{\sin\Theta}{\sqrt{2}}\beta_2 & \frac{\sin\Theta}{\sqrt{2}}\beta_1 & \cos\Theta\,\beta_3 \\ 0 & \sin\Theta\,\beta_3 & \cos\Theta\,\beta_2 \end{pmatrix} \qquad \text{(III.12)}$$

where Θ is an arbitrary parameter and α_i, β_i are defined in II.5.

There are many such solutions with up to nine free parameters (including α_i, β_i). We have sufficient freedom to fit the up and down quark masses. Note, we have the standard SU_5 result, that the lepton mass matrix is identical to the down quark matrix at M_{GUT}. This is clearly a problem. Moreover, clearly the finiteness conditions are, unfortunately not powerful enough to explain the family hierarchy.

IV. Top quark mass

We now want to present the one loop results for α_G, M_G, $\sin^2\theta$, m_b/m_τ and m_t. The model has three relevant scales M_G, m, and M_W. Between M_G and m we have two pair of color triplet Higgs and one pair of Higgs doublets; below m we have only one Higgs doublet pair. In both cases, we have three families of quarks and leptons and the $SU_3 \times SU_2 \times U_1$ gauge multiplets. We find

$$\begin{aligned} M_G &= 1.9 \times 10^{17}\ \text{GeV} \quad , \\ \alpha_G &= .047 \quad , \\ \sin^2\theta &= .21 \quad . \end{aligned} \qquad \text{(IV.1)}$$

using $\alpha_3 = .1$, $\alpha^{-1} = 128$ at M_W.

In general, if we obtain a family hierarchy, i.e., $m_t \gg m_c \gg m_u$ and $m_b \gg m_s \gg m_d$, then up to corrections of order $(m_c/m_t)^2$ or $(m_s/m_b)^2$, we have

$$\lambda_t^2 = \frac{8}{5} g^2 ,$$

$$\lambda_b^2 = \lambda_\tau^2 = \frac{6}{5} g^2 , \tag{IV.2}$$

at M_G. $\lambda_t, \lambda_b, \lambda_\tau$ are Yukawa couplings for top, bottom, and tau, respectively. In order to obtain $\lambda_t, \lambda_b, \lambda_\tau$ at M_W, we use the one loop renormalization group equations given in ref 4. Between M_G and m we must use the low energy superspace potential, including the color triplet Higgs fields $H_3, \bar{H}_3$:

$$W \supset \lambda_t H Q \bar{T} + \lambda_b \bar{H} Q \bar{B} + \lambda_\tau \bar{H} L \bar{\tau}$$

$$+ \lambda_1 H_3 \bar{T}\bar{\tau} + \frac{\lambda_2}{2} H_3 Q_\alpha Q_\beta \varepsilon^{\alpha\beta} + \lambda_3 \bar{H}_3 \bar{T}\bar{B} + \lambda_4 \bar{H}_3 Q_\alpha L^\alpha . \tag{IV.3}$$

The boundary conditions at M_G are

$$\lambda_t = \lambda_1 = \lambda_2 = \sqrt{\frac{8}{5}}\, g ,$$

$$\lambda_b = \lambda_\tau = \lambda_3 = \lambda_4 = \sqrt{\frac{6}{5}}\, g . \tag{IV.4}$$

$$g = \sqrt{4\pi\alpha_G} .$$

We then obtain $\lambda_b, \lambda_\tau, \lambda_t$, at M_W.

The masses are given by the expressions:

$$m_b = \lambda_b \langle \bar{H}_0 \rangle ,$$

$$m_\tau = \lambda_\tau \langle \bar{H}_0 \rangle ,$$

$$m_t = \lambda_t \langle H_0 \rangle . \tag{IV.5}$$

The ratio $m_b/m_\tau\Big|_{M_W} = \frac{\lambda_b}{\lambda_\tau}\Big|_{M_W}$ is completely determined. However, in order to determine m_b or m_t we must know $\langle\bar{H}_0\rangle$ and $\langle H_0\rangle$. These are, in principle, determined by running the soft SUSY breaking scalar masses Eq. (II.12) from M_G to M_W. However, since we have not yet done this analysis, we use as input

$$M_W^2 = \frac{g_2^2}{2}(\langle H_0\rangle^2 + \langle\bar{H}_0\rangle^2) \quad ,$$

$$\frac{m_\tau}{\lambda_\tau} = \langle\bar{H}_0\rangle \quad . \tag{IV.6}$$

We find

$$\frac{m_b}{m_\tau}\Big|_{M_W} = 1.8 \quad , \tag{IV.7}$$

and

$$m_t = 155 \pm 10\% \text{ GeV} \quad . \tag{IV.8}$$

Using $m_b(m_b) = 4.25$ GeV[7] and renormalizing to $m_b(M_W)$, we find the experimental result

$$\frac{m_b}{m_\tau}\Big|_{M_W,\ \exp} = 1.8 \tag{IV.9}$$

in excellent agreement with (IV.7).

Finally the 10% error in (IV.8) takes into account thresholds we have ignored at M_G and M_W. We note that the renormalization group equations have an infra-red fixed point at $m_t \sim 180$ GeV. The result (IV.8) is strongly affected by the running between M_G and m.

V. Generalized two-loop finite SU_5 GUT

In this section we shall discuss the question, how general are the results obtained in section IV? We consider the more general notation

$$\bar{5}_x \quad , \quad x = 1, \ldots, 7$$

$$5_\alpha \quad , \quad \alpha = 1, \ldots, 4$$

$$10_i \quad , \quad i = 1,2,3 \quad . \tag{V.1}$$

The most general superspace potential with these states (respecting SU_5) is given by

$$W = \lambda_{x\alpha}\bar{5}_x 24 5_\alpha + \bar{g}_{ixy} 10_i \bar{5}_x \bar{5}_y + g_{ij\alpha} 10_i 10_j 5_\alpha$$
$$+ M_{x\alpha}\bar{5}_x 5_\alpha + q\mathrm{Tr}(24^3) + M\mathrm{Tr}(24^2) \quad . \tag{V.2}$$

The last two terms in (V.2), once again, have a minimum which breaks SU_5 to $SU_3 \times SU_2 \times U_1$. This gives us the $\bar{5}5$ mass matrix

$$\begin{array}{c|c} & 5_1 \;\cdots\; 5_4 \\ \begin{array}{c}\bar{5}_1 \\ \vdots \\ \bar{5}_4 \\ \vdots \\ \bar{5}_7\end{array} & \left(\; \langle 24\rangle\lambda_{x\alpha} + \bar{M}_{x\alpha} \; \right) \end{array} \quad . \tag{V.3}$$

We can then always rotate the states $\bar{5}_x$ to $\tilde{\bar{5}}_x$ (which includes $\bar{5}_\alpha, \alpha=1,\ldots,4$ and $\bar{5}_i, i=1,\ldots,3$) defined by the matrix

$$\begin{array}{c c} & 5_\beta \\ \bar{5}_\alpha & \left(\begin{array}{c} \langle 24\rangle\tilde{\lambda}_{\alpha\beta} + \tilde{M}_{\alpha\beta} \\ \\ 0 \end{array} \right) \\ \bar{5}_i & \end{array} \quad . \tag{V.4}$$

We then obtain

$$\begin{aligned} W = \; & \bar{5}_\alpha(\tilde{\lambda}_{\alpha\beta}24 + \tilde{M}_{\alpha\beta})5_\beta \\ & + \bar{5}_i(\tilde{\lambda}_{i\beta}24 + \tilde{M}_{i\beta})5_\beta \\ & + g_{i\alpha\beta}10_i\bar{5}_\alpha\bar{5}_\beta \\ & + \bar{g}_{ijk}10_i\bar{5}_j\bar{5}_k \\ & + \tilde{\bar{g}}_{ij\alpha}10_i\bar{5}_j\bar{5}_\alpha \\ & + g_{ij\alpha}10_i10_j5_\alpha \; , \end{aligned}$$

$$\text{where } \tilde{\bar{g}}_{ij\alpha} \equiv 2\bar{g}_{ij\alpha} \; . \tag{V.5}$$

Several terms must be deleted since they are dimension four baryon violating operators. We therefore set

$$\tilde{\lambda}_{i\beta},\ \tilde{M}_{i\beta},\ \bar{g}_{ijk},\ \bar{g}_{i\alpha\beta} \tag{V.6}$$

to zero. We now have

$$W = \bar{5}_\alpha(\tilde{\lambda}_{\alpha\beta}24 + \tilde{M}_{\alpha\beta})5_\beta + \tilde{\bar{g}}_{ij\alpha}10_i\bar{5}_j\bar{5}_\alpha + g_{ij\alpha}10_i10_j5_\alpha \; , \tag{V.7}$$

where $\bar{5}_\alpha, 5_\alpha$ are Higgs fields and $10_i, \bar{5}_i$ are the three families of quarks and leptons.

The finiteness conditions (Eq. I.3) are now given by

$$\bar{5}_\alpha \ : \ \frac{24}{5}\tilde{\lambda}^*_{\alpha\gamma}\tilde{\lambda}_{\beta\gamma} + 4\tilde{\bar{g}}^*_{ij\alpha}\tilde{\bar{g}}_{ij\beta} = 2\delta_{\alpha\beta}\frac{12}{5}g^2 \ ,$$

$$\bar{5}_i \ : \ 4\tilde{\bar{g}}^*_{ki\alpha}\tilde{\bar{g}}_{kj\alpha} = 2\delta_{ij}\frac{12}{5}g^2 \ ,$$

(V.8)

$$5_\alpha \ : \ \frac{24}{5}\tilde{\lambda}^*_{\gamma\alpha}\tilde{\lambda}_{\gamma\beta} + 3g^*_{ij\alpha}g_{ij\beta} = \frac{24}{5}\delta_{\alpha\beta}g^2 \ ,$$

$$10_i \ : \ 2\tilde{\bar{g}}^*_{ik\alpha}\tilde{\bar{g}}_{jk\alpha} + 3g^*_{ik\alpha}g_{jk\alpha} = 2\delta_{ij}\frac{18}{5}g^2 \ .$$

By taking the appropriate traces we find

$$\mathrm{Tr}\ \bar{5}_i \ : \ \sum_{ki\alpha}|\tilde{\bar{g}}_{ki\alpha}|^2 = \frac{18}{5}g^2 \ ,$$

$$\mathrm{Tr}\ \bar{5}_\alpha \ : \ \sum_{\alpha\beta}|\tilde{\lambda}_{\alpha\beta}|^2 = g^2 \ ,$$

(V.9)

$$\mathrm{Tr}\ 5_\alpha \ : \ \sum_{ij\alpha}|g_{ij\alpha}|^2 = \frac{24}{5}g^2 \ ,$$

$$\mathrm{Tr}10_i \ : \ \text{consistent} \ .$$

Thus, the four expressions in (V.8) are self-consistent. Note a theory with two pairs of Higgs fields and four families of quarks and leptons (satisfying I.2) cannot be made realistic. In this case the same equations (V.8) apply with now $\alpha,\beta,\gamma = 1,2$ and $i,j,k = 1,\ldots,4$. The trace conditions (V.9) are however, inconsistent in this case and the theory cannot be made finite.

Let us now assume that we have only one light pair of Higgs doublets in the low energy theory. Let $\alpha = 1$ be that light Higgs. We then obtain the analogous equations to (IV.2) for the quark and lepton Yukawa couplings to the light Higgs. We find

$$|\tilde{\bar{g}}_{ij1}|^2 = \frac{6}{5} g^2 \left(1 - \frac{|\tilde{\lambda}_{1\gamma}|^2}{g^2}\right) \sim \lambda_b^2 ,$$

$$|g_{ij1}|^2 = \frac{8}{5} g^2 \left(1 - \frac{|\tilde{\lambda}_{\gamma 1}|^2}{g^2}\right) \sim \lambda_t^2 . \tag{V.10}$$

where $\frac{|\tilde{\lambda}_{1\gamma}|^2}{g^2} \lesssim 1$ and $\frac{|\tilde{\lambda}_{\gamma 1}|^2}{g^2} \leq 1$ [using (V.9)]. Thus λ_t will in general be smaller than $\sqrt{\frac{8}{5}}\, g$. Moreover, λ_τ will in general start smaller than $\sqrt{\frac{6}{5}}\, g$; thus $\langle \bar{H}_0 \rangle$ will in general be larger and $\langle H_0 \rangle$ smaller. Both effects will thus tend to decrease the value of m_t. As a result, the value $m_t = 155$ GeV is probably an "upper bound". Upper bound is in quotes since in general there may not be any color triplet Higgs at $m \sim 10^{10}$ GeV. These had a big effect on the running; tending to lower λ_t.

Finally if there are no intermediate mass Higgs triplets, then the GUT scale will in general be smaller. Thus $M_{GUT} = 1.9 \times 10^{17}$ GeV may be an "upper bound".

In conclusion, we have presented a realistic two-loop finite SUSY GUT. Finiteness conditions are seen to incorporate Higgs Yukawa couplings into the overall scheme; i.e., they are no longer completely arbitrary. In a particular model the top quark mass is 155 GeV ± 10%. However, a general analysis finds that there is still an enormous amount of freedom in the Higgs Yukawa couplings and 155 GeV may only be an upper bound for m_t. Finally, the constraints of finiteness give no clue to the origin of a family hierarchy.

References

1. S. Mandelstam, Proc. of the XXl International Conference on High Energy Physics, Paris (1982); A. Parkes and P. West, Phys. Lett. B122, 365 (1983); Nucl. Phys. B222, 269 (1983); A. Namazie, A. Salam and J. Strathdee, Phys. Rev. D28, 1481 (1983); J. G. Taylor, Phys. Lett. 121B, 386 (1983); J. J. van der Bij and Y. P Yao, Phys. Lett. 125B, 171 (1983); S. Rajpoot, J. G. Taylor and M. Zaimi, Phys. Lett. 127B, 347 (1983).
2. M. T. Grisaru and W. Siegel, Nucl. Phys. B210, 29 (1982); L. Girardello and M. T. Grisaru, Nucl. Phys. B194, 65 (1982); P. S. Howe, K. S. Stelle and P. K. Townsend, Nucl. Phys. B214, 513 (1983); A. Parkes and P. West, Phys. Lett. 127B, 35 (1983), J. M. Frere, L. Mezincescu and Y. P. Yao, Phys. Rev. D29, 1196 (1984).
3. A. Parkes and P. West, Phys. Lett. 138B, 99 (1984); P. West, Phys. Lett. 137B, 371 (1984); D. R. T. Jones and L. Mezincescu, Phys. Lett. 136B, 242 (1984); 138B, 293 (1984).
4. D. R. T. Jones and S. Raby, to be published in Phys. Lett B; S. Hamidi and J. H. Schwarz, CALT-68-1159, CAL-TECH (1984).
5. D. R. T. Jones, L. Mezincescu and Y. P. Yao, Univ. of Michigan preprint (1984).
6. J. Bjorkman, D. R. T. Jones and S. Raby, in preparation.
7. J. Gasser and H. Leutwyler, Phys. Rep. 87 (1982).

Table I

Representation	$T(R)$	$C_2(R)$
24	5	5
5	$\frac{1}{2}$	$\frac{12}{5}$
$\bar{5}$	$\frac{1}{2}$	$\frac{12}{5}$
10	$\frac{3}{2}$	$\frac{18}{5}$
$\overline{10}$	$\frac{3}{2}$	$\frac{18}{5}$
45	12	$\frac{32}{5}$
50	$\frac{35}{2}$	$\frac{42}{5}$
75	25	8

$d_R C_2(R) \equiv d_{Adj} T(R)$

with d_R, the dimension of the representation R.

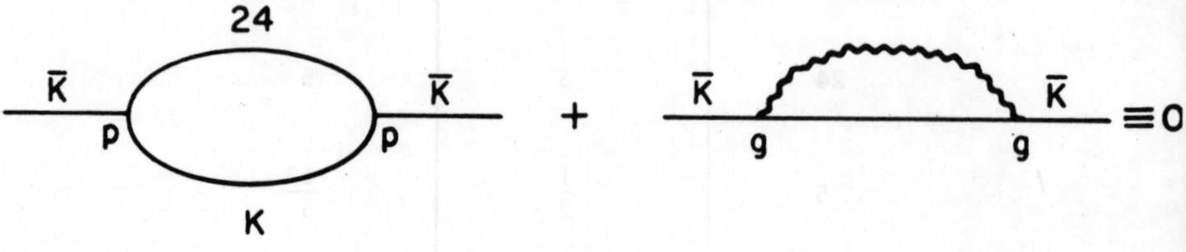

Fig. 1

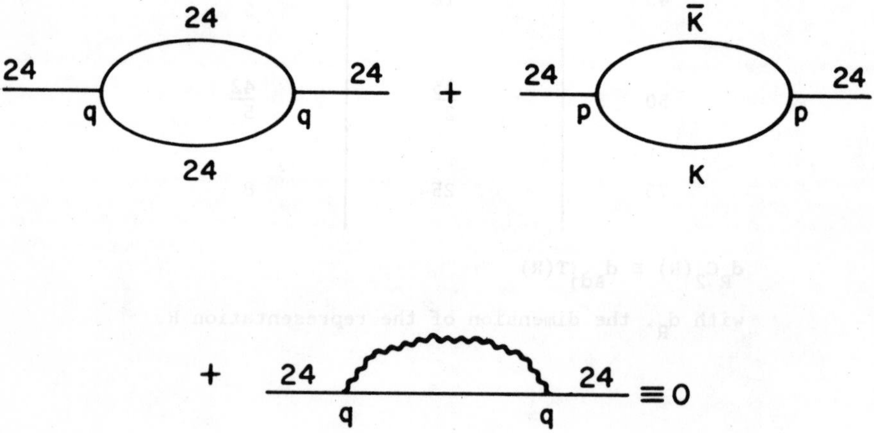

Fig. 2

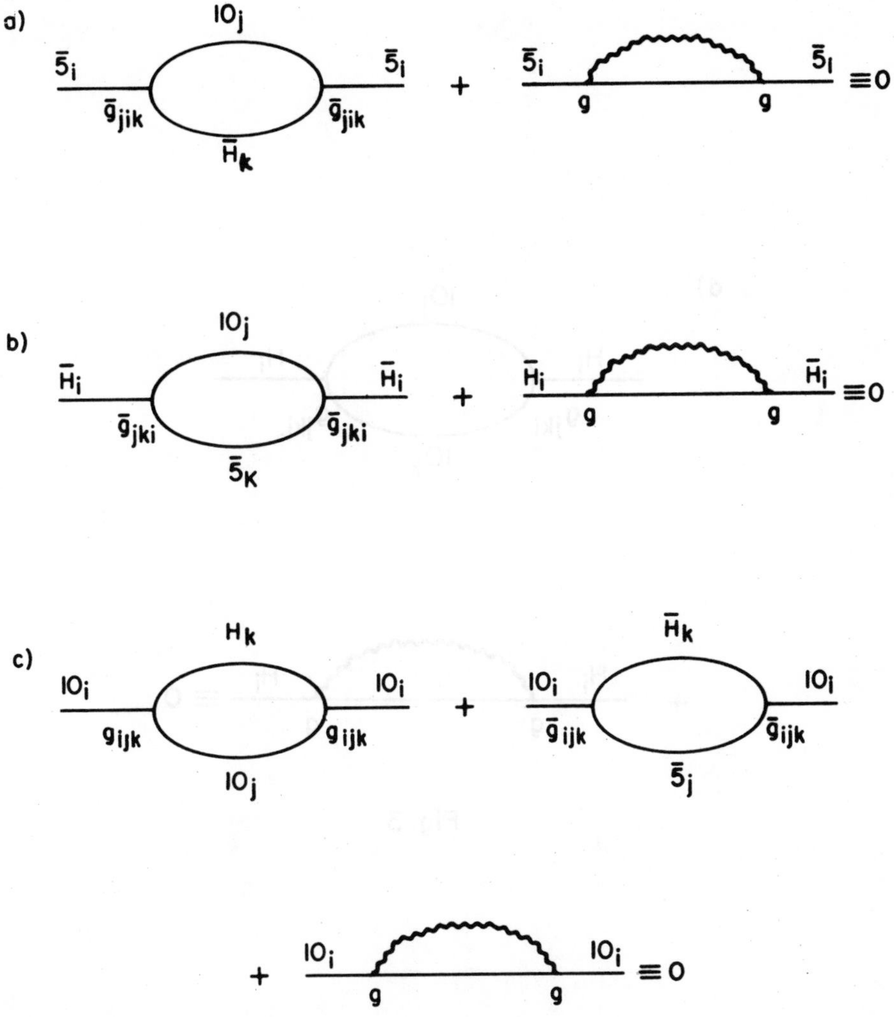

a)
10_j
$\bar{5}_i$
$\bar{5}_i$
$\bar{g}_{jik}$
$\bar{g}_{jik}$
$\bar{H}_k$
+
$\bar{5}_i$
$\bar{5}_i$
g
g
$\equiv 0$
b)
10_j
$\bar{H}_i$
$\bar{H}_i$
$\bar{g}_{jki}$
$\bar{g}_{jki}$
$\bar{5}_K$
+
$\bar{H}_i$
$\bar{H}_i$
g
g
$\equiv 0$
c)
H_k
10_i
10_i
g_{ijk}
g_{ijk}
10_j
+
$\bar{H}_k$
10_i
10_i
$\bar{g}_{ijk}$
$\bar{g}_{ijk}$
$\bar{5}_j$
+
10_i
10_i
g
g
$\equiv 0$

Fig. 3

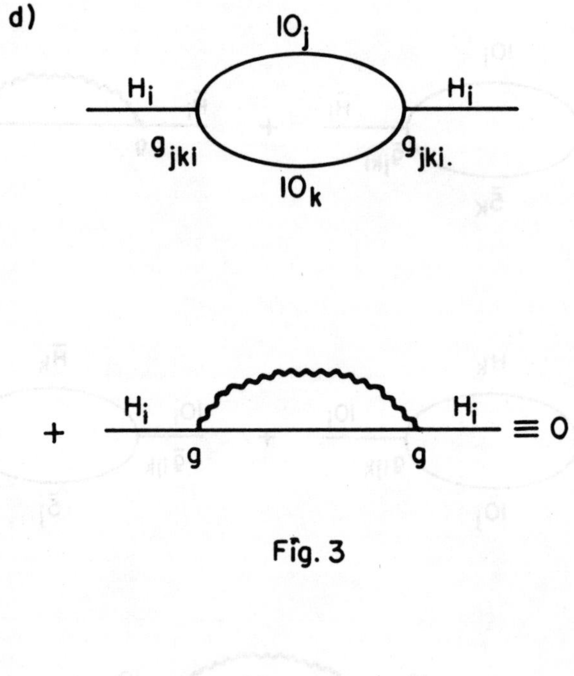

Fig. 3

SUPERSYMMETRIC GAUGE THEORY MODELS

G.G. Ross
University of Oxford
Theoretical Physics Department
1 Keble Road
Oxford
ENGLAND

1.1 Introduction

The standard model, based on the gauge group SU(3) x SU(2) x U(1), has been very successful in describing the strong, weak and electromagnetic interactions. However, it falls short of what one might hope for in the ultimate theory: particles are assigned by hand to multiplets of the gauge group and there are a large number of parameters, gauge couplings and masses, needed to specify the theory. Grand Unified Theories (GUTs) seek to improve this situation by embedding the gauge group in a simple grand unified group with a single gauge coupling constant. The multiplet structure for a given particle spin can also be simplified in GUTs, but there is still no symmetry relating different spins. Only the vector bosons are uniquely specified, by the local gauge principle, to belong to a definite representation of the gauge group (the adjoint). Supersymmetry is the only symmetry known which can relate different spins, and one may hope it supplies the missing ingredient for constructing the ultimate theory by relating the representation content of scalars and fermions to those of vectors and by relating the scalar and Yakawa couplings to the gauge coupling. Moreover, local supersymmetry necessarily includes the theory of gravity, leading to the unification of gravity with the other interactions.

Unfortunately, as we have heard at this meeting, these supergravity theories do not have the correct spectrum to describe the known particles. For example, even the largest (N=8) supergravity includes only the gauge group SU(3) x U(1) x U(1) and not the standard model. Thus supergravity theories may be the fundamental theory, but only at a more fundamental level at which the W^s bosons, for example, are composite.

Thus , in building supersymmetric models, we must give up the hope that the extended supersymmetry models contain the spectrum of the observed states in a single irreducible representation and build models based on the direct product structure G x[N extended supersymmetry]. At first sight, this appears to lose many of the potential benefits of supersymmetry, and indeed in this type of theory there may remain many arbitrary parameters. However, such models can solve the hierarchy problem common to Grand Unified Theories, and for this reason they have been extensively studied. In these lectures, we will consider the simplest such theories based on the group G x[N=1 supersymmetry]. Higher N theories may be possible, but they all require minor fermions and, since there is no evidence yet for such states, it seems best to start with a structure capable of accommodating the known chiral fermion structure.

1.2 Rules for construction of N=1 supersymmetric Lagrangians

It is straightforward to construct Lagrangian densities which are globally supersymmetric using the superspace rules introduced by Dr. Stelle[1]. The gauge invariant supersymmetric matter kinetic energy term is of the form

$$L_{km} = [\Phi^+ e^{2gV} \Phi]_D \tag{1}$$

where D denotes the $\Theta\Theta\bar{\Theta}\bar{\Theta}$ projection. Φ is a (left handed) chiral superfield

$$\Phi = \phi + \sqrt{2}\, \psi\, \Theta + F\Theta\Theta \tag{2}$$

with complex scalar component ϕ, two component Weyl fermion field ψ and auxiliary field F. V is a vector supermultiplet defined by

$$V \equiv \{-\Theta\, \sigma^\mu\, \bar{\Theta}\, v^a_\mu + i\Theta\Theta\bar{\Theta}\, \bar{\lambda}^a - i\bar{\Theta}\bar{\Theta}\Theta\, \lambda^a + \tfrac{1}{2}\, \Theta\Theta\bar{\Theta}\bar{\Theta}\, D^a\}T^a \tag{3}$$

where v^a_μ, λ^a are gauge bosons and their fermion partners the gauginos. The index a refers to the local gauge group, the fields v^a_μ and λ^a belonging to its adjoint representation. The matrices T^a are representations of the group acting on the matter fields Φ.

Eq.(1) uniquely specifies the couplings of the gauge bosons and their gaugino partners to matter fields once their representation is specified. For example, if the gauge group is SU(2) with $\binom{\nu}{e}$ a left-handed fermion doublet component of eq.(2), the interaction with the charged gauge field, W^+ is given by the usual form

$$\frac{g}{2\sqrt{2}}\, \bar{e}\, \gamma_\mu (1-\gamma_5)\nu\, W^{\mu-} + \text{h.c.} \tag{4}$$

while the gaugino coupling contains the piece

$$\frac{g}{\sqrt{2}}\, (\bar{e}\, \tilde{\nu}\, \lambda^-_W + \tilde{e}^*\, \nu\, \bar{\lambda}_W) + \text{h.c.} \tag{5}$$

where $\tilde{\nu}$ and $\tilde{e}$ are the scalar neutrino and electron respectively, and λ_W is the W gaugino, the Wino. We will give a more complete form for the standard model in the next section. Thus, to define a supersymmetric gauge theory it is necessary to choose the gauge group and the matter representations. This specifies the gauge boson and gaugino couplings. In addition, we must specify the non-gauge interactions between the components of the chiral (matter) supermultiplets. The form of these interactions follows from the allowed supersymmetric interaction term which may be written[1] as

$$L^{int} : [P(\Phi)]_F + \text{h.c.} \tag{6}$$

where F denotes the $\Theta\Theta$ projection and the superpotential, P, is a general function of the chiral superfields Φ_i, but <u>not</u> of their hermitian conjugates. If L^{int} is to have maximum field dimension of 4, necessary for a renormalisable theory, $P(\Phi)$ should be at most cubic in Φ. In terms of the component fields, L^{int} may be written as[1) $L^{int} = L^{Fermion} + L^{scalar}$, where

$$L^{Fermion} = \sum_{i,j} \left.\frac{\partial^2 P}{\partial\Phi_i\,\partial\Phi_j}\right|_{\Phi=\phi} \psi_i\psi_j$$

$$L^{scalar} = \sum_i \left|\frac{\partial P}{\partial\Phi_i}\right|^2_{\Phi=\phi} + \sum_{\alpha,i} |g_\alpha\, \phi_i^+\, T^{\alpha,i}\, \phi_i|^2 \tag{7}$$

where $T^{\alpha,i}$ are representations of the gauge group, α, acting on the representation ϕ_i and g_α is the associated gauge coupling. The last term in L^{scalar} comes from eliminating the (non-propagating) D auxiliary field in eq.(3) while the first term in L^{scalar} comes from eliminating the F auxiliary field in eq.(2). They are often referred to as D and F terms respectively.

Once P is specified, the model is complete and in the next section I will discuss reasonable forms for P which can generate realistic fermion masses through $L^{Fermion}$ and acceptable patterns of spontaneous symmetry breaking via L^{scalar}.

<u>1.3 A supersymmetric version of the standard model</u>(2,3)

The standard model starts with the gauge group SU(3) x SU(2) x U(1). The vector supermultiplet transforms as the adjoint under this group giving the particle content shown in Table 1. Unfortunately, we cannot identify any of the new fermions, the gauginos, with the quarks and leptons as the latter do not transform as the adjoint. We are therefore forced to introduce chiral supermultiplets to accommodate the quarks and leptons as in Table 1. (Note that, for a theory with renormalisable gauge interactions the vector supermultiplets must be in the adjoint of the gauge group). Once again we are forced to double the number of states this time introducing (complex) scalar fields to partner the known fermions. Finally, we must include Higgs scalar fields, needed to spontaneously break the gauge group SU(3) x SU(2) x U(1) to SU(3) x $U(1)_{em}$. Our original hope was that the Higgs sector would be simplified, by assigning the Higgs scalar to the same supermultiplets as the known fermions. An obvious possibility is to identify the Higgs SU(2) doublet as a partner of a lepton doublet. However, this is not possible, for such an assignment in supersymmetry does not give an acceptable pattern of fermion masses. The reason is that supersymmetry restricts the possible forms of Yukawa couplings and the couplings necessary to give down quarks and charged leptons a mass are not present. To see this, note that Yukawa couplings are simple described in terms of the superpotential P via eq.(7). The Yukawa couplings needed to give all charged fermions a mass has a superpotential of the form

$$P = \sum_{j,i=d,s,b} m_{ij}^{(d)} \varepsilon^{\alpha\beta} \psi_{i,a}^{(q)} H_{2\beta} \psi_j^{(d^c)} + \sum_{i,j=u,c,t} m_{ij}^{(u)} \varepsilon^{\alpha\beta} \psi_{i,\alpha}^{(q)} H_{1\beta} \psi_j^{(u^c)}$$

$$+ \sum_{i=e,\mu,\tau} m_i \, \varepsilon^{\alpha\beta} \psi_{i,\alpha}^{(1)} \, H_{2\beta} \psi_i^{(1^c)} \tag{8}$$

where $H_{1\alpha}$ and $H_{2\alpha}$ are chiral supermultiplets transforming as doublets under SU(2), but with U(1) charge $\pm\frac{1}{2}$, respectively,

so that their charge states are $\begin{bmatrix} H_1^+ \\ H_1^0 \end{bmatrix}$ and $\begin{bmatrix} H_2^0 \\ H_2^- \end{bmatrix}$ respectively.

The indices i and j are family indices and $m_{ij}^{(d)}$, $m_{ij}^{(u)}$ are the mass matrices for the up and down quark masses. The lepton supermultiplet doublets, $\psi_{i,\beta}^{(1)}$, have the correct charges to be identified with $H_{2\beta}$. In the non-supersymmetric standard model $H_{2\beta} = \varepsilon_{\beta\gamma} H_1^{c\gamma}$, but, in the supersymmetric case, the rules of section (1.2), require that P only be formed using products of (left handed chiral) supermultiplets and not their (right-handed chiral) conjugates. Thus, in eq.(8), $H_{1\beta}$ must be identified with a completely new chiral supermultiplet. In addition, more states are needed, for $H_{1\beta}$ contains new, charged, Weyl fermions and we must add, further, charged fermions to allow for the construction of Dirac masses to ensure that the final theory has no massless charged states (we cannot give charged fermions a Majorana mass without violating charge conservation) and to avoid anomalies. The simplest solution is to introduce another new SU(2) doublet chiral superfield which is usually identified with $H_{2\beta}$. In constructing simple grand unified generalisation of the standard model this, in fact, is the only possibility, for, if $H_{2\beta}$ is identified with a lepton doublet, proton decay proceeds too fast because the colour triplet components, grand unified multiplet partners of the leptons, are necessarily light and mediate proton decay (see section 2.2).

Thus the final multiplet structure for a supersymmetric version of the standard model includes two new chiral supermultiplets whose scalar partners are to be identified with the Higgs scalars needed to break the SU(3) x SU(2) x U(1) to SU(3) x $U(1)_{em}$ and to give all charged fermions a mass. The full multiplet structure is given in Table (1).

Although the SU(3) x SU(2) x U(1) x N=1 supersymmetry structure fails to simplify the multiplet structure of the original model (indeed it more than doubles the spectrum!) it does have a redeeming property that has caused it to be studied intensively, recently, as a possible theory for the strong, weak and electromagnetic interactions - it solves the hierarchy problem. In the next sections, we will construct the Lagrangian for this theory and show how this solution is achieved.

Table 1 Multiplet structure for the minimal supersymmetric SU(3) x SU(2) x U(1) theory

Vector Supermultiplets			Spin J
V_G	$g_\mu^{a=1...8}$	Gluons	1
	$\tilde{g}^{a=1...8}$	Gluinos	$\frac{1}{2}$
V_W	$W_\mu^{\pm}$, Z_μ	W, Z bosons	1
	$\tilde{W}^{\pm}$, $\tilde{Z}$	Winos, Zino	$\frac{1}{2}$
V_γ	A_μ	Photon	1
	$\tilde{A}$	Photino	$\frac{1}{2}$

Chiral Supermultiplets			Spin J
$\psi_{i,\alpha}^{(q)}$	q_L, $q_L^{(c)}$	Quarks	$\frac{1}{2}$
$\psi_i^{u^c}$, $\psi_i^{d^c}$	$\tilde{q}$, $\tilde{q}^c$	Scalar quarks	0
$\psi_{i,d}^{(1)}$	ℓ_L, $\ell_L^{(c)}$	Leptons	$\frac{1}{2}$
$\psi_i^{1^c}$	$\tilde{\ell}$, $\tilde{\ell}^c$	Scalar leptons	0
	H_1, H_2	Fermionic Higgs	$\frac{1}{2}$
	H_1, H_2	Higgs doublets	0

1.4 <u>The SU(3) x SU(2) x U(1) supersymmetric Lagrangian</u>

Once the transformation properties of the supermultiplets under the gauge group are specified and the superpotential is given, the Lagrangian density may be immediately constructed using the result of section (1.2)

$$\begin{aligned} L^G = & -\frac{1}{4}\,\mathrm{Tr}\{W^{\mu\nu}W_{\mu\nu}\} - i\mathrm{Tr}\{\lambda\sigma^m D_m\bar{\lambda}\} \\ & + \sum_j \phi_j D_m D^m \phi_j^* + i\sum_j D_m\bar{\psi}_j\bar{\sigma}^m\psi_j \\ & - \frac{i}{\sqrt{2}}\sum_{j,a} g_a(\phi_j\tau^a\bar{\psi}_j - \phi_j^*\tau^a\psi_j)\lambda^a \\ & - \frac{1}{8}\sum_a\left|g_a\sum_j \phi^+{}_g\tau^a\phi_j\right|^2 \end{aligned} \tag{9}$$

In this, λ represents the gaugino, the trace implies a sum over all the gauge indices, a, of SU(3) x SU(2) x U(1) and the sum over j is over all the chiral superfields of Table 1. D_m are the usual gauge covariant derivatives. For convenience, we split L^G up into the usual (non-supersymmetric) kinetic energy term, L_{Kin}, and a term, L^G_{Yuk}, describing the new Yukawa interactions induced by the supersymmetric form of the gauge interactions and L^D the scalar interactions obtained by eliminating the D auxiliary fields. We have

$$\begin{aligned} L_{Kin} = & -\frac{1}{4}\,\mathrm{Tr}\{W^{\mu\nu}W_{\mu\nu}\} - i\mathrm{Tr}\{\lambda\sigma^m D_m\bar{\lambda}\} \\ & + \sum_j \phi_j D_m D^m \phi_j^* + i\sum_j D_m\bar{\psi}_j\bar{\sigma}^m\psi_j \end{aligned} \tag{10}$$

The covariant derivatives generate couplings between the gauge bosons and the other fields in the theory. The self couplings of the gauge bosons and the couplings to the quark and lepton fields are the usual ones of the standard model. The new couplings involve Higgsinos, gauginos and quarks and sleptons. For example the Higgsino, Wino and slepton couplings are[3]

$$\begin{aligned} L'_{Kin} = & (e/\cos\theta_W)\frac{1}{2}B_\mu(\bar{\tilde{H}}_1^+\bar{\sigma}^\mu\tilde{H}_1^+ + \bar{\tilde{H}}_1^0\bar{\sigma}^\mu\tilde{H}_1^0 \\ & - \bar{\tilde{H}}_2^-\bar{\sigma}^\mu\tilde{H}_2^- - \bar{\tilde{H}}_2^0\bar{\sigma}^\mu\tilde{H}_2^0 - \bar{\tilde{\nu}}\bar{\sigma}^\mu\nu - \bar{\tilde{e}}^-\bar{\sigma}^\mu\tilde{e}^- + 2\,\bar{\tilde{e}}^+\bar{\sigma}^\mu\tilde{e}^+) \\ & + (e/\sin\theta_W)W_\mu^+\{\frac{1}{\sqrt{2}}(\bar{\tilde{H}}_1^+\bar{\sigma}^\mu\tilde{H}_1^0 + \bar{\tilde{H}}_2^0\bar{\sigma}^\mu\tilde{H}_2^-) \end{aligned}$$

$$+ (-\bar{\tilde{W}}^{+}\bar{\sigma}^{\mu}\tilde{W}^{o} + \bar{\tilde{W}}^{o}\bar{\sigma}^{\mu}\tilde{W}^{-}) \quad + \text{h.c.} + \frac{}{\sqrt{2}}\,\tilde{\nu}^{*}\bar{\sigma}^{\mu}\tilde{e}^{-}\} + \text{h.c.}$$

$$+ (e/\sin\theta_W)W^{3}_{\mu}\{\tfrac{1}{2}(\bar{\tilde{H}}_1{}^{+}\bar{\sigma}^{\mu}\tilde{H}_1{}^{+} - \bar{\tilde{H}}_1{}^{0}\bar{\sigma}^{\mu}\tilde{H}_1{}^{0}$$

$$+ \bar{\tilde{H}}_2{}^{o}\bar{\sigma}^{\mu}\tilde{H}_2{}^{o} - \bar{\tilde{H}}_2{}^{-}\bar{\sigma}\tilde{H}_2{}^{-}) + (\bar{\tilde{W}}^{+}\bar{\sigma}^{\mu}\tilde{W}^{+} - \bar{\tilde{W}}^{-}\bar{\sigma}^{\mu}\tilde{W}^{-})\}$$

$$+ \frac{1}{2}(\tilde{\nu}^{*}\bar{\sigma}^{\mu}\tilde{\nu} - \tilde{\bar{e}}^{*}\bar{\sigma}^{\mu}\tilde{\bar{e}}^{*})\} \tag{11}$$

The remaining piece of eq.(9) is

$$L^{G}_{Yuk} = (e/\cos\theta_W)\; i\,\frac{\sqrt{2}}{2}\; H_1^{*+}\tilde{B}^{o}\tilde{H}_1^{+} + H_1^{*o}\tilde{B}^{o}\tilde{H}_1{}^{o}$$

$$- H_2^{*o}\tilde{B}^{o}\tilde{H}_2{}^{o} - H_2^{*-}\tilde{B}^{o}\tilde{H}_2{}^{-} - \tilde{\nu}^{*}\,\tilde{B}^{o}\nu - \tilde{\bar{e}}^{*}\tilde{B}^{o}\bar{e} + 2\,\tilde{e}_{+}^{*}\tilde{B}^{o}e_{+}\} + \text{h.c.}$$

$$+ e/\sin\theta_W)i\;\{H_1^{*+}\tilde{W}^{+}\tilde{H}_1{}^{o} + H_1^{*o}\tilde{W}^{-}\tilde{H}_1{}^{+} + H_2^{*o}\tilde{W}^{+}\tilde{H}_2{}^{-} + H_2^{*-}\tilde{W}^{-}\tilde{H}_2{}^{o}$$

$$+ \tilde{\nu}^{*}\tilde{W}^{+}e^{-} + \tilde{\bar{e}}\tilde{W}^{-}\bar{\nu}\} + \text{h.c.}$$

$$+ (e/\sin\theta_w)\,\frac{i}{\sqrt{2}}\,\{H_1^{*+}\tilde{W}^{o}\tilde{H}_1{}^{+} - H_1^{*o}\tilde{W}^{o}\tilde{H}_1{}^{o} + H_2^{*o}\tilde{W}^{o}\tilde{H}_2{}^{-} + H_2^{*-}\tilde{W}^{-o}\tilde{H}_2^{-}$$

$$+ \tilde{\nu}^{*}\tilde{W}^{o}\nu - \tilde{\bar{e}}^{*}\tilde{W}^{o}e^{-}\} + \text{h.c.} \tag{12}$$

The Yukawa couplings may be read immediately from the superpotential using the form of eq.(7). They are of the form given in the standard model, except that the Higgs, H_1, couples only to the right handed up quarks and H_2 couples only to the right handed down quarks and charged leptons. The scalar interactions in L_{int} also follow from P using eq.(7).

The two component form for the interactions, used here, is simply related to the more familiar four component form by noting, in the Weyl basis, that

$$\lambda\sigma^{\mu}\psi = \Lambda\gamma^{\mu}\left(\frac{1-\gamma_5}{2}\right)\Psi \tag{13}$$

where λ and ψ are 2 component Weyl spinors (see Appendix A) related to the usual four component Dirac spinors Λ and Ψ by

$$\Psi = \begin{bmatrix} \psi \\ \text{any} \end{bmatrix} \qquad \Lambda = \begin{bmatrix} \lambda \\ \text{any} \end{bmatrix} \tag{14}$$

We may also write this vertex in the form

$$\bar{\lambda}\bar{\sigma}^{\mu}\psi = -\bar{\Lambda}\,\gamma^{\mu}\left(\frac{1+\gamma_5}{2}\right)\Psi' \tag{15}$$

where

$$\Psi' = \begin{bmatrix} \text{any} \\ - \\ \psi \end{bmatrix}, \qquad \Lambda' = \begin{bmatrix} \text{any} \\ - \\ \lambda \end{bmatrix} \tag{16}$$

(This follows from the result $\bar{\lambda}\bar{\sigma}^{\mu}\psi = -\psi\sigma^{\mu}\bar{\lambda}$ which applies to anticommuting operators).

For scalar couplings $\lambda\psi = \psi\lambda$ so

$$\lambda\psi = \bar{\Lambda}'_R\Psi_L = \bar{\Psi}'_R\Lambda_L \tag{17}$$

The multiplet structure and interactions discussed above are common to all attempts to generalise the standard model. The main uncertainty in the model proves to be the mechanisms driving spontaneous symmetry breaking of the SU(2)xU(1) and of supersymmetry and we will spend some time discussing the alternatives that have been explored. Before doing this, however, let us consider the symmetries of the standard model.

1.5 Symmetries of the standard SUSY model

In the non-supersymmetric standard model there are various symmetries which occur. Baryon and lepton numbers are automatically conserved, as is lepton type (e, μ, τ). Strangeness is conserved by the strong and electromagnetic interactions and, due to the GIM mechanism, by the neutral weak currents at tree level. C and P are conserved by the strong and electromagnetic interactions, but not by the weak. CP is violated for three or more generations. Neutrinos have zero mass.

Let us consider the situation in the supersymmetric case. This time, because there are new supersymmetric partners carrying the same quantum numbers as their partners, there are possibilities for symmetry violation.

Baryon and lepton numbers are conserved by the supersymmetric Lagrangian (cf. eqs.(7)-(12)). However, unlike the standard model, this does not automatically follow[(4)(5)] for we have omitted terms in the Lagrangian which would have violated B and L. For example the terms

$$(LH)_F,\ (QQU)_F \tag{18}$$

are SU(3)xSU(2)xU(1) symmetric and supersymmetric. Thus they could be added to the original Lagrangian, but as they violate Lepton and baryon numbers, respectively, the were excluded from the original Lagrangian. An interesting property of supersymmetric theories is that if a term is not included in the original superpotential, it will not be generated in any order of perturbation theory[(6)], so that, even if there is no symmetry forbidding terms such as in eq.(18), they will not occur if not present at tree level. This remarkable fact can give rise to 'supernatural' or accidental symmetries. However, for many people, this is an unsatisfactory origin for a symmetry as basic as lepton or baryon number and, as we will discuss shortly, terms such as in eq.(18) can be forbidden by R symmetries.

In the standard model strangeness is violated because the quark mass matrix splits the degeneracy between d and s quarks and so the charged currents, when expressed in terms of the quark mass eigenstates, have a strangeness changing component. In the lepton sector, because neutrinos are massless, the charged currents do not change τ, μ or e lepton number, as the neutrino current eigenstates can be taken as the physical neutrinos leaving the charged currents lepton flavour diagonal. In the supersymmetric version of the standard model strangeness may be violated in the squark sector giving rise to new sources of strangeness violation. In the lepton sector, there is an even more interesting change for the sneutrinos are not massless and, when the charged currents are expressed in terms of the neutrino mass eigenstates, they have a lepton flavour violating component, which can induce processes such as $\mu \to e\gamma$, $\mu \to 3e$, etc. Finally, the squark and slepton sectors may introduce new sources of CP violation as discussed below.

1.6 Mixing angles and CP violation in the supersymmetric sector

Let us discuss the structure of the charged and neutral currents in some more detail[7]. Expressed in terms of "current" quark eigenstates the charged weak current is diagonal and is simply given by

$$L_{cc} = \frac{g_2}{\sqrt{2}} \tilde{u}_i^{+} \gamma_\mu \tilde{d}_i W_\mu^{+} + \text{h.c.} \tag{19}$$

In terms of the mass eigenstates, $\tilde{u}_i^m$ and $\tilde{d}_i^m$, the current eigenstates are related by unitary matrices $\tilde{X}, \tilde{Y}$

$$\tilde{u}_i^m = \tilde{X}_{ij} \tilde{u}_j, \quad \tilde{d}_i^m = \tilde{Y}_{ij} \tilde{d}_j \tag{20}$$

and hence the charged current becomes

$$L_{cc} = \frac{g_2}{\sqrt{2}} \tilde{u}_{K_i}^{m+} \gamma^\mu \tilde{U}_{KMij} \tilde{d}_{Lj}^m W_\mu^{+} + \text{h.c.} \tag{21}$$

where

$$\tilde{U}_{KM} = \tilde{X}\tilde{Y}^{+} \tag{22}$$

and, as $\tilde{X}$ and $\tilde{Y}$ are unitary matrices, so too is $\tilde{U}_{KM}$. Since supersymmetry is broken there is no need for the squark mass eigenstates to correspond to the quark eigenstates and so $\tilde{U}_{KM}$ need not be equal to U_{KM}. $\tilde{U}_{KM}$ may be written in the same form as the Kobayashi Maskawa mixing matrix, U_{KM}, in the standard model, but with different angles and phase. The neutral currents coupled to the Z, the γ or the gluons remain invariant under this rotation - the supersymmetric generalisation of the GIM mechanism. This is because they always involve the combination $\tilde{X}\tilde{X}^{+}$ or $\tilde{Y}\tilde{Y}^{+}$.

The situation is quite different for interactions involving the gauginos. Consider, for example, the interaction of gauginos which is (suppressing colour indices on the quarks and squarks)

$$L_{\tilde{g}} = \sqrt{2} i g_3 \bar{\lambda}^{\alpha} [u_{L_i}^{\sim +} T^{\alpha} u_{L_i} + \tilde{d}_{L_i}^{+} T^{\alpha} d_{L_i}] + \text{h.c.} \tag{23}$$

The interaction between mass eigenstates is

$$L_{g} = \sqrt{2} i g_3 \bar{\lambda}^{\alpha} [u_{L_i}^{m+} V_L^u T_{ij}^{\alpha} u_L^m + d_L^{+m} V_L^d T_{ij}^{\alpha} d_{l_j}^m] + \text{h.c.} \tag{24}$$

where

$$V_L^u = \tilde{X}X^+ \quad \text{etc.} \tag{25}$$

and i,j are family indices. There is a similar form for the right handed components.

V^u and V^d can again be written in the standard Kobayashi Maskawa form, but in general involving new mixing angles and phases. Their values depend on the mechanism giving squarks and sleptons mass, a subject discussed further in sections (3.2-3.4). In the minimal version of the model[8]

$$\tilde{U}_{KM} \simeq I$$

$$V_L^u \simeq V_R^u \simeq I$$

$$V_L^d \simeq U_{KM} \qquad V_R^d \approx I \tag{26}$$

where U_{KM} is the Kobayashi Maskawa mixing matrix of the standard model. Thus gluinos will change flavour principally where coupled to the left handed down doublets of quarks.

The lepton sector may similarly be analysed with analogous new mixing angles needed when expressing the current in terms of mass eigenstates.

In addition to the CP violation introduced in the mixing matrix, there are further new sources of CP violation arising from the gaugino mass terms (see section (1.8)). To be consistent with observed limits on CP violation for the neutron dipole electric moment, the imaginary part of the gluino mass must satisfy[9].

$$\mathrm{Im}[m_{\tilde{g}}] < 10^{-3} \frac{m^3_{3/2}}{(1000\mathrm{GeV})^2} \tag{27}$$

In non-minimally coupled supergravity models, the gauginos acquire mass of $O(m_{3/2})$ at tree level and eq. (27) gives a constraint which may be difficult to satisfy.

1.7 Flavour changing neutral currents[7,10]

The supersymmetric partners we have introduced, contribute, via loops, to flavour changing neutral currents involving light hadrons and leptons. For example, the graph of Fig.(1a) gives (assuming $m_{\tilde{q}} > m_{\tilde{W}}$) the $\Delta S=2$ operator contributing to the K_L-K_S mass difference:

$$\frac{g_2{}^4}{256\pi^2 \, m_{\tilde{q}}^6} \, U^+_{KM_{s_i}} \, \Delta m^2_{\tilde{q}_i} \, U_{KM_{id}} \, (\bar{s}_L \gamma_\mu d_L)^2 \tag{28}$$

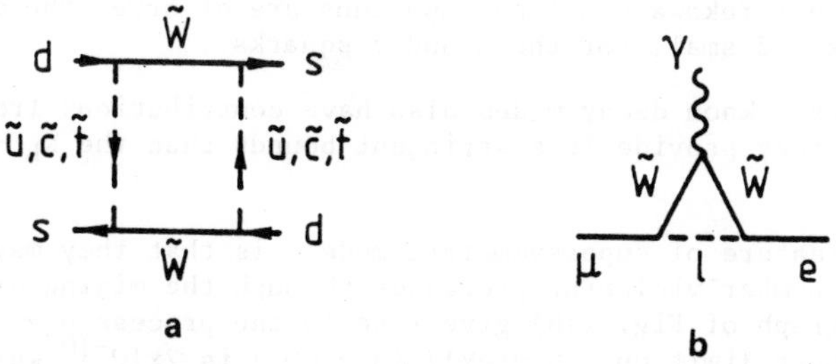

Fig. 1(a) Graph contributing to the $\Delta S=2$, $\Delta Q=0$ amplitude via supersymmetric virtual states. (b) Graph generating $\mu \to e\gamma$

$m_{\tilde{q}}$ is the mean up squark mass and $\Delta m^2_{\tilde{q}i} \equiv m_{\tilde{q}i} - m_{\tilde{q}}{}^2$ is the up squark mass difference. The experimental measurement of the K_L-K_s mass difference puts strong constraints on the size of this term, and assuming the mixing angles of the $\tilde{u}$ and $\tilde{d}$ squarks are smaller or comparable to the Cabibbo angle, gives the bound

$$\frac{\Delta \tilde{m}^2_{q_{u,c}}}{m_{\tilde{q}}^2} < 0\left(\frac{1}{30}\right) \frac{(m_{\tilde{q}}^2)}{M_W{}^2} \tag{29}$$

An even stronger bound comes from the contribution of Fig.(1a) to the imaginary part of the K_L, K_s mass difference (assuming phases of order one)

$$\frac{\Delta \tilde{m}^2_{q_{u,c}}}{m^2_{\tilde{q}}} < 0(10\text{-}3)\ \frac{(m^2_{\tilde{q}})}{M_W^2} \qquad (30)$$

For the top squark the bound may be weaker because we expect the associate mixing angle to be smaller.

In fig. (1a) we obtain non-zero contributions if we replace the $W^{\pm}$ by $\tilde{W}^o$, $\tilde{B}^o$ and $\tilde{g}$ exchange with $\tilde{d}$ and $\tilde{s}$ squark exchange along the horizontal for, as discussed in section (1.6), in this case the GIM mechanism does not prevent flavour changing neutral currents. Evaluation of these graphs lead to similar constraints for the $\tilde{d}$ and $\tilde{s}$ squark mass differences, slightly weakened by the additional mixing angle factors.

For values of $m_{\tilde{q}} \lesssim M_W$ these bounds appear very strong constraints, but, for the models introduced above, they are all easily satisfied because squarks get a large, flavour independent, mass from the supersymmetry breaking mechanism. The mass differences between them are introduced by their Yukawa couplings and thus are of order the quark mass differences and small for the u and c squarks.

The other rare kaon decay modes also have contributions from the new states, but they provide less stringent bounds than the $K_L - K_S$ systems.(10)

A novel feature of supersymmetric models is that they may generate lepton number violating processes through the mixing of the sleptons. The graph of Fig. (1b) give rise to the process $\mu \rightarrow e\gamma$. Currently the upper limit on $(\mu \rightarrow e\gamma)/\Gamma(\mu \rightarrow e\nu\bar{\nu})$ is $2\text{x}10^{-10}$ which leads to the limit

$$\frac{1}{(m^2_{\tilde{l}} \text{ or } m^2_{\tilde{B}})}\ \frac{\Delta m^2_{\tilde{l}}}{m^2_{\tilde{l}}} < 0(10^{-7})\text{GeV}^{-2} \qquad (31)$$

Once again this condition can be met in the models above, but, because slepton masses are usually less than squark masses, it may be that the rate $\mu \rightarrow e\gamma$ is close to the current limit.

1.7 R symmetry and R parity (11)

In addition to the usual symmetries, supersymmetric theories may have new "R" symmetries. We note first that the supersymmetric Lagrangian was formed by either a $\Theta\Theta\bar{\Theta}\bar{\Theta}$ projection, eq.(1), or a $\Theta\Theta$ projection, eq.(6). The first is invariant under the "R symmetry" characterised by the angle β given by

$$\Theta \to e^{-i\beta}\Theta$$

$$P(\Phi) \to e^{2i\beta}P(\Phi) \tag{32}$$

$$V \to V$$

The form of the superpotential P determines whether there is an R symmetry for a general β, other than the discrete values $\beta = 0,\pi$. The value $\beta = \pi$ is always a symmetry of the Lagrangian and corresponds to fermion number conservation since (cf. eqs.(2) and (3)) under such a rotation fermion fermion fields change sign. L will be invariant under a more general R symmetry, characterised by β, provided we can find a choice of phase rotations for the chiral fields Φ_i

$$\Phi_i \to e^{in_i\beta}\,\Phi_i \tag{33}$$

under which all components of $P(\phi)$ have the transformation of eq.(32) The transformation of the component fields under R symmetry are[11]

$$V_\mu{}^a \to V_\mu{}^a \;;\; \lambda_\alpha{}^a \to e^{i\beta}\,\lambda_\alpha{}^a \;;\; D^a \to D^a$$

$$\phi^i \to e^{in_i\beta}\,\phi^i \;;\; \psi^i \to e^{i\gamma_5(n_i-1)\beta}\,\psi^i \;;\; F \to e^{i(n_i-2)\beta}F \tag{34}$$

Clearly R symmetry forbids a gaugino mass term of the form $\lambda^a\lambda^a$. This must be broken if gluinos are to acquire a mass consistent with current bounds and in the next sections we discuss how this comes about. However, even if R symmetry is broken there may remain unbroken discrete symmetries. For example, the superpotential of eq.(8) is invariant under "matter parity"

$$\psi^{(i)} \to -\,\psi^{(i)} \;;\; i = q,\ u^c,\ d^c,\ \ell,\ \ell^c$$

$$H_{1,2} \to H_{1,2} \tag{35}$$

Such a symmetry is clearly desirable as it forbids terms such as in eq.(18) which could violate lepton and baryon number and for this reason almost all supersymmetric models are matter parity invariant. Combining matter parity and the residual, unbroken, R symmetry with $\beta = \pi$ gives a symmetry known as "R parity" under which

$$\nu_\mu,\ q,\ \ell,\ H^o \xrightarrow{R_\pi} \nu_\mu,\ q,\ \ell,\ H^o_{1,2}$$

$$\lambda,\ \tilde{q},\ \tilde{\ell},\ \tilde{H}_{1,2} \xrightarrow{R_\pi} -\lambda,\ -\tilde{q},\ -\tilde{\ell},\ -\tilde{H}^o_{1,2}$$

Thus all the new supersymmetric states are odd under R parity. This has a profound effect on the phenomenology of such theories for then the new states will only be produced and decay in pairs. However, R parity is by no means an essential ingredient of supersymmetric theories and can be broken[5,12] by the judicious inclusion of terms such as are given in eq.(18) or, perhaps more plausibly, spontaneously by a scalar neutrino, eg. $\tilde{\nu}_\tau$, acquiring a vacuum expectation value (vev). We will briefly discuss the differences expected in section (1.9).

1.8 Symmetry breaking in supersymmetric models

In order to build a phenomenologically acceptable model, it is necessary to break both the electroweak gauge symmetry, SU(3)xSU(2) SU(3)xSU(2)xU(1) $\rightarrow$ SU(3)xU(1)$_{em}$, and the supersymmetry. The latter is necessary to split the degeneracy between squarks and quarks, between sleptons and leptons and between gauginos and gauge bosons.

Electroweak breaking proceeds via the Higgs scalars, H_1^o and H_2^o acquiring vevs. We will discuss in section (3.6) how this comes about; for the purposes of discussing the phenomenology we can assume they do.

There are some novel features in the spontaneous breaking of supersymmetry compared to the spontaneous breaking of Lie groups[1]. It is straightforward to calculate the Hamiltonian, H, in terms of the supersymmetry generators Q_α and $Q_{\dot{\alpha}}$

$$H = P_o = \tfrac{1}{4} \sum_{\alpha,\dot{\alpha}=1,2}^{2} (Q_\alpha^{\ 2} + Q_{\dot{\alpha}}^{\ 2}) \tag{37}$$

Thus the energy of a state is positive semidefinite. This immediately implies that global supersymmetry is spontaneously broken if, and only if, the ground state of the theory has non-zero energy since the broken case requires $Q_{\alpha,\dot{\alpha}}|o\rangle \neq 0$ for some $\alpha,\dot{\alpha}$. From eq.(7), we see H can be non-zero either through the F or the D term acquiring non-zero vevs[13]. The latter requires an Abelian factor in the group (in addition to the U(1) of the standard model) and cannot be Grand Unified in a simple group. For this reason, most models use the F term to break supersymmetry. It is straightforward to construct superpotentials involving new chiral superfields which lead to a non-zero F term. To construct realistic supersymmetric models, it is necessary then to couple these fields and the supersymmetric breaking to the conventional sector of Table 1. Various models have been constructed in which this is done by Yukawa, gauge or gravitational interactions. The latter has been exhautively explored recently[14,15,16] and we will concentrate on it here. Gravitationally induced masses are typically or order $E_{vac}^{\ 2}/M_{planck}$ where $M_{planck}(10^{19}$ GeV) is the Planck mass and E_{vac} is the vacuum potential energy induced by the supersymmetry breaking sector. For significant effects the supersymmetry breaking scale, E_{vac}, must be large, of order 10^{11} GeV.

If supersymmetry is made a local symmetry, one automatically obtains gravitational interactions for the supersymmetry algebra includes the Poincare group. The (N=1) locally supersymmetric gauge theory modifies the Lagrangian given by eq.(7) in the globally supersymmetric case. The potential in the minimally coupled case, may be written as[14]

$$V(\phi,\bar{\phi}) = e^{\sum_i|\phi_i|^2/M^2}\{\sum_j\left|\frac{\partial P}{\partial\phi_j} + \frac{\phi_j^* P}{M^2}\right|^2 - \frac{3|P|^2}{M^2}\} + \tfrac{1}{2}\sum_\alpha D_\alpha^* D^\alpha \qquad (38)$$

where ϕ_j are the chiral superfields in the theory, P is the usual superpotential, D the usual D term, and M is related to the Planck mass

$$M = \frac{1}{\sqrt{8\pi}} M_{Planck} = 2.4\times10^{18}\ \text{GeV}$$

In the limit $M \to \infty$, V reduced to the globally supersymmetric form, eq.(7).

Supersymmetry is broken if

$$[e^{\sum_i|\phi_i|^2/M^2}\sum_j\left|\frac{\partial P}{\partial\phi_j} + \frac{\phi_j^* P}{M^2}\right|^2 + \tfrac{1}{2}\sum_\alpha D_\alpha^* D^\alpha] \equiv \mu^4 \text{ is non-zero, but, due to} \qquad (39)$$

the extra term in eq.(38), and in contrast to the globally supersymmetric case, it is possible for this to happen while V remains zero. It is usual, in building supergravity models, to exploit this possibility and keep V zero because this corresponds to the desirable case of zéro cosmological constant. If this is done the P itself must have non-zero vacuum expectation value. This is related to the gravitino mass

$$m_{3/2} = \frac{\langle P\rangle}{M^2}\exp\left(\tfrac{1}{2}\frac{\sum|\langle\phi_i\rangle|^2}{M^2}\right) \qquad (40)$$

where $\langle P\rangle$ and $\langle\phi_i\rangle$ are the vacuum expectation values of P and ϕ_i

Now we can ask about symmetry breaking in the gauge nonsinglet sector. The interesting thing is that there are terms in eq.(38) which, at tree level, give a common mass to the scalar fields. Expanding the term $\left|\frac{\partial P}{\partial\phi_j} + \phi_j^*\frac{P}{M^2}\right|$, we find, using eq.(40) the scalar mass term $m_{3/2}|\phi_j|^2$. From eq.(38) we can see that, for zero cosmological constant, $\sqrt{3}\,\frac{\langle P\rangle}{M} = \mu^2$, where μ is the supersymmetry breaking scale. Thus $m_{3/2} \approx \frac{\mu^2}{M}$ and gravitational corrections, as anticipated above, give a common (supersymmetry breaking) mass to the scalar fields of magnitude $m_{3/2}$.

To be more specific in giving the predictions of locally supersymmetric N=1 theories, we need to specify the supersymmetry breaking mechanism. We will discuss the simplest class of model in which supersymmetry is broken in the gauge singlet sector and is only communicated to the gauge nonsinglet sector by gravitational effects[16].

The choice of a suitable supersymmetry breaking potential is rather arbitrary, particularly in the heavy sector. One simple form that has been analysed in detail is the superpotential

$$P = p(Z) + g(y) \tag{41}$$

where p (Z) involves only singlet fields, Z, and g (y) involves the gauge nonsinglet fields of the standard model. Let us assume that the form of p and g are chosen so that, at the minimum of the potential, some fields acquire a vacuum expectation value of the form

$$\langle Z_i \rangle = b_i M$$

$$\left\langle \frac{\partial p}{\partial Z_i} \right\rangle = a_i \mu M$$

$$\langle p \rangle = M^2$$

$$\langle y_a \rangle = 0 \tag{42}$$

The condition the cosmological constant vanish at the minimum is

$$\sum_i |a_i + b_i|^2 = 3 \tag{43}$$

The effective low energy potential is found by keeping those terms nonvanishing in the limit $M \to \infty$

$$V = |\tilde{g}_{,i}|^2 + m_{3/2}{}^2 |y_{,i}|^2 + m_{3/2}{}^2 (A\tilde{g} + \text{h.c.}) + \tfrac{1}{2} D_\alpha D^\alpha \tag{44}$$

where

$$m_{3/2} = \exp(\tfrac{1}{2}\sum_i |b_i|^2)\mu$$

$$\tilde{g} = \exp(\tfrac{1}{2}\sum_i |b_i|^2) g$$

and

$$A = b_i^*(a_i + b_i) \tag{45}$$

The scalar potential of eq.(44) shows that the gravitational corrections couple supersymmetry breaking from the singlet "hidden" sector to the nonsinglet sector in a simple manner, through soft supersymmetry breaking terms of $O(m_{3/2})$. Every scalar field obtains a common mass $m_{3/2}$ and there are further soft terms contributing to the scalar potential proportional to the superpotential of the nonsinglet sector, with all fields replaced by their scalar partners. The matter fermions do not acquire a mass from gravitational corrections. Gauginos can acquire a mass of order $m_{3/2}$ if the gauge kinetic energy is non-minimal[14]. It is not related to the scalar masses and can be much smaller.

The simplest example of a suitable superpotential (the Polonyi potential[17]) is to choose for p(Z)

$$h = \lambda Z + \Delta \qquad (46)$$

In this case $\left|\frac{\partial P}{\partial Z} + \frac{Z^* P}{M^2}\right|$ is non-zero so supersymmetry is broken with

$$A = 3 - \sqrt{3} \qquad (47)$$

in eq.(44).

To summarise, the supergravity potential contains terms that give mass to all supersymmetric partners of quarks and leptons, and to the Higgs scalars. Gauginos may also have a mass at tree level of $O(m_{3/2})$ or, if for some reason this is absent, they will be generated in radiative order[9], as the constant term in eq.(46) violates the R symmetry forbidding their masses (see section (1.7))

1.9 Phenomenology of the minimal model

The most direct test of supersymmetry is to observe directly one of the many new states predicted by the theory. In section 1.3, we introduced the multiplets needed to build the basic SU(3)xSU(2)xU(1) [SU(3)xSU(2)xU(1)] x (N=1 supersymmetry) model. The various Grand Unified versions of the theory all have this low energy structure, with the possible addition of a light singlet field. In fact most of the supersymmetric models that have been considered, whether global or local, have the same low energy spectrum. What differs between these models is the pattern of masses. In particular, different models have as the lightest supersymmetric particle (LSP) the photino, the gravitino of the Higgsinos. In most models there is a multiplicative R parity conserved, where R is +1 for conventional hadrons and -1 for the new supersymmetric states (see section (1.7)). In models with an unbroken R parity, the new states may only be produced in pairs and, once produced, a new supersymmetric state will ultimately decay into the lightest such state which will be stable. Their decay patterns will thus depend sensitively on the identity of the lightest new state. As we have discussed, it is also possible to construct models in which the R parity is broken, allowing for single production of the new states and their decay into convention states only.

For the case of theories with a conserved R parity, the phenomenology, for various choices of LSP, has been extensively discussed[2,3]. The $R_\pi$odd particles have definite couplings to conventional states (cf. section (1.4)) and it is straightforward to discuss their production and decay. For example, squarks and sleptons can be produced via $e^+e^- \to \tilde{q}\tilde{q}, \tilde{\ell}\tilde{\ell}$. The sleptons are expected to decay via

$$\tilde{\ell} \to x + \ell$$

where x may be the photino, gravitino or higgsino depending on their masses. The current experimental bound on the process $e^+e^- \to \ell^+\ell^- + 2x$ places a lower limit of about 20 GeV on the slepton mass and the equivalent processes involving squarks yields $m_{\tilde{q}} > 15$ GeV, the difference being that squarks may decay into gluinos giving a more complicated final state. We do not have time here for a full discussion of the supersymmetric phenomenology so let me concentrate on a topic not widely discussed namely the different phenomenology one might expect in models without an R_π parity[12]. For definiteness, and since it is in any case most interesting, suppose R parity is broken spontaneously through the tan sneutrino acquiring a vev, $\langle \nu_\tau \rangle \neq 0$. In this case, the physical τ^- is a mixture of the original τ^- together with the wino w^- and Higgsino H_2^-. In realistic models, it turns on the τ_L^- mixes predominantly with H_2^- while the τ_R^+ is mainly unmixed. Also the physical ν_τ is a mixture of the current ν_τ and H_2^o and there is a new neutrino-like state ν_N. Remarkably, the forward backward symmetry in $e^+e^- \to \tau^+\tau^-$ and the τ lifetime are unchanged. However, since τ and ν_τ are mixtures of $R_\pi$even and $R_\pi$odd states, new supersymmetric states may decay singly. For example squarks (sleptons) may decay into quarks (leptons) plus τ or ν_τ, and the photino can decay into $\nu_\tau(e^+e^-$ or $\mu^+\mu^-$ or $\tau^+\tau^-$ or $q\bar{q}$ or $\nu_e\nu_e$ etc.)

Single production of $R_\pi$odd states is also possible. For example, the graphs of Fig. 2 would allow for the single production of gluinos or squarks. The production rate is suppressed by Yukawa couplings, but for a top quark of O(40 GeV) the suppression factor is not small and the production of $q\tilde{q}\nu_\tau$, for example, could be appreciable.

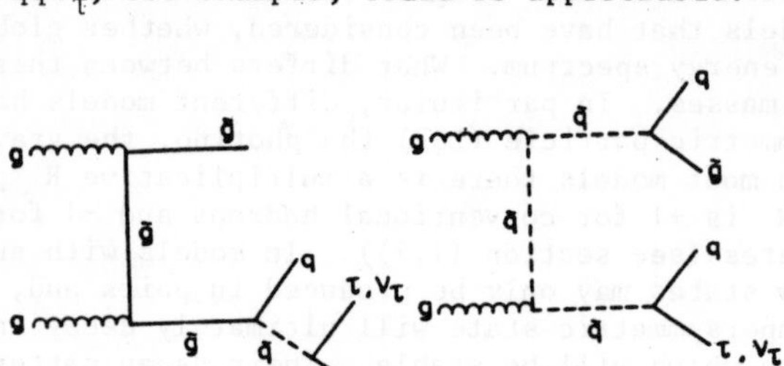

Fig. 2 Typical graphs leading to single production of R_π odd states (a) gluino production (b) squark production

2.1 Grand Unification and Super-Grand Unification

The basic idea of grand unification is that the standard model, based on the gauge group SU(3) x SU(2) x U(1), is part of a larger semisimple gauge group G with a single gauge coupling constant (or, possibly, a product of such semisimple groups with a discrete symmetry relating these groups)[18]. G is spontaneously at a scale, M_x, to the standard model giving, to order $\frac{\mu^2}{M_x^2}$, where μ^2 is a typical hadronic mass scale, the standard low energy phenomenology. The best known example is that invented by Georgi and Glashow with G = SU(5) with the breaking pattern

$$G = SU(5) \underset{M_x}{\rightarrow} SU(3) \times SU(2) \times U(1) \underset{M_W}{\rightarrow} SU(3) \times U(1)_{em} \tag{48}$$

The multiplet structure needed for the simplest version of SU(5) is shown in Fig. 3. It is minimal in the sense that the fewest number of scalar representations are used, although it should be said that there is no symmetry principle requiring this and there are reasons to suppose that the minimal theory is already sick because of the predicted fermion mass spectrum.

PARTICLE CONTENT	SPIN
$A_\mu^{a=1--24} = \underline{24}$ $W_\mu^\pm Z_\mu \gamma_\mu\ \bar{A}_\mu^{a=1...8},\ x_\mu^{2/3},\ y_\mu^{1/2}$	J = 1
+ 2 Generations	
$\begin{bmatrix} \bar{d} \\ \bar{d} \\ \bar{d} \\ e^- \\ \nu_e \end{bmatrix}_L = \underline{\bar{5}} \qquad \begin{bmatrix} 0 & \bar{u} & -\bar{u} & u & d \\ -\bar{u} & 0 & \bar{u} & u & d \\ \bar{u} & -\bar{u} & 0 & u & d \\ -u & -u & -u & 0 & e^+ \\ -d & -d & -d & -e^+ & 0 \end{bmatrix} \frac{1}{\sqrt{2}} = \underline{10}$	J = ½
$\bar{H} = \begin{bmatrix} \phi^{1/3} \\ \phi^{1/3} \\ \phi^{1/3} \\ \phi^- \\ \phi^0 \end{bmatrix} = \underline{\bar{5}} \quad + \quad \Sigma_j^i = \underline{24}$	J = 0

Fig. 3 Particle content of minimal SU(5) theory.

One problem arises in implementing the breaking schemes such as is given in eq.(48). To inhibit proton decay, the scale M_x is often very large ($0(10^{15}$GeV) in SU(5)), much larger than the electroweak scale $M_W = 0(10^2$GeV). Unless one requires unnatural cancellations on the theory radiative corrections will give $\frac{M_W}{M_X} \geq 0(\alpha_G)$, where α_G is the grand unified fine structure constant which is typically $> 0(10^{-2})$. This is clearly inconsistent with the required ratio $\frac{M_W}{M_X} \leq 0(10^{-13})$ giving rise to the hierarchy problem of GUTs. The best solution is to increase the symmetry of the system, the new symmetry preventing the feed through of the large mass scale, M_x, to the electroweak scale, M_w. The only symmetry which can do this is supersymmetry giving rise to supersymmetric grand unified theories (or SUSY GUTs). The SUSY Guts constructed so far are based on the direct product structure $G^{SUSY} = G \times$ [N extended supersymmetry]. In this lecture I will discuss the construction and phenomenology of the simplest SUSY GUT.

2.2 A supersymmetric version of SU(5)[20]

To construct an SU(5) x [N=1 supersymmetry] SUSY GUT we need to assign the states of Table 1 to N=1 supermultipets. Following the reasoning of section (1.3) leads immediately to the multiplet structure of Table 2.

Table 2: Supermultiplets used in SU(5) SUSY GUT a=1,...N_g is a family index and α,β are group indices.

Role	Notation	SU(5) Representation Content
Matter	$\psi_a^\alpha, \chi_a{}^{\alpha\beta}$	$N_g \times (\bar{5} + 10)$ (chiral supermultiplet)
Higgs	H_1^α, H_2	$(\bar{5} + 5)$ (chiral supermultiplet)
	Σ_α^β	24
Vector	V^α	24 (vector supermultiplet)

The Yukawa couplings needed to give all charged fermions a mass has a superpotential of the form

$$P_{5_M} = \frac{1}{\sqrt{2}} M_{ij}{}^{(d)} \psi_{i\alpha} \chi^{\alpha\beta} H_{2\beta} - \frac{1}{4} M_{ij}{}^{(u)} \varepsilon_{\alpha\beta\gamma\delta\rho} \chi_i^{\alpha\beta} \chi_j^{\gamma\delta} H_1^\rho \tag{49}$$

where H_1 and H_2 are (left handed) chiral superfields transforming as 5 and $\bar{5}$, respectively, under SU(5) and i and j are family indices. It is necessary that H_1 and H_2 be distinct chiral supermultiplets, and not hermitian conjugates, for supersymmetry does not allow us to build the superpotential with a chiral superfield and its conjugate. It is also impossible to identify H_2, transforming as a $\bar{5}$, with one of the $\bar{5}^s$ introduced to describe the quark and lepton secotr. In addition to the reasons given in section (1.3), there is an even more pressing one for the colour triplet components of H_2 mediate proton decay and must have very large mass ($\gtrsim 0(10^{10}$ GeV)), if the proton is not to decay too quickly. If we identify the doublet components of H_2 with the sneutrino, selection members of a $\bar{5}$, then the triplet components will be partners of the down antisquarks. Since, as we will shortly show, supermultiplet splitting in a gauge nonsinglet representation is < $0(1\text{TeV}^2 \times \frac{\pi}{\alpha})$, the triplet components will mediate proton decay far too fast. Consequently, to accommodate the necessary Higgs scalars, we must choose two new chiral supermultiplets H_1 and H_2 transforming as a 5 and $\bar{5}$ respectively. We must also ensure that the H_1 and H_2 multiplets split so that their triplet components are heavy ($\gtrsim 10^{10}$ GeV)^2 while leaving their doublets light, 0(1TeV). We postpone a discussion of the mechanisms generating this multiplet splitting until section (3.5). In addition to the multiplets introduced above, it is necessary to add an ajoint chiral supermultiplet, Σ, to accommodate the adjoint of Higgs scalars necessary to break SU(5) to SU(3)xSU(2)xU(1).

The final multiplet choice for our SU(5)x N=1 supersymmetry model is given in Table 2. It will, of course, be necessary to break the supersymmetry, and this requires further fields. We will discuss this later.

2.3 The Hierarchy problem in supersymmetry

The hierarchy problem is the difficulty in keeping a low electroweak scale, M_W when grand unification occurs at a scale $M_X \gg M_W$ (19). Although we have not yet discussed the mechanism responsible for electroweak breaking, we can still ask whether a low scale for M_W will be stable against the radiative corrections.

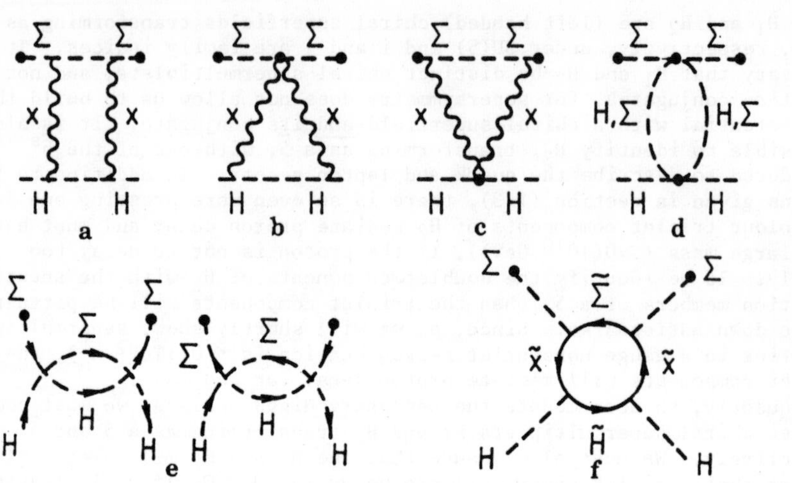

Fig. 4 Graphs contribution to Higgs doublet masses in supersymmetric SU(5).

The dangerous terms are of the form $\lambda H^{+}H\ Tr(\Sigma^{+}\Sigma)$ which generate a large doublet mass. We may avoid complications concerning the renormalisation conditions by performing a finite calculation relating the Higgs mass at two momentum scales

$$m_H^2(p^2) = m_H^2(M_X^2) + [\lambda(p^2) - \lambda(M_X^2)] < Tr(\Sigma^{+}\Sigma) > \tag{50}$$

where the graphs of Fig. (5) generate a coupling, , which is logarithmically dependent on the momentum scale. Since the vev $<Tr(\Sigma^{+}\Sigma)>$, is of order M_X^2 the condition that $m_H^2(1\ Tev^2)$ should be $< 0(1\ TeV^2)$ (so M_W is < 100 GeV) can only be achieved if $m_H^2(M_X^2)$ is $0(M_X^2)$ and there is a delicate cancellation of the terms on the right hand side. This is the hierarchy problem - why should physics at the microscopic scale be complicated ($m_H(M_X) \simeq 10^{15}$ GeV) in order to achieve simplicity at the macroscopic scale ($m_H(1\ TeV) \simeq 0$)?

The remarkable thing is that λ is zero in supersymmetric theories. This is a manifestation of the nonrenormalisation theorem(6) which says terms in the superpotential are not radiatively corrected and which comes about through a cancellation of the graphs involving boson and fermion loops. Thus supersymmetric theories avoid the hierarchy problem (an improvement on ordinary non-supersymmetric SU(5) for which λ is non -zero, because there are no cancelling fermion contributions).

2.4 Supersymmetry breaking and the hierarchy problem

If supersymmetry were unbroken, there would be scalar leptons and scalar quarks degenerate with the chraged leptons and quarks as well as gauginos degenerate with the gauge bosons. No such states have been observed, so supersymmetry must be broken. However, once the degeneracy between such states is broken, the cancellation between the graphs of Fig. 4 is spoilt by the difference in masses in the propagators. Evaulating the graphs of Fig. 4 gives

$$m_H^2(\mu^2) = m_H^2(M_X^2) + \frac{\alpha}{\pi}(m_X^2 - m_X^2) \tag{51}$$

Now, the condition that $m_H(\mu)$ should naturally be less than O(1 TeV), at $\mu \approx 1$ TeV, puts a bound on the mass splitting with a vector supermultiplet

$$(\tilde{m}_X^2 - m_X^2) < O\left(\frac{(1\ \mathrm{TeV})^2\pi}{\alpha}\right) \tag{52}$$

Similarly, if the mass splitting between supersymmetric partners in a chiral supermultiplet is Δm_H^2, the condition that the contribution from the loops similar to Fig. 4 do not contribute unacceptably to m_H is

$$\Delta m_H^2 < O\left(\frac{(1\ \mathrm{TeV})^2\pi}{\beta}\right) \tag{53}$$

where β is the coupling of the chiral supermultiplet to the Higgs doublet sector.

Thus, although supersymmetry can solve the hierarchy problem, it does so only if the masses of the super partners are so low that they are accessible to experimental discorvery. For this reason, there has been considerable effort to try to construct viable theories, and to determine the expected spectrum of states and their phenemonology.

2.5 Phenomenology of SUSY-GUT^s

Grand unification requires that the three couplings constants of the standard model be related to a single, grand unified coupling. In SU(5) the relation is

$$g_3 = g_2 = \sqrt{\frac{5}{3}}\, g_1 = g_5 \tag{54}$$

The couplings, however, depend on the scale, μ^2, at which they are measured for radiative corrections, such as shown in Fig. 5 cause them to "run".

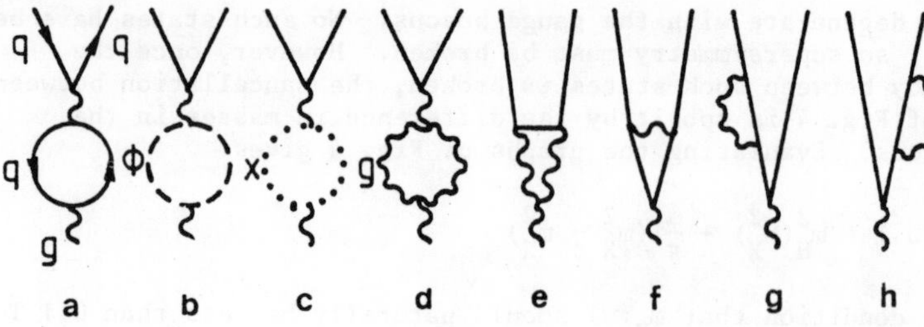

Fig. 5 Diagrams contributing to the running of gauge couplings: solid lines are fermions, wavy lines gauge bosons, dashed lines scalars and dotted lines ghosts.

In leading order the renormalisation group summation gives for the running couplings

$$\alpha_i^{-1}(\mu^2) \equiv \frac{g_i^2(\mu^2)^{-1}}{4\pi} = \alpha_i^{-1}(M_X) + \frac{1}{6\pi} b_i \ln \frac{M_X}{\mu} \tag{55}$$

where $\alpha_i^{-1}(M_X)$ are given by eq.(1) and the β_i are obtained from the graphs of Fig. 5

$$b_3 = 33 - 4N_g \quad ; \; b_2 = 22 - 4N_g \quad ; \; b_1 = 4N_g \tag{56}$$

Plotting α_i v/s μ^2 gives the graph of Fig. 6.

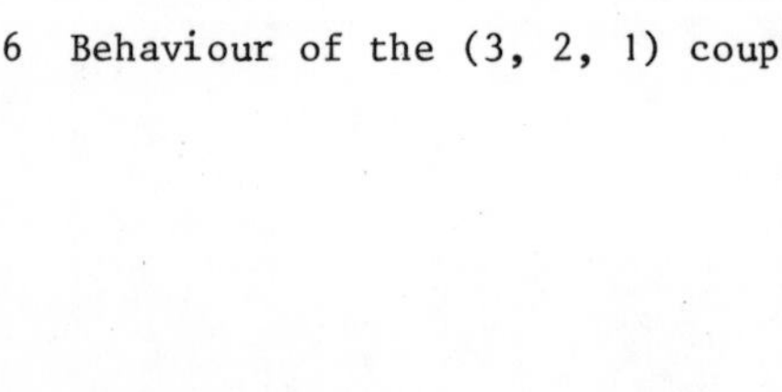

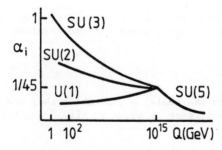

Fig. 6 Behaviour of the (3, 2, 1) couplings with energy

Remarkably, it is found that the values of $\alpha(\mu)$ as measured in the laboratory are consistent with eq.(54) provided[21]

$$M_X = 2.4^{+2.8}_{-1.6} \times 10^{14}\ \text{GeV} \quad \text{for} \quad \Lambda_{\overline{MS}} = 0.16^{+0.1}_{-0.08}\ \text{GeV}$$

$$\sin^2\tilde{\Theta}_W(80) = \frac{g_2^2}{g_1^2 + g_2^2}(80) = 0.212^{-0.004}_{+0.006} \tag{57}$$

(These results came from a more exact two loop calculation of the running couplings). The weak angle $\sin^2\tilde{\Theta}_W$ is evaluated in the MS regularisation scheme (important when calculating beyond the leading order) and for comparison the experimental value for the same quantity is

$$\sin^2\tilde{\Theta}_W(80)\Big|_{Expt} = 0.215 \pm 0.015 \tag{58}$$

in remarkable agreement with SU(5) prediction.

What happens in the SU(5) SUSY-GUT. Because it involves new light supersymmetric particles (see Table 1) it is necessary to recompute the values for the grand unified mass and $\sin^2{}_W$. The initial ratios of the couplings, eq.(54), are unchanged since we still have the same assignment of states to representations of SU(5). However, the radiative corrections will differ for now, in Fig. 5, we must include loops containing the new states. The values of b_i are easily calculated, including the new light supersymmetric partners of the quarks, leptons and SU(3) x SU(2) x U(1) gauge bosons in the graphs of Fig. 5. This gives[22]

$$b_3 = 27 - 6N_g \quad ; \; b_2 = 18 - 6N_g \quad ; \; b_1 = 6N_g \tag{59}$$

Including two loop effects[23] gives the predictions (for Λ_{MS} = 150 Mev)

$$M_X = 7.7 \times 10^{15}\ \text{GeV}$$

$$\sin^2\Theta_W = 0.233 \tag{60}$$

The reason M_X has increased is principally because the octet of gluinos reduces b_3 in eq.(59) relative to the SU(5) value in eq.(56) thus slowing the evolution of α_3 and postponing the crossover of α_3 and α_2 (cf. Fig. 6). It is possible to change the prediction for M_X by changing the rate of evolution of α_2; for example adding two further light Higgs SU(2) singlets (mimicing the structure of a lepton family) brings M_X back to the usual SU(5) value and lowers $\sin^2\Theta_W$ by 0.215[22]. Both values for $\sin^2\Theta_W$ are consistent with the experimental value of eq. (58).

Quark and lepton masses are predicted as in SU(5) to have the value at the scale $M_X = m_b = m_\tau$; $m_s = m_\mu$; $m_d = m_e$ at $\mu = M_X$.

Radiative corrections change this at low scales and, surprisingly, the predictions are very similar for SUSY SU(5) and ordinary SU(5)[23, 24]

$$m_b \approx 3m_\tau \; ; \; m_s \approx 3m_\mu \; ; \; m_d \approx 3m_s \quad \text{at} \quad \mu \approx 1 \text{ GeV}.$$

The first relation is in good agreement with experiment but the latter two are not and imply there must additional structure beyond the minimal SU(5) introduced above.

2.6 Nucleon decay

The most dramatic prediction of SU(5) and most GUT^s is that the proton will decay via the graphs of Fig. 7a, through the exchange of X or Y bosons, the new gauge bosons in SU(5) in addition to those in SU(3) x SU(2) x U(1). Evaluation of these graphs gives the dominant contribution

$$\frac{1}{4} L_{B=1} = \frac{g^2}{8M_X^2} \varepsilon_{ijk} \bar{u}^{c\,k}_L \gamma_\mu u^j_L \; \bar{e}^+_L \gamma^\mu d^i_L \tag{61}$$

To compute the estimated proton decay lifetime one must estimate the overlap of the quark states appearing in eq.(61) with the asymptotic hadronic states, the proton etc. This gives for the dominant decay mode (including radiative corrections)[25]

$$\tau_p \to e^+\pi^o = 5\text{x}10^{29\pm1.7} \text{ years} \tag{62}$$

a result inconsistent with current limits.

In supersymmetric SU(5) one might think that the proton lifetime will be invisibly long for, from eq.(61), $\tau_p \, \alpha M_X^4$ and M_X in SUSY SU(5) is (cf. eq.(57) and (60)) $\approx$ 40 times the SU(5) value. However, in SUSY SU(5) there are new processes involving squarks which mediate nucleon decay[4],[26]. These are shown in Fig. (7b) (plus graphs with d and s interchanged on external states).

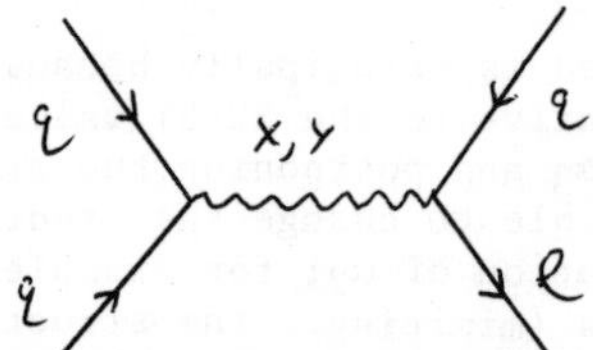

(a)

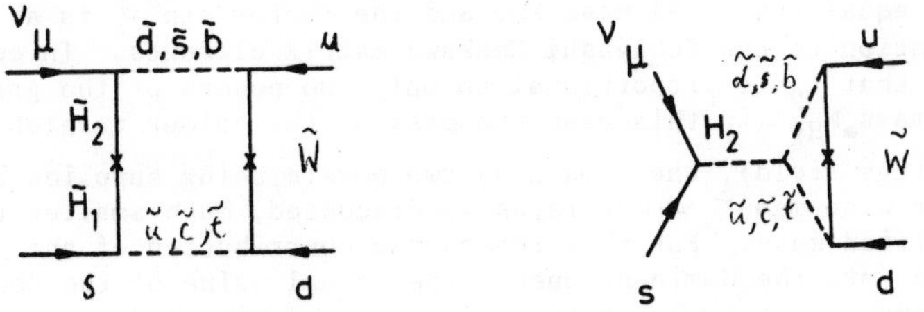

(b)

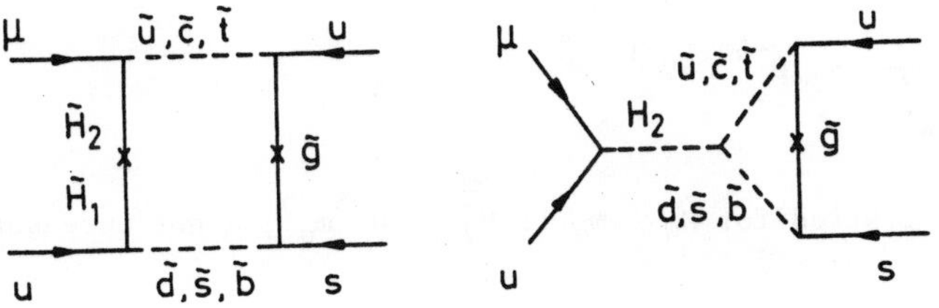

(c)

Fig. 7 Graphs contributing to proton decay (a) in GUTs. (b),(c) In SUSY-GUTs. The x denotes a mass insertion.

Consider first the graphs of Fig. 7b. Because the graphs involve Higgsino exchange whose couplings are proportional to masses (cf. eqs. (7) and (8) the dominant graph involves a strange quark and hence the expected proton decay mode in SUSY SU(5) is $p \to K^+\nu$ rather than $p \to e^+\pi^o$ in SU(5). Evaluating these graphs and estimating the quark matrix elements gives(27)

$$\tau^{-1}_{p\to K^+\bar{\nu}} = 4\left|\frac{m_c\, m_p\, m_p^2}{2v_1 v_2 M_{H_X}}\right|^2 b^{o2}\sin^4\Theta_c \times 10^{27}\ \text{years}^{-1}$$

where

$$b^o = \frac{g_2^2 M_2^{\tilde{W}}}{512\pi^2}\left(\frac{1}{M_{\tilde{q}}^2 \text{ or } M_{\tilde{W}}^2}\right) \tag{63}$$

Here $M_2^{\tilde{W}}$ is the supersymmetry breaking component of the Wino mass which need not equal its total mass M_W, and the factor $\sin^4\Theta_c$ is an approximation to the Kobayashi Maskawa matrix elements. In eq.(16) it is clear that t_p is proportional to only two powers of the grand unified mass M_{H_X} (in this case the mass of the colour triplet component of the Higgs field), the remaining two powers being supplied by the squark or Wino mass^2 which is, as we discussed, much smaller than the grand unified mass. For this reason the contribution of the graphs of Fig. 7b,c are the dominant ones. The actual value of the decay lifetime depends sensitively on the values of the wino and squark masses. For a lifetime $\tau_p \geq 10^{31}$ years consistent with current limits

$$b^o \approx \frac{1}{3} \times 10^{-8}\ \mathrm{GeV}^{-1}$$

which implies

$$\frac{M_2^{\tilde{W}}}{(m_{\tilde{q}}^2 \text{ or } M_{\tilde{W}}^2)} < 10^{-5}\ \mathrm{GeV}^{-1} \qquad (64)$$

which can be satisfied for $M_{\tilde{W}} < m_{\tilde{q}}$ if $M_2^{\tilde{W}} < 10^{-5} m_q{}^2$, a not unreasonable restriction.

What is the dominant decay mode in SUSY SU(5)? It is easy to check that the μ decay modes in the Wino dressed graphs are suppressed by a factor of $\frac{m_u}{m_c \sin\Theta}$ because the H_1 coupling is proportional to $m^{(u)}$ and μ decay is associated always with an external $Q = \frac{2}{3}$ quark leading to a m_u factor whereas $\bar{\nu}$ decay is associated with an internal $Q = \frac{2}{3}$ quark which can be the charm squark with a m_c factor. The gluino dressed graphs Fig. 7(c) can contribute to μ decay modes but, if the gluino does not change flavour and the squarks are degenerate the graphs cancel identically. However, the gluino can change flavour (see section 1.6) As a result the graphs of Fig. 7(c) contribute. The dominant contribution to $p \to \mu^+ K^o$ comes from an internal top squark with a factor m in the second of Fig. 7(c). Evaluating this diagram gives the relative rate[26]

$$\frac{\tau_p(\text{gluino exchange})}{\tau_p\ (\text{wino exchange})} \approx \left| \frac{m_c M_2^{\tilde{W}}\ \alpha_2}{m_t \sin^2\Theta_c \sin^2\tilde{\Theta}_c M_2 g^2 \alpha_3} \right|^2 \qquad (65)$$

where we have written $\sin^2\tilde{\Theta}_c = (U^+\tilde{U})_{23}$. Model dependent estimates suggest $\sin^2\tilde{\Theta}_c \lesssim \sin^2\Theta_c$ and $M_2{}^g/M_2{}^W \equiv \alpha_3/\alpha_2$. Using this we find the gluino exchange graphs may be dominant for $m_t \geq 40$ GeV. In this case, Figs.(7c) and (7d) suggest the dominant SUSY GUT nucleon decay modes will be $p \to \mu^+ K^o$, νK^+ and $n \to \nu K^o$

The absolute prediction for the proton lifetime through gluino dressed graphs depends on numerous poorly known parameters and is therefore itself poorly determined. With $M(\tilde{g}) = 100$ GeV, $m_{\tilde{u}} = 1$TeV and mixing angles $\tilde{U} = O(U)$ for m_t =40 GeV the proton lifetime is[28]

$$\tau_p = 6.10^{31} \text{ years} \tag{66}$$

This choice of parameters seems quite reasonable within the context of supersymmetric models and the result holds out some hope that proton decay may be visible to the experiments now running.

Supersymmetry can generate proton decay through new operators involving squarks and sleptons. These new states can also contribute to many other processes, for example in section (1.7) we discussed their role in strangeness violation and lepton-flavours violation. More general studies have been made of their role in n-$\bar{n}$ oscillation, ν masses etc., and we encourage the interested reader to refer to the extensive literature on the subject[29].

3.1 Supersymmetric models

We turn now to a discussion of specific SUSY-GUT models. In order to construct realistic supersymmetric models, it is necessary to introduce supersymmetry breaking to split the new states from their supersymmetric partners. Thus we must couple the supersymmetry breaking mechanism (either F type or D type) (see section (1.8)) to the light sector involving SU(3) x SU(2) x U(1) gauge nonsinglet fields. There are several different ways to do this and each leads to a different type of model.

(i) Couplings via gauge interactions. For D type breaking, the U(1) Abelian gauge group interactions couple the symmetry breaking to the various fields in the theory, and the supersymmetry breaking in a given supermultiplet is dependent on its U(1) charge.

Symmetry breaking models using F type breaking may be built using gauge non-singlet fields in the O'Raifeartaigh potential and then gauge interactions will, in higher order, couple the symmetry breaking to the light sector[30].

(ii) Coupling via Yukawa interactions. F type models with singlet fields, may have additional Yukawa couplings to gauge non-singlet fields. The supersymmetry breaking in a given supermultiplet is then proportional to its effective Yukawa coupling to the singlet fields[31].

(iii) Coupling via gravity [14,15,16] F (or D) type models, with supersymmetry breaking only in the gauge singlet sector, will still induce supersymmetry breaking in the non-singlet sector via gravitational corrections. This provides an attractive alternative to method (ii) as no, ad hoc, Yukawa coupling between the sectors need be introduced. Such gravitationally induced masses are typically of order

(E^2_{vac}/M_{Planck}), where M_{Planck} (10^{19} GeV) in the Planck mass. For significant effects, the supersymmetry breaking scale, E_{vac}, must be large, or order 10^{11} GeV. Of course, gravitational corrections may also be important in models with supersymmetry breaking in the non-singlet sector.

We will consider three types of model which illustrate the points discussed above.

3.2 F type models with Yukawa coupling[31]

Supersymmetry is broken if the F term is nonzero. This may be achieved by adding to the SU(5) theory discussed above three gauge singlet fields A,B,C with the superpotential (known as the O'Raifeartaigh potential)

$$P = \lambda_1(A^2 - M^2)B + \lambda_2 A^2 C \tag{66}$$

Using eq.(7) gives

$$V = |\lambda_1(A^2-M^2)|^2 + |\lambda_2 A^2|^2 + |2\lambda_1 B + 2\lambda_2 C|^2 \tag{67}$$

For no value of A is V zero, so supersymmetry is broken, but only in the A, B, C singlet sector at this stage.

To couple supersymmetry breaking to the gauge nonsinglet sector, we add to the superpotential a term

$$P_\phi = \lambda_3 \phi_1{}^a \phi_{3a}\ A + \lambda_4 AM^2 \tag{68}$$

where ϕ_1 and ϕ_2 transform as 5 and $\bar{5}$ under SU(5). The last term is to break R invariance, defined in section (1.7), which, if exact, forbids gaugino masses. It is easy to check that ϕ_1 and ϕ_2 do not acquire a vacuum expectation value. At tree level ϕ_1 and ϕ_2 do not couple to F_B or F_C, so they do not feel the supersymmetry breaking. However, beyond tree level, terms of the form $[\phi_1{}^+\phi_1 C^+ C]_D$ are generated via the graph of Fig. (8) which gives a scalar, supersymmetry breaking, mass

$$m_\phi{}^2 \approx \frac{\lambda_3{}^2}{8\pi^2}\frac{\langle F_C\rangle^2}{\mu^2} = \frac{\lambda_3{}^2}{32\pi^2}\mu^2 \quad (\text{for } \lambda_3 \ll \lambda_1, \lambda_2) \tag{69}$$

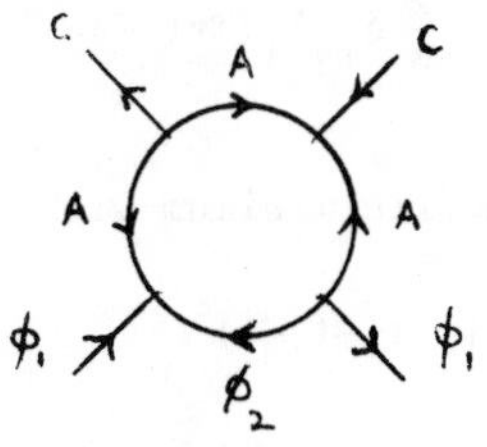

Fig. 8 Supergraph contributing to m_{ϕ_1}

Similarly, there may be contributions to masses of all the other gauge nonsinglet fields in the SU(5) theory occurring through their gauge couplings to ϕ_1 and ϕ_2 and hence to the supersymmetry breaking sector. This gives squarks, sleptons, Higgs scalars and gauginos a mass, just a as is required. On the other hand, quarks, leptons and gauge bosons remain massless at this stage because they require SU(2) x U(1) breaking to be non-zero. Thus it is naturally explained why we have not yet seen the new supersymmetric states - they are heavy $\simeq$ 0(1 Tev).

3.3 F type model with gauge couplings - the geometric hierarchy(30)

It is straightforward to build a version of the O'Raifeartaigh potential with gauge nonsinglet fields. We introduce the chiral supermultiplets Y and **Σ**, adjoints under SU(5), together with a singlet field X. The superpotential is

$$P = \lambda_1 \mathrm{Tr}(Y\Sigma^2) + \lambda_2(\mathrm{Tr}\Sigma^2 - \mu^2)X \tag{70}$$

The potential at tree level is

$$V = |F_X|^2 + \sum_{ij} |F_{Y_{ij}}|^2 + \sum_{ij} |F_{\Sigma_{ij}}|^2 + \frac{1}{2}\mathrm{Tr}|D|^2 \tag{71}$$

where

$$D = -g([\Sigma^*,\Sigma] + [Y^*,Y] + \text{quark, lepton and Higgs terms}) \tag{72}$$

F_X, F_Y and F_Σ are given (for a general SU(N)) group by

$$-F_Y^* = \lambda_2(\mathrm{Tr}\Sigma^2 - \mu^2)$$

$$-F^*_{Y_{ij}} = \lambda_1((\Sigma)^2_{ij} - \frac{1}{N}\delta_{ij}(\mathrm{Tr}\Sigma^2)) \tag{73}$$

$$-F^*_{\Sigma_i} = (2\lambda_1 Y\Sigma + 2\lambda_2 X\Sigma)_{ij} - \frac{2}{N}\delta_{ij}\lambda_1(\mathrm{Tr}(Y\Sigma)) \tag{73}$$

For SU(6) there is a supersymmetric minimum with

$$<\Sigma> \quad \frac{\mu}{\sqrt{6}} \times \text{Diagonal } (1,1,1,-1,-1,-1) \tag{74}$$

For SU(5) it is not possible to achieve the supersymmetric minimum, but the minimum requires $<\Sigma>$ be as close to the identity as possible

$$<\Sigma> = V \times \text{Diagonal } (2,2,2,-3,-3) \tag{75}$$

$$\text{with} \quad V = \frac{\lambda_2\mu}{(\lambda_1^2 + 3Q\lambda_2^2)^{\frac{1}{2}}} \tag{76}$$

Also at the minimum

$$<Y> = \frac{H_2}{\lambda_1} X_o \times \text{Diagonal } (2,2,2,-3,-3) \tag{77}$$

$$<X> = X_o$$

where X_o is undetermined at tree level. In contrast to the potential of eq.(10.94), SU(5) is uniquely broken in this scheme to SU(3)xSU(2)xU(1).

Radiative corrections to the effective potential give, for large X_o [30]

$$V_{1\ \mathrm{loop}} = \frac{\lambda_1^2\lambda_2^2\mu^4}{(\lambda_1^2 + 30\lambda_2^2)}\left[1 + \frac{\lambda_2^2}{\lambda_2^2 + \frac{\lambda_1}{30}}\left(\frac{29\lambda_1^2 - 50g^2}{80\pi^2}\right)\ell n\frac{X_o}{\mu^2}\right] \tag{78}$$

For $50g^2 > 29\mu_1^2$ this requires X_o should be larger than μ. The minimum of V can be found be resumming the large logs, using the renormalisation group, and thus X_o is determined in radiative order usually with $X_o \gg \mu$.

The symmetry breaking pattern in this model is easy to estimate qualitatively. The Σ field couples through the terms of eq. (71) to the non-vanishing F terms F_Y and F_X. Thus all components of Σ, apart from the hypercharge component which does not couple to these F terms, have supersymmetry breaking masses of order μ. On the other hand the potential V of eq. (71) generates supersymmetric mass terms for the Σ superfield of order $\lambda_2\lambda_o \simeq M_G$. This may be much larger than μ.

Supersymmetry in other sectors is communicated by radiative corrections. Let us denote by $\tilde{X}$ the combination of chiral superfields X and Y which has non-zero F term. Its fermion component is the goldstino, which in local supersymmetry becomes a component of the gravitino. The goldstino couplings to a multiplet are proportional to the supersymmetry breaking in that multiplet, so, to estimate the supersymmetry breaking in a given supermultiplet, it is sufficient to look for the possible terms generating goldstino couplings to that supermultiplet. There are only two possibilities for a radiatively induced coupling

$$[\tilde{X}^*\tilde{X}\phi_i\phi_j]$$

$$[\tilde{X}^*\tilde{X}\phi_i\phi_j^*]_D \qquad (79)$$

where ϕ_i and ϕ_j are general chiral superfields. These can generate scalar masses or order $\langle F_{\tilde{X}}\rangle^2/M^2$ where M is a mass term generated on evaluating the Feynman diagram. As these terms are given by graphs involving the Σ field we know these terms vanish as $M_\Sigma \to \infty$. A careful analysis shows that the mass parameter M^2 must be quadratic in M and so the scalar masses^2 are of order $\langle F_{\tilde{X}}\rangle^2/M_\Sigma{}^2$. From eq. (73) we see that this is $O(\mu^4/M_\Sigma{}^2)$. Thus multiplets not directly coupled to the $\tilde{x}$ field will obtain a supersymmetry breaking mass of at most $O(\mu^2/M_\Sigma)$. Since M_Σ is to be identified with the Grand Unified mass ($\approx 10^{16}$ GeV), this requires that the supersymmetry breaking scale μ^2 should be very large. Allowing for the fact that coupling constants are also needed, a scale for μ of $O((10^9 - 10^{11})$ GeV) is necessary to generate supersymmetry breaking masses of the order of 10^2 GeV in the light sector. This geometric relation between masses has led to the name geometric hierarchy for this type of model.

Squarks, slepton and Higgs scalars can get mass via gauge boson and gaugino exchanges. They give masses of order $\frac{\alpha}{2\pi}\frac{\mu^2}{M_G}$, where α is the relevant coupling; they are larger for coloured states than for uncoloured states. We note particularly that masses driven by gauge interactions are the same for squarks or sleptons of different families. The corrections to this come from graphs involving quarks and leptons and are proportional to their masses m_i. Evaluation of these graphs gives only a small mass splitting $\delta\mu_i^2$ between squarks and sleptons, in agreement with the bounds of section (1.7).

$$\frac{\delta\mu_i^2}{\mu_i^2} \approx m_i/(\mu^2/M_G) \qquad (80)$$

This feature is true also in the F type models discussed above.

3.4 Models with supersymmetry breaking communicated by gravity[14,15,16]

A particularly economical class of models have been constructed in which supersymmetry breaking is communicated via gravitational couplings in a manner analagous to that discussed in section (1.8) for the standard model. From eq. (44) we see that gravitational effects couple the supersymmetry breaking in the gauge singlet sector to the gauge nonsinglet sector giving a commong mass, $m_{3/2}$, at tree level to the scalar fields. If we allow non-minimal forms for the kinetic energy (ie. derivative terms of dimension greater than four suppressed by inverse powers of the Planck mass) gauginos may also acquire a common mass of $O(m_{3/2})$ at tree level. Even if they do not, gaugino masses will be generated in radiative order[9] as R symmetry is violated by the constant term in eq. (46).

3.5 SU(5) breaking and multiplet splitting

SU(5) may be broken to SU(3) x SU(2) x U(1) through an adjoint of scalar fields, Σ, acquiring a vev as in eq. (75). An example of a superpotential giving rise to this form is given in eq. (70); other (simpler) forms may easily be constructed[32]. One problem we must tackle is the need to split the Higgs multiplets, H_1 and H_2, so that the colour triplet components should be superheavy, while the doublet components remain light $\leqslant O(M_W)$. This is necessary to inhibit proton decay, while allowing for SU(2) x U(1) symmetry breaking (see section (2.2)).

The simplest way to achieve the splitting is by introducing the superpotential[20] (I is the unit matrix)

$$P_{5H} = H\,(m'I+\gamma_{24}\Sigma)H_2 \tag{81}$$

This gives terms in the low energy potential of the form

$$V = |H_1(m'I+\gamma_{24}\Sigma)|^2 + |(m'I+\gamma_{24}\Sigma)H_2|^2 \tag{82}$$

From this we find Higgs masses

$$m_{H_D}^{\;2} = m' - m$$

$$m_{H_Y}^{\;2} = m' + \frac{2}{3}m \tag{83}$$

where $m = -3\gamma_{24}v_3$, and subscripts D and T refer

to SU(2) doublet and SU(3) colour triplet components. If m' = m then the doublets remain massless while the triplets acquire a mass of order M_X. This certainly achieves the aim of splitting the Higgs multiplets although there is no symmetry giving the relation m = m'.

Do radiative corrections spoil the result? We have already noted the property of supersymmetry theories that the parameters in the superpotential are not remormalised[6]. Thus when supersymmetry is exact (as it is at the scale M_X), there will only be wave function renormalisation affecting the scalar masses. Consequently m_{H_D} remains zero beyond tree level, if m = m', up to supersymmetry violating effects which we will discuss later.

This illutstrates the first method for splitting the H and $\bar{H}$ multiplets, namely fixing it by hand at a scale M_X and relying on the non-renormalisation theorems to maintain the splitting at low scales. A second way has been suggested which has the advantage of explaining why m' = m initially. In this method the condition m = m' is necessary to minimise the potential energy[33]. Consider the modified form of eq.(71)

$$P_{5H} = H_1(\gamma_1 ZI+\gamma_{24}\Sigma)H_2 \tag{84}$$

where Z is an SU(5) singlet superfield. If this is the only time Z is mentioned in the superpotential the potential will only involve Z through the terms

$$V = |H_1(\gamma_1 ZI+\gamma_{24}\Sigma)|^2 + |(\gamma_1 ZI+\gamma_{24}\Sigma)H_2|^2 \tag{85}$$

Now, if SU(2) is broken through H_1 and H_2 acquiring vevs (how this happens we will discuss later) minimisation of V with respect to the vev of Z gives

$$\gamma_1 \langle Z\rangle - m = 0 \tag{86}$$

Clearly $\gamma_1\langle Z\rangle$ play the same role as m' in eq. (83) so the above equation requires m' = m as is needed to split the multiplets. For obvious reasons this method has become known as the "sliding singlet" technique.

The sliding singlet scheme is elegant and economical, but works only if the vacuum expectation value of the field Z is determined solely by the terms in eq. (85). This can be easily upset by radiative corrections and it turns out that the solution only works if the scale, μ, of supersymmetry breaking is small ($\ll 10^{11}$ GeV).

In models involving a large supersymmetry breaking scale μ, such as the Geometric Hierarchy or supergravity models discussed above, it is necessary to adopt another solution. The most popular method is known as the missing doublet scheme[34]. This idea relies on choosing a new Higgs representation whose SU(3) x SU(2) x U(1) content does not include a SU(2) doublet, SU(3) singlet. As a result its Yukawa couplings to H_1 and H_2 cannot give their doublet components a mass but can give one to their triplet components. For example, the SU(3)xSU(2) content of a 50 under SU(5) is

$$\underline{50} = (8,2) + (6,3) + (\bar{6},1) + (3,2) + (\bar{3},1) + (1,1) \quad (87)$$

It has, as desired, no (1,2) components. We couple two 50s, θ and $\bar{\theta}$ (together these have no anomalies) to H_1 and H_2 via the superpotential

$$P_{MD} = b\theta\Sigma H_1 + b'\bar{\theta}\Sigma H_2 + \tilde{M}\,\bar{\theta}\theta \quad (88)$$

where Σ is here a 75, needed instead of a 24 to construct the θ,H mixing terms. The field Σ will develop vacuum expectation value Mx/g breaking SU(5) uniquely to SU(3) x SU(2) x U(1). Then P_{MD} gives mass terms involving H_1 and H_2

$$P_{MD} = \theta_3(bM/g)H_{2_T} + \bar{\theta}_3(cM/g)H_{2_T} + \tilde{M}\tilde{\theta}_3\theta_3 \quad (89)$$

where the subscripts refer to the SU(3) transformation properties. From this it follows that the triplets acquire a mass $\simeq \frac{M_X^2}{\tilde{M}}$ while the doublets reamin massless.
remain massless. With $\tilde{M} >> M_X$ the triplets may be much lighter than the vector bosons of the theory.

3.6 Electroweak breaking and the problem of mass scales

One of the main motivations for building N=1 supersymmetric models was to solve the hierarchy problem to allow the electroweak breaking scale to coexist with the grand unified scale. As we have seen, the models introduced above can do this, having light Higgs doublets, H_1 and H_2, which, if they acquire vacuum expectation values, will generate electroweak breaking. It is easy to construct a superpotential which will force H_1 and H_2 to acquire vevs, but only if we are prepared to add a mass parameter of order 300 GeV to the theory. This raises the question of where such a scale should come from. In our theory, including gravity, there are four fundamental mass scales: M_{Planck}, M_X, $m_{3/2}$ and M_W. Obviously it would be aesthetically better to relate these scales, and build a theory with only one fundamental mass scale. It turns out that, in many models, electroweak breaking and M_W may be related either to $m_{3/2}$ (a measure of the supersymmetry breaking scale) or directly to M_{Planck}, and in this section we will concentrate on examples of these models.

At scales much below the Grand Unified mass the scalar potential, involving the light Higgs fields has the form[2]

$$V(H_1,H_2) = m_1^2|H_1|^2 + m_2^2|H_2|^2 + m_3^2 H_{1_\alpha} H_{2_\beta} \varepsilon^{\alpha\beta}$$

$$+ \frac{g_2^2 + g_1^2}{8}(|H_1|^4 + |H_2|^4) + \frac{(g_2^2 - g_1^2)}{4}|H_1|^2|H_2|^2 - \frac{g_2^2}{2}|\varepsilon^{\alpha\beta}H_{1_\alpha}H_{2_\beta}|^2 \quad (90)$$

where the quartic terms are D terms and we have allowed for possible mass terms. Electroweak breaking occurs if this potential has a minimum away from the origin. For example, this will happen if m_1^2 becomes negative.

Let us consider, first, models of the type of section (3.2), in which gauge nonsinglet scalar fields acquire a symmetry breaking mass through radiative coupling to the symmetry breaking sector. Scalar particles get a contribution to their mass in radiative order of the form

$$m^2 \approx \frac{\alpha}{\pi}(m_F^2 - m_B^2) \quad (91)$$

where m_F and m_B are the masses of the fermion and boson partners in the virtual loop. If, for example, the exchanged particles are gluons and gauginos the contribution will be positive for the gluons are massless while the gluinos must acquire a supersymmetry breaking mass. The squarks will obtain a larger radiative mass than sleptons or Higgs scalars through their larger gauge couplings. On the other hand, the Higgs scalar H_1 couples via a large Yukawa coupling to the top quark and squark. In this case the contribution of eq. (81) will be negative for the top squark is much heavier than the top quark. If the Yukawa coupling is large enough this term dominates the H_1^2 mass^2, and H_1 will acquire a vev[35], Minimising eq. (90) shows that H_2 will also acquire a vev along its neutral direction (ie. charge conservation is unbroken). The scale of these veves is given by the magnitude of m_1^2 which is clearly related to the supersymmetry breaking scale and hence to $m_{3/2}$. It is remarkable that it is only the Higgs multiplets that acquire a negative mass squared, the fact that $\alpha_3 > \alpha_2 > \alpha_1$ ensuring that SU(3) x SU(2) x U(1) is broken to SU(3) x $U(1)_{em}$, for the top squark gets a large positive contribution to its mass from graphs involving gluons and gluinos and thus does not become negative even when the negative contribution from the loops involving the large top Yukawa coupling are included.

In supergravity models the masses in eq. (90) arise at tree level through gravitational coupling to the supersymmetry breaking sector. They are given in eq. (44). If $\tilde{g}$ is cubic in the light fields we have

$$|g_i \pm m_{3/2} y_i|^2 = |g_i|^2 + m_{3/2}^2 |y_i|^2 \pm 3m_{3/2}(g+g^*) \tag{92}$$

Thus V of eq. (44) is positive for $A < 3$ and is minimised if all light fields have zero vacuum expectation values leaving SU(2) x U(1) unbroken at tree level. For $A > 3$ there is a lower minimum with non-zero vevs and in this case SU(2) x U(1) may be broken at tree level. As in our first example M_W is related to $m_{3/2}$ in this approach.

However, the global minimum is not the desired one, breaking charge and lepton number. For some values of A the desired minimum (a local one) is stable and so this method of electroweak breaking may be a possibility, but it looks somewhat contrived.

However, even for $A < 3$, as in the case for the simple Polonyi potential, eq.(90), once we include radiative corrections there is the possibility for SU(2)xU(1) breaking[36]. From eq.(87) we see that all scalar masses are equal to $m_{3/2}^2$ at tree level. Presumably, this result applies at the Planck scale. Radiative corrections due to the renormalisable gauge and Yukawa interactions will split this equality and, in continuing down to low scales of $O(M_W)$, they contribute terms

$$\propto [\ln(\frac{M_{Planck}}{M_W})]^n.$$

These terms may be resumed using the renormalisation group analysis, leading to running masses $m_i^2(H_i^2)$ in eq.(90). For the same reason as was discussed in our first example, these radiative corrections may selectively drive the Higgs scalar mass^2 negative triggering SU(2)xU(1) breaking. Minimisation of the potential eq.(90) gives two possibilities. In the first the combination of $m_i^2(H_i^2)H_i^2$ and H_1^4 terms lead to a minimisation with $\langle H_i \rangle^2 = O(m_i^2)$. Since the scale of m_i is set by $m_{3/2}$ this gives a solution with M_W related to $m_{3/2}$. In the second possibility the H_i^4 terms are anomalously small through a cancellation. The minimum is then close to the point at which $m_i^2(H_i^2)$ goes negative. Since $m_i^2(H_i^2)$ evolves logarithmically from the scale of the Planck mass, this point will be directly related to the Planck mass times an exponentially small factor. This is known as dimensional transmutation[37].

Thus these models all have the electroweak breaking scale as a dependent mass scale. However there still remain three independent mass parameters M_{Planck}, M_X and $m_{3/2}$. Can these be related? Since, in supersymmetry models, M_X is usually very close to M_{Planck} they perhaps should not be considered as independent mass scales, and may be related by the mechanism giving rise to the effective N=1 theory valid below the Planck scale. However, the supersymmetry breaking scale μ, or the related gravitino mass $m_{3/2}$, are much smaller and it is clearly desirable to relate it to the other scales. Two mechanisms have been proposed. The first[38] applies to a class of models in which, in the globally supersymmetric limit, supersymmetry is unbroken but in achieving the supersymmetric minimum a field, B, acquires an infinite vev. Including the gravitational strength couplings of eq.(38), the vev of B is limited by the Planck mass and consequently the potential energy is non zero, but suppressed from its natural scale by powers of (M_X/M_{Planck}). As a result the scale of superheavy breaking is very small.

We may illustrate this by a very simple example involving singlet fields X Y and B with superpotential

$$P = \lambda_1 (X\,Y - M^2) B + \lambda_2\, X^3 \tag{93}$$

The globally supersymmetric potential following from this is

$$V = |\lambda_1 (XY - M_X{}^2)|^2 + |\lambda_1\, XB|^2 + |\lambda_1 YB + 3\lambda_2\, X^2|^2 \tag{94}$$

which has a supersymmetric minimum at

$$\langle XY\rangle = M_X{}^2$$

$$\langle X\rangle \;= 0 \tag{95}$$

This minimum requires $\langle Y\rangle = \infty$, a physically unreasonable value for we expect large gravitational corrections of the form Y^{n+4}/M^n. Indeed in the N=1 locally supersymmetric case the potential has the form

$$V = e^{(|X|^2 + |Y|^2 + |B|^2)/M} \quad \{|\lambda_1 (XY - M_X{}^2) + B^* P/M^2|^2$$

$$+ |\lambda_1\, X\, B + Y^* P/M^2|^2 + |\lambda_1\, YB + 3\lambda_2\, X^2 + X^* P/M^2|^2$$

$$- 3|P|^2/M^2\} \tag{96}$$

For large $\langle Y\rangle$, V has then the form

$$V = e^{Y^2/M^2} \{ \left| - \frac{3}{\langle Y\rangle^4} + \frac{1}{M^2 \langle Y\rangle^2} \right|$$

$$- \frac{3}{M^2} \left| \frac{1}{\langle Y\rangle^3} \right|^2 \} \; |\lambda_2 M^6|^2 \tag{97}$$

This has a minimum with $\langle Y\rangle^2 = 6M^2$. Thus gravitational effects limit the vev of $\langle Y\rangle$ and prevent the supersymmetric minimum eq.(95), being reached. As anticipated, the deviation from this minimum is suppressed by powers of (M_X/M) but the surprise is that for the very simple potential given above at the minimum, each term of eq.(97) is of the order $\lambda_2^2 \frac{M_X^{12}}{M^8}$ giving a supersymmetry breaking scale

$$M_{SUSY} = 0(\sqrt{\lambda_2}\, \frac{M_X^3}{M^2}) \tag{98}$$

and gravitino mass

$$m_{3/2} = M_{Planck}\, \lambda_2 (\frac{M_X}{M})^6 \tag{99}$$

For a rather moderate value for the initial ratio, $(\frac{M_X}{M})$ of $0(10^{-2.5})$, a gravitino mass of $0(10^2 \mathrm{GeV})$ is easily obtained. Although this model has been built using gauge singlet fields X,Y,B the idea can easily be generalised to gauge non-singlets fields[39] in which case M_X can be related to the GUT breaking scale.

The model presented above is an example of how $m_{3/2}$ may be made much smaller than M_{Planck} through gravitational corrections present at tree level in the N=1 supergravity theory. The reason it works is because without gravitational corrections the potential, eq.(94), is flat in the Y direction at minimum ($m_Y = 0$) and so small gravitational corrections can play an important role in fixing the Y vev. It is possible to invent models[40] in which a singlet field, Y, is massless even in the presence of tree level gravitational corrections. In these models the vev of Y is completely undetermined at tree level as are the supersymmetry breaking scale and $m_{3/2}$ which are related to $\langle\langle Y\rangle\rangle$. It is only in radiative order that the vev of Y and the supersymmetry breaking scale occur. However, radiative corrections a only logarithmically with scale and one finds $m_{3/2}$ is related to the Planck mass times an exponentially small factor giving rise to a natural hierarchy as desired. The beauty of this approach is that the original flatness of the potential in the Y direction can be shown to arise as a result of a symmetry which is a general property of many supergravity models rather than, as in our first example, as a consequence of a particular choice of superpotential. However, these models achieve their small $m_{3/2}$ through the vev of Y becoming exponentially larger than the Planck scale and I find it hard to accept that gravitational radiative corrections which can induce terms of the form Y^{n+4}/M^n will not destabilise the solution. There is a further, cosmological, problem with both approaches discussed above which follows directly from their central ingredient - a very light field. This field couples only with gravitational strength to other light states and consequently decays at very late times in the evolution of theuniverse.

It behaves as matter not radiation for much of its life and during this time its energy density relative to the radiation background grows like R. When it decays it dumps unacceptably large amounts of entropy into the observable universe - the so-called Polonyi problem(41). It remains to be seen whether this problem can be circumvented.

To summarise, ignoring this cosmological problem, locally supersymmetric theories can solve the problem of mass scales and we can build theories with only one fundamental mass scale M_{Planck} in which both the electroweak breaking scale M_W and the supersymmetry breaking scale $m_{3/2}$ are given in terms of M_{Planck} x naturally small factors.

References

(1) K. Stelle, these proceedings

(2) For more detailed reviews see:
P. Fayet and S. Ferrara, Phys. Rep. 32C (1977) 259.
H. Nilles, Phys. Rep. 110 (1984) 1.

(3) The phenomenological aspects of the supersymmetric standard model are covered in H. Haber and G.L. Kane, Phys. Rep. (to be published).
J. Ellis, Supersymmetric GUT^s, Lectures presented at the Advanced Study Institute on Quarks, Leptons and Beyond, Munich CERN preprint Ref. TH 3802-CERN: see also Cornell Conference proceedings 1983)

(4) S. Weinberg, Phys. Rev. D26 (1982).

(5) L.J. Hall, M. Suzuki, Nucl. Phys. B231, 419 (1984).
M. Bow ick M. Chase and P. Ramond, Phys. Lett. 128B (1983) 185.
I.H. Lee, Phys. Lett. 138B (1984) 121.

(6) P. West, Nucl. Phys. B106 (1976) 219.
M. Grisaru, M. Rocek and W. Siegal, Nucl. Phys. B159 (1979) 429.

(7) R. Barbieri and R. Gatto, Phys. Lett. 110B (1982) 211.
M.J. Duncan, Nucl. Phys. B224 (1983) 289; ibid B221 (1983) 285.
M. Suzuki, Phys. Lett. 115B (1982) 40.
J. Donaghue, H. Nillesand D. Wyler, Phys. Lett. 135B (1984) 423.
A. Bouquet, J. Kaplan and C.A. Savoy, Univ. of Paris preprint PAR LPTHE 8425 (1984).
J.P. Derendinger and C.A. Savoy Nucl. Phys. B237 (1984) 307.

(9) R. Barbieri, L. Girardello and A. Masiero, Phys. Lett. 127B (1983) 429; W. Buchmuller and D. Wyler, Phys. Lett 121B (1983) 321;
R. Arnowitt, A.H. Chamseddine and P. Nath, Phys. Lett 50 (1983) 232; S. Weinberg, ibid, 50 (1983) 387.
D.V. Nanopoulos and M. Srednicki, Phys. Lett. 128B (1983) 61.
F. Del Aguila, M. Gavela, J. Grifols and A. Mendez, Phys. Lett. 126B (1983) 71.

(10) J. Ellis and D.V. Nanopoulos, Phys. Lett. 110B (1982) 44.

T. Inami and C.S. Lim, Nucl. Phys. B207 (1982) 533.

(11) For an introduction to R symmetries see also Fayet and Ferrara, ref.2.

(12) G.G. Ross and J.W.F. Valle, Rutherford Appleton Lab. preprint RAL-84-102.
J. Ellis, G. Gelmini, C. Jarlskog, G.G. Ross and J.W.F. Valle, Rutherford Appleton Lab. preprint RAL-84-085.

(13) P. Fayet and J. Iliopoulos, Phys. Lett. 51B (1974) 461.
L. O'Raifeartaigh, Nucl. Phys. B96 (1975) 331.
P. Fayet, Phys. Lett. 58B (1975) 67.

(14) E. Cremmer, B. Julia, J. Scherk, P. van Nieuwenhuizen, S. Ferrara and L. Giradello, Phys. Lett. 79B (1978) 231; Nucl. Phys. B147 (1979) 105.
E. Cremmer, S. Ferrara, L. Giradello and A. van Proyen, Phys. Lett. 116B (1982) 231; Nucl. Phys. B212 (1983) 413.
J. Bagger and E. Witten, Phys. Lett. 115B (1982) 202.

(15) L. Ibanez, Phys. Lett. 118B (1982) 73.
R. Barbieri, S. Ferrara and C.A. Savoy, Phys. Lett. 119B (1982) 343
A. Chamseddine, P. Nath , and R. Arnowitt, Phys. Rev. Lett. 49 (1982) 970
P. Nath , R. Arnovitt and A.P. Chamseddine, Phys. Lett. 121B (1983) 33.
J. Ellis, D.V. Nanopoulos and K. Tamvakis, Phys. Lett. 121B (1983) 123.
L. Hall, J. Lykken and S. Weinberg, Phys. Rev. D27 (1983) 2359.

(16) H.P. Nilles, T. Sredricki and D. Wyler, Phys. Lett. 120B (1982) 346.

(17) J. Polonyi, Univ. of Budapest report No. KFKI-1977-93 (1977).

(18) J.C. Pati and A. Salam, Phys. Rev. D8 (1973) 1246.
H. Georgi and S.L. Glashow, Phys. Rev. Lett. 32 (1974) 438.

(19) E. Gildener, Phys. Rev. D14 (1976) 1667; Phys. Lett. 92B (1980) 111.
E. Gildener and S. Weinberg, Phys. Rev. D15 (1976) 3333.
G. 't Hooft, Proceedings of the advanced study institute, Cargese 1979 Eds. G. 't Hooft et al. (Plenum Press NY 1980)

(20) N. Sakai, A.F. Phys. CH (1981) 153.
S. Dimopoulos and H. Georgi, Nucl. Phys. B193 (1981) 150.

(21) H. Georgi, H.R. Quinn and S. Weinberg, Phys. Rev. Lett. 33 (1974) 451.
J. Ellis, M.K. Gaillard, D.V. Nanopoulos and S. Rudaz, Nucl. Phys. B176 (1980) 61.
T. Goldman and D.A. Ross, Phys. Lett. 843 (1979) 208.

C.H. Llewellyn Smith, G.G. Ross and J. Wheater, Nucl. Phys. B177 (1981) 263.
S. Weinberg, Phys. Lett. 91B (1980) 51.
I. Hall, Nucl. Phys. B178 (1981) 75.
P. Binetruy and T. Schucker, Nucl. Phys. B178 (1981) 293.

(22) L. Ibáñez and G.G. Ross, Phys. Lett. 105B (1981) 439.
J. Ellis, D.V. Nanopoulos and S. Rudaz, Nucl. Phys. B202 (1982) 43.

(23) M.B. Einhorn and D.R.T. Jones, Nucl. Phys. B196 (1982) 475.
W. Marciano and G. Senjanovich, Phys. Rev. D25 (1982) 3092.

(24) D. Nanopoulos and D.A. Ross, Phys. Lett. 118B (1982) 99.

(25) For reviews see P. Langacker, Phys. Reports 72 (1981) 185.
W. Marciano, BNL preprint 33415 (1983)

(26) N. Sakai and T. Yanagida, Nucl. Phys. B197 (1982) 83.
S. Dimopoulos, S. Ruby and F. Wilczek, Phys. Lett. 112B (1982) 133.

(27) J. Ellis et al. ref. 21.
P. Salati and C. Wallet, Nucl. Phys. B209 (1982) 389.
J. Ellis, J.S. Hagelin D.V. Nanopoulos and K. Tamvakis, Phys. Lett. 124B (1983) 464.
S. Chadha and M. Daniel, Nucl. Phys. B229 (1983) 105.

(28) S. Chadha, G. Coughlan, M. Daniel and G.G. Ross, Physics Lett. (in press) 1984.

(29) J. Ellis, ref(3); G.G. Ross, "Grand Unified Theories", The Benjamin Cummings Publishing Co. (1984); Proceedings of the XVII International Conference on ν physics, Noordkirching (1984).

(30) S. Dimopoulos and S. Raby, Los Alamos report (1982).
E. Witten, Phys. Lett. 105B (1981) 267.
J. Polchinski and L. Susskind, Phys. Rev. 26D (1982) 3661.

(31) J. ELlis, L.E. Ibañez and G.G. Ross, Phys. Lett. 113B (1982) 283. Nucl. Phys. B221 (1983) 29.
L.E. Ibañez and G.G. Ross, Phys. Lett. 110B (1982) 215.

(32) See Georgi and Dimopoulos (ref. 20), also Nilles (ref.2).

(33) E. Witten ref.(30).
L.E. Ibañez and G.G. Ross, Phys. Lett. 105B (1981) 439.

(34) B. Grinstein, Nucl. Phys. B206 (1982) 387.
A. Masiero, D.V. Nanopoulos, K. Tamvakis and T. Yanagida, Phys. Lett. 115B (1982) 380.

(35) L.E. Ibáñez and G.G. Ross, Phys. Lett. 110B (1982) 215.

L. Alvarez-Gaumé, M. Claudson and M.B. Wise, Nucl. Phys. B207. (1982) 96.
C.R. Nappi and B.A. Ovrut, Phys. Lett. 113B (1982).
K. Inoue, A. Kakuto, H. Komatsu and S. Takeshita, Prog. Theor. Phys. 68 (1982) 927.

(36) L.E. Ibañez, Phys. Lett. 118B (1982) 73.
J. Ellis, D.V. Nanopoulos and K. Tamvakis, Phys. Lett·121B (1983)
L. Alvarez-Gaume, J. Polchinski and M.B. Wise, Nucl. Phys. B221 (1983) 495.
S.K. Jones and G.G. Ross, Phys. Lett. 135B (1984) 69.

(37) S. Coleman and E. Weinberg, Phys. Rev. D7 (1973) 1888.

(38) C.J. Oakley and G.G. Ross, Phys. Lett. 125B (1983) 59.

(39) L. Ibanez and G.G. Ross, Phys. Lett. 131B (1983) 335.
B.A. Ovrut and S. Raby, Phys. Lett. 134B (1984) 51.

(40) J. Ellis, A.B. Lahanas, D.V. Nanopoulos and K.A. Mamvakis, Phys. Lett. 134B (1984) 429.
J. Ellis, C. Konnas and D.V. Nanopoulos, Nucl. Phys. B241 (1984) 406.

(41) G. Coughlan, W. Fischler, E. Kolb, S. Raby and G.G. Ross, Phys. Lett. 131B (1983) 59.

SUPERSYMMETRIC INFLATIONARY COSMOLOGY*

P. Ramond**
Physics Department, University of Florida,
Gainesville, Florida 32611

ABSTRACT

We review the general requirements imposed on cosmology by particle physics. In particular, we discuss a model which includes Inflation, Supersymmetry and Grand Unification. We show how this model can be used to put an upper bound on the proton lifetime.

The standard cosmological model [1] for the formation of the elements has proven increasingly successful with the explosion of astrophysical data over the last two decades. According to it, gravity is described by the homogeneous, isotopic metric of Friedman, Robertson and Walker, it contains one time dependent scale parameter R(t) and can occur in three varieties corresponding to an open, closed or critical expansion of the universe. Einstein's equations then demand that matter be described as a perfect cosmic fluid with energy density ρ(t) and pressure p(t). There is no reason to doubt the adequacy of this picture from the time of cosmological nucleosynthesis to the present, provided one includes density fluctuation δρ to account for the observed inhomogeneities we are a part of. At that time, the assumption that the matter part was dominated by a heat bath at a temperature T of the order of Mev (corresponding to the p-n mass difference and inversely proportional to R) seems to be successful. The question of interest is to determine what happened before that time, and in order to answer this question one needs information as to the type of matter and its behavior at those early times; for this, one turns to particle physics. In fact the second marriage of particle physics and cosmology was due to the observation [2] that Grand Unification provided a ready-made mechanism for generating baryon number asymmetry. Armed with a specific model of particle physics at high energy, one can start exploring the earlier epochs of the universe. No model with sufficiently high credibility exists but we

*Invited talk at the Eighth Johns Hopkins Workshop, June 1984.

**Supported in part by the U. S. Department of Energy under contract No. DE-AS-05-81-ER40008.

will assume that it incorporates supersymmetry as a way of preserving very small scale ratios in the presence of radiative corrections. We call such a model Supersymmetric Quantum Interacting Dynamics (SQID). Before discussing specific SQIDs let us recall some elementary facts about the FRW description of cosmology.

Einstein's equations reduce to

$$\left(\frac{\dot{R}}{R}\right)^2 + \frac{k}{R^2} = \frac{1}{3M^2}\rho(t), \qquad (1)$$

where $M = 2.4 \times 10^{18}$ GeV, and we define it to be one Planck (P), and $k = 0,\pm 1$ depending on the eventual fate of R(t). The matter density $\rho(t)$ is govered by

$$\dot{\rho}(t) + 3\left(\frac{\dot{R}}{R}\right)(\rho+p) = 0, \qquad (2)$$

where p is the pressure. These equations then have to be augmented by giving the equation of state of matter, i.e. by expressing ρ as a function of p. We summarize some interesting cases:

a) Relativistic matter $\rho = \frac{1}{3}p$ gives $\rho \sim R^{-4}$;

$H \equiv \left(\frac{\dot{R}}{R}\right) \sim t^{-1}$; $R(t) \sim t^{1/2}$.

b) Non Relativistic matter $\rho \gg p$ gives $\rho \sim R^{-3}$;

$H \sim t^{-1}$; $R(t) \sim t^{2/3}$.

c) Cosmological matter $\rho = -p =$ constant;

it gives $H \simeq \frac{1}{3M^2}\rho_0$ $\quad R(t) \sim e^{Ht}$.

One can envisage other interesting cases, such as "stiff matter" for which $\rho = p$, but popular particle physics scenarios seem to produce only cases a), b) and c).

Although matter, at least from the time of cosmological, nucleosynthesis is described by thermal considerations, it must be emphasized that this is so only because at those times the interaction rate among particle species is in general much faster than the expansion rate of the universe, thus allowing for the particles to share their energy and for the notion of temperature to emerge. Thus one has to continuously check whether

$$H \lessgtr \langle n\sigma v\rangle \qquad (3)$$

is satisfied or not, where σ is the cross section, n the density and v the velocity. We will see that it is not inconceivable to have ρ(t) at some earlier time dominated by particles which do not have time to achieve equilibrium due to the weakness of their interactions.

Having assembled these tools, we can now start to describe the demands on the right-hand side of Einstein's equation set both by cosmology and by particle physics.

At the time of, say, cosmological nucleosynthesis, various parameters must be given initial values in order to successfully describe the universe today. In particular the observed isotropy and homogeneity of the detected microwave background and the observed value of the mass density suggests that there was an enormous amount of entropy [3] in the very early universe. The problem is how to explain its origin by using a credible particle physics model. Such a large initial entropy would help explain why the observed universe is so close to criticality and yet so old; namely, if you define the critical density, ρ_c, via Hubble's constant as

$$H^2 = \frac{1}{3M^2}\rho_c, \qquad (4)$$

the observed luminous density ρ_ℓ is seen to be

$$\Omega_\ell \equiv \frac{\rho_\ell}{\rho_c} \sim 10^{-2},$$

while dark matter around which luminous matter congregates is estimated to be

$$\Omega_d \equiv \frac{\rho_d}{\rho_c} \sim 2 \times 10^{-1}.$$

Now this value is very close to 1 (in cosmology, errors are usually quoted in the exponent), and the evolution equation (1) shows that given the "age of the universe", Ω would have been to be equal to 1 in one part in 60(!) given adiabaticity. However a severe loss of adiabaticity, such as a large entropy release would take care of this peculiar initial condition.

One finds that a period of inflation [3,4], with constant energy density with its consequent exponential increase of the scale parameter, $R(t) \sim e^{H_0 t}$ "solves" this type of problem provided that $H_0\delta t \gtrsim 60$. However such a scenario makes the k/R^2 term in equation (1) totally negligible. Thus inflation predicts that today Ω should be equal to 1, and yet it is measured to be at most 0.3! The nature of the missing matter is not clear at present (if it exists at all), but there is no lack of candidates from particle physics models.

At this stage, we should mention one great advantage of the inflationary scenario: it provides the model builder with a natural mechanism to get rid of embarrassing particles. All that one has to do is to arrange for the unwanted particle not to be produced after inflation takes place, via the so-called reheating process. This trick, as we shall see, is used at least twice, once for monopoles, once for gravitinos. This type of procedure was known to the ancients (re: Gilgamesh and/or Noah's ark).

Let us now describe sensible requirements on the SQID. First of all the scale of GUT breaking, M_x, is bounded by the absence of proton decay to be greater than 10^{15} GeV or 1 milliPlanck. Such a breaking produces magnetic monopoles, which do not appear to be plentiful in nature. Since it seems that monopoles cannot be destroyed except by antimonopoles, one had better arrange for inflation to take place <u>after</u> GUT breaking. In this way, the monopole density will be diluted by the exponential phase (inflation). However in the reheating after inflation the germs of baryogenesis must appear, and this poses very severe constraints on the various models.

Finally it is desirable to include supersymmetry (Susy "pour les intîmes") in order to protect the weak interaction scale [5]. This implies a typical splitting within a supersymmetric multiplet (Tev). If μ is the SUSY breaking scale, then the gravitino mass $m_{\tilde{G}}$ is given by

$$m_{\tilde{G}} \sim \frac{\mu^2}{M} \sim \text{Tev} \tag{5}$$

i.e. $\mu \sim 10^{10}$ GeV, or 10 nanoPlanck. The gravitino is a particle which is itself embarrassing. For this particular mass, it would decay after cosmological nucleosynthesis. Its decay would have several negative consequences [6] - one it will produce more photons, thus reheating and then diluting the baryon asymmetry, second its decay products would change the abundance of elements laboriously formed earlier. For instance in the decay $\tilde{G} \to \gamma + \tilde{\gamma}$ ($\tilde{\gamma}$ is the photino), photons capable of dissociating deuterium would be produced. On the other hand in the decay into a gluon-gluino pair, $\bar{p}$ would eventually result, which could change ^{4}He into D. The only way out seems to be to limit the original abundance of gravitinos. Thus can be done by appealing to inflation. Then one has to worry only about gravitinos produced either by reheating or by the decay of particles whose interactions break supersymmetry. The first constraint puts a limit to the temperature of the reheated bath at $T_R \lesssim 10^8 - 10^{12}$ GeV, depending on who you read. The second constraint (called the Entropy Crisis of Supersymmetry) has to be examined for each model of SUSY breaking.

Thus we will need three mechanisms for our various constraints: in temporal order 1) GUT breaking at a scale M_x ($\sim$mP), Inflation at a scale Δ, and SUSY breaking at a scale

μ ($\sim$ 10 nP). We now present our model [7]. We will use the formalism of N=1 supergravity to describe it.

Starting from a chiral superfield $\Phi_i(x,\theta)$, each containing a complex spinless boson and a Weyl fermion, one describes their non gauge interactions by means of a superpotential $P(\Phi)$, from which the ordinary potential is derived [8]

$$V(\phi_i) = e^{-\frac{|\phi_i|^2}{M^2}} \{\sum_i |\frac{\partial P}{\partial \Phi_i} + \frac{\phi_i^* P}{M^2}|^2 - \frac{3}{M^2}|P|^2\}, \tag{6}$$

where ϕ_i are the spinless components of Φ_i. This form is valid when the scalar fields have the usual kinetic terms. This potential breaks SUSY if and only if

$$|\frac{\partial P}{\partial \Phi} + \frac{\phi^* P}{M^2}| \neq 0 \text{ at minimum.} \tag{7}$$

Now if one starts from a superpotential made up of the sum of several disconnected parts, the structure of (6) forces interactions between them of $\mathcal{O}(\frac{1}{M^2})$. The philosophy of hidden sector physics is to have different sectors which interact with one another only through gravitational strength interactions.

Our model for the superpotential is then

$$P = I + G + S, \tag{8}$$

where I describes inflation, S describes supersymmetry breaking, and G describes the GUT sector. Thus in this picture fields in a given sector will not be in thermal equilibrium with fields of another sector until very late times when the expansion rate is very slow. This is because they only interact gravitationally with one another. [Interestingly, in a contracting universe, particles would be thrown together and equilibrium would be easier with time, but this does not apply to our universe, at least not yet.]

The I part of the superpotential is now built according to the following tenets - use only one superfield containing one complex scalar field, the inflaton, and its partner the inflatino - at minimum its potential does not generate any cosmological term and does not break supersymmetry, i.e.

$$|\frac{\partial I}{\partial \phi} + \frac{\phi^* I}{M^2}|^2 - \frac{3}{M^2}|I|^2 = 0 \quad \text{at minimum,} \tag{9}$$

and

$$\left|\frac{\partial I}{\partial \phi} + \frac{\phi^* I}{M^2}\right| = 0 \qquad \text{at minimum.} \tag{10}$$

These together imply that both I and $\frac{\partial I}{\partial \phi}$ are zero at minimum. Hence the simplest expression for I obeying these criteria is

$$I = \frac{\Delta^2}{M} (\Phi - \phi_0)^2 , \tag{11}$$

where Δ is an unknown parameter with dimension of mass and ϕ_0 is the vacuum value of the inflaton. We require further that I inflates, i.e. $\frac{\partial V}{\partial \phi} = 0$ at the origin. In this case we find that ϕ_0 is fixed to be M. Hence we set

$$I = \frac{\Delta^2}{M} (\Phi - M)^2. \tag{12}$$

As an added bonus we find that $\frac{\partial^2 V}{\partial \phi^2} = 0$ at the origin as well. In fact near the origin

$$V_I(\phi) \approx \Delta^4 [1 - (\phi/M)^3 + ..]; \quad \phi/M \ll 1, \tag{13}$$

where ϕ is the inflaton field. The inflaton potential has a minimum at ϕ=M, where the mass of the inflaton field is

$$m_\phi = \frac{\Delta^2}{M} . \tag{14}$$

This very simple form for the inflaton sector leads to the following scenario. The inflaton has only gravitational strength self interactions. One can check that except for temperatures near the Planck mass it never is in thermal equilibrium. It produces about 10^8 e-folds of inflation. At $t_* \sim m_\phi^{-1}$ the inflaton field remembers its mass and inflation stops; the evolution becomes matter dominated until a time $t_R \sim \Gamma_\phi^{-1}$ when the inflaton decays. It decays universally since it is hidden from all the other sectors. Its decay rate is given by

$$\Gamma_\phi \sim \frac{\Delta^6}{M^5} , \tag{15}$$

corresponding to a reheat temperature

$$T_R \sim (M\Gamma_\phi)^{1/2} \sim \frac{\Delta^3}{M^2} . \tag{16}$$

Now the hitherto unknown parameter Δ is fixed by the needed scale of density fluctuations; these are given by [9]

$$\frac{\delta\rho}{\rho} \simeq \frac{H^2}{\dot{\phi}}, \tag{17}$$

evaluated at a time during inflation when the fluctuation leaves the horizon. In our model one can estimate this expression for fluctuation relevant to galactic sizes to be

$$\frac{\delta\rho}{\rho} \simeq 1.5 \times 10^4 \left(\frac{\Delta}{M}\right)^2, \tag{18}$$

by using the approximation (13) for the potential, which is certainly valid in this regime. Thus in order to get the desired value of $\frac{\delta\rho}{\rho} \simeq 10^{-4}$, we must fix

$$\Delta \simeq 10^{-4}\ P \approx .1 \text{ milliPlanck}. \tag{19}$$

There are no more free parameters left in this sector. Let us now turn to the other sectors and see if contradictions can be found.

We take the SUSY breaking superpotential to be of the O'Raifeartaigh [10] type, i.e.

$$S = A[\lambda B^2 + \mu^2] + \mu BC + q, \tag{20}$$

where A is the superfield whose F-term breaks the supersymmetry. Its scalar component is called, the "O'Raifearton" and its spinor component is "eaten" by the gravitino to provide the helicity ± 1/2 states. The O'Raifearton acquires a mass at the one loop level

$$m_A \simeq \alpha_\lambda \mu \; ; \quad \alpha_\lambda = \frac{\lambda^2}{4\pi}, \tag{21}$$

and its preferred decay channel is into gravitinos

$$A \to \tilde{G}\tilde{G}\,, \quad \Gamma_A \sim \frac{\alpha_\lambda^3}{(4\pi)^2}\, m_A \tag{22}$$

Note that with our value of Δ, the inflaton is (barely) kinematically forbidden to decay into the O'Raifearton, since

$$m_\phi \lesssim m_A \tag{23}$$

Alternatively, we could say that (23) puts an upper bound on Δ, specifically $\Delta \lesssim 10^{-3.5}$ P, perfectly consistent with the value (19) obtained from fluctuations.

Next one checks that as long as α_λ is reasonably small the decay of the inflaton into O'Raifearton by one loop effects can also be suppressed. Thus in our model, there is essentially no direct production of gravitinos by inflaton decay. However, there is a more worrisome source of O'Raifeartons which comes about by means of the induced interactions between I and S: after inflation, the A field is left with a non-zero potential energy. This potential energy is then eventually translated into gravitinos. However one can check that with our values for μ and Δ this produces only a negligible gravitino abundance at the time of nucleosynthesis. Further since our reheat temperature is, for our value of Δ, low i.e. $\sim 10^6$ GeV, we completely avoid the gravitino problem.

There remains, however, one hurdle to overcome, and it is that of baryogenesis [2]. One wants the reheating process to include particles whose interactions can lead to baryon asymmetry; yet the inflaton mass is fixed by Δ [$M_\phi \sim 10^{10}$ GeV] - it follows that the Higgs triplets generated in inflaton decay cannot have masses higher than that, otherwise baryogenesis could not occur. On the other hand such low mass for the triplets will cause in principle fast proton decay. Thus we are in the enviable (but dangerous) situation of obtaining an <u>upper</u> bound for proton decay [11] from the above cosmological considerations!

Because the inflaton mass is bounded in our model, we have to consider Higgs triplets with masses $\sim (10^{10}$ GeV). Also because of the low reheat temperature (10^6 GeV), we have to rely on a non-equilibrium scenario for producing baryon asymmetry. Fortunately the inflaton is never in equilibrium, so we can check that its preferential decay modes will occur into Higgs triplets. We have then [12]

$$n_{B-\overline{B}} \simeq n_H \delta B \approx n_\phi \delta B, \tag{24}$$

where δB is the specific baryon number asymmetry, n_H is the triplet density, n_ϕ the inflaton density. Also then

$$n_\gamma T_R \approx m_\phi n_\phi, \tag{25}$$

so that

$$\frac{n_{B-\overline{B}}}{n_\gamma} \simeq \frac{\Delta}{M} \delta B \ . \tag{26}$$

There remains to see if we can arrive at GUTs models which produce the required δB while not causing too fast a proton decay rate.

In a supersymmetric GUT, the presence of Higgs triplets with mass $\sim (10^{10}$ GeV) [13] can cause fast proton decay by means of the quark chiral operators (dimension five). Specifically given the

Yukawa couplings

$$\phi_{\bar{5}}\phi_{10}H_1 + \phi_{10}\phi_{10}H_2 , \tag{27}$$

where H_1 and H_2 contain the Higgs color triplets, the following interactions are induced

a) $\phi_{\bar{5}}\phi_{10}\bar{\phi}_{\bar{5}}\bar{\phi}_{10}$, mediated by H_1

b) $\phi_{10}\phi_{10}\bar{\phi}_{10}\bar{\phi}_{10}$, mediated by H_2

and

c) $\phi_{\bar{5}}\phi_{10}\phi_{10}\phi_{10}$, mediated by H_1 and H_2

exchange, provided H_1 and H_2 mix. The last interaction contributes to proton decay in supersymmetric theories as an interference term, and for $H_1,H_2 \sim (10^{10}$ GeV) with $\mathcal{O}(1)$ mixing, will cause fast proton decay, in contradiction with experiment. The only hope is to suppress the mixing between H_1 and H_2, while relying on straight H_1 and H_2 exchange to give proton decay. A careful analysis of these decay rates already exists in the literature [14]. One finds that if H_1 is lighter than H_2, the preferred decay modes are $p \to e^+\pi^0$, $e^+\omega$ and our limit from the inflaton mass gives [11]

$$\tau_p(p\to e^+\pi^0(\omega)) \lesssim 10^{34} \text{ years;} \tag{28}$$

it is therefore uninteresting since known backgrounds make this upper limit unattainable. However, if H_2 is lighter than H_1 the limit becomes more interesting, namely [11]

$$\tau_p(p \to \mu^+K^0) \lesssim 10^{31} \text{ years.} \tag{29}$$

The next question involves the actual mechanism by which the baryon asymmetry is produced. In our model we have assumed that the $H_1 - H_2$ mixing term is suppressed. Such models exist [11] but they seem to involve a large number of Higgs particles.

At any rate we hope that we have shown that with a definite cosmological model for inflation, the existence of baryogenesis and the absence of proton decay poses severe constraints on any SQID one might want to propose.

Acknowledgement

The author thanks the Lewes Center for Physics for providing a suitable environment for writing this paper.

COPYRIGHT JFCARTIER 1984
It's been over 700 years
and we haven't observed a single event.
We will have to make it BIGGER.

References

(1) See one of the many excellent books on the subject; e.g. S. Weinberg, Gravitation and Cosmology, New York, Wiley 1972.

(2) M. Yoshimura, Phys. Rev. Lett. 41, 281 (1978), (E) 42, 746 (1978). A. D. Sakharov, Zh. Eksp. Teor. Fiz. Pis'ma 5, 32 (1967). E. Kolb, M. Turner in Ann. Rev. Nucl. Part. Sci. 1983, 33:645.

(3) A. Guth, Phys. Rev. D23, 347 (1981).

(4) A. D. Linde, Phys. Lett. 108B, 389 (1982); A. Albrecht and P. J. Steinhardt, Phys. Rev. Lett. 48, 1220 (1982). P. J. Steinhardt and M. S. Turner, Phys. Rev. D29 2126 (1984).

(5) L. E. Ibañez, Phys. Lett. 118B, 73 (1982); R. Barbieri, S. Ferrara and C. A. Savoy, Phys. Lett. 119B (1982) 343; P. Nath, R. Arnowitt and A. H. Chamseddine, Phys. Rev. Lett. 49 (1982) 970; H. P. Nilles, M. Srednicki and D. Wyler, CERN preprint TH-3432 (1982); S. Ferrara, D. V. Nanopoulos and C. A. Savoy, Phys. Lett. 123B (1983) 214; L. Hall, J. Lykken and S. Weinberg, University of Texas preprint UTTG-1-83 (1983);
L. E. Ibañez and G. G. Ross, Phys. Lett. 131B (1983) 335.

(6) For the latest see M. Yu. Khlopov and A. D. Linde, Phys. Lett. 138B, 265 (1984); John Ellis, Jihn E. Kim and D. V. Nanopoulos, CERN preprint 3839 (1984).

(7) R. Holman, P. Ramond and G. G. Ross, Phys. Lett. 137B, 343 (1984); G. D. Coughlan, R. Holman, P. Ramond, G. G. Ross, University of Florida preprint, 1984; R. Holman, University of Florida preprint UFTP-84-6 (1984).

(8) E. Cremmer et al., Phys. Lett. 79B, 231 (1978); Nucl. Phys. B147, 105 (1979); Phys. Lett. 116B, 231 (1982); Nucl. Phys. B212, 413 (1982).

(9) See for instance J. M. Bardeen, P. J. Steinhardt and M. S. Turner, Phys. Rev. D28, 679 (1983) and references contained within.

(10) L. O'Raifeartaigh, Nucl. Phys. B96, 331 (1975).

(11) G. Coughlan, R. Holman, P. Ramond and G. G. Ross, in preparation.

(12) L. F. Abbott, E. Fahri, and M. B. Wise, Phys. Lett. 117B, 29 (1982).

(13) B. Grinstein, Nucl. Phys. B206, 387 (1982); S. Dimopoulos and F. Wilczek, ICTP UM-HE81-71 (1982); A. Masiero et al., Phys. Lett. 117B, 380 (1982).

(14) P. Salati and J. C. Wallet, Nucl. Phys. B209, 389 (1982).

The Jordan formulation of quantum mechanics : a review

P. K. Townsend
DAMPT . University of Cambridge . U. K.

In the standard formulation of quantum mechanics an observable is represented by a (possibly infinite dimensional) Hermitian matrix. But if a and b are Hermitian the matrix product ab is not necessarily Hermitian. Consider instead the Jordan product,

$$a \circ b = \frac{(ab+ba)}{2} \qquad (1)$$

which is Hermitian if a and b are. With respect to the Jordan product Hermitian matrices form a closed algebra that is obviously commutative, but is nonassociative. However the Jordan product does have the property that

$$a \circ (a^2 \circ b) = (a^2 \circ (a \circ b) \qquad (2)$$

where $a^2 \equiv a \circ a$ =the matrix product of two a's . If a and b are genuine matrices, i. e. associative with respect to the matrix product, this relation is an identity and is known as the Jordan identity.

Hermitian matrices also have the additional property that

$$a \circ a + b \circ b = 0 \Rightarrow a = b = 0 \qquad (3)$$

An algebra with this property is said to be formally real. Now given reality it can be shown that the Jordan identity is equivalent to power associativity. As we have remarked, $a \circ a$ is unambiguously a^2. The product $a \circ a \circ a$ is also unambigously a^3 because of commutativity, but $a \circ a \circ a \circ a$ is potentially ambiguous as it could be $(a \circ a) \circ (a \circ a)$ or $a \circ (a \circ (a \circ a))$. Power associativity ensures that they are the same, and more generally that

$$a^n \circ a^m = a^{m+n} \qquad (4)$$

for all positive integers m and n. This property allows us to define unambiguously functions of an observable as power series expansions. Physically we expect that if α is the value of a in some state then $f(\alpha)$ will be the value of f(a) in that state. This is the physical motivation for power associativity and hence for the Jordan identity.

Of course, as long as we continue to regard the Jordan product as derived from the matrix product we have no need of motivations for identities such as (2) they are simply true. But we can define the Jordan product directly in terms of the Jordan identity. More precisely, we define a Jordan algebra as a vector space with a commutative bilinear product satisfying the Jordan identity. To qualify as an algebra of observables we require in addition that there be a unit element (because the unit matrix is Hermitian) and that the algebra be formally real.[1] In this way we arrive at the axioms

1. $a \circ b = b \circ a$ (commutativity)

2. $a \circ (a^2 \circ b) = a^2 \circ (a \circ b)$ (Jordan identity)

3. $a^2 + b^2 = 0 \Rightarrow a=b=0$ (reality)

4. $\exists\, 1$ s.t. $1 \circ a = a \circ 1 = a$ (existence of 1)

Although Hermitian matrices certainly constitute an example of such an algebra with the Jordan product as in (1), it is by no means obvious that they exhaust the possibilities. To investigate this point we need a classification of formally real unital Jordan algebras. This task is greatly simplified by the theorem that all non-simple Jordan algebras are direct sums of simple algebras. In the finite dimensional case the classification of <u>simple</u> formally real unital Jordan algebras was achieved in 1934[2]. There are five classes which are

1. $J(Q)$

2. $H_m^{\mathbb{R}}$

3. $H_m^{\mathbb{C}}$

4. $H_m^{\mathbb{H}}$

5. $H_3^{\mathbb{O}}$

Every element a of a Jordan algebra can be written as

$$a = a^I e_I \tag{5}$$

where $(e_I) = (1, e_i)$, $i=1,2,\ldots,n$, is a basis for the algebra. The basis can be chosen such that

$$e_i \circ e_j = \delta_{ij} 1 + T_{ijk} e_k \tag{6}$$

In general, an arbitrary nonsingular matrix would replace δ_{ij} but reality implies that this may be transformed to δ_{ij} by a change of basis. Similarly, one can take the structure constants T_{ijk} to be totally symmetric in ijk without loss of generality.

The type 1 algebras are the simplest because T_{ijk} vanishes. They are the generalizations of the algebra of σ matrices under anticommutation and are therefore sub-algebras of the Clifford algebras of positive definite quadratic forms. The type 2, 3, and 4 algebras are realizable as Hermitian matrices over the real numbers, complex numbers, and quaternions respectively, with the Jordan product again being the anticommutator. By contrast, the single type 5 algebra is exceptional because it has no such matrix realization. It can be realized by 3x3 Hermitian "matrices" over the octonions, but these are not matrices in the usual sense because they are not associative with respect to the matrix product. The exceptionality of $H_3^{\emptyset}$ can be proved by means of the "Glennie identity"[3], which is eighth order in elements of the algebra. If the elements are ordinary matrices and the Jordan product the anticommutator then this is indeed an identity. But it is not satisfied by $H_3^{\emptyset}$.

The classification of infinite dimensional Jordan algebras has recently been acomplished. It appears that all infinite dimensional simple Jordan algebras are extensions of types 1 to 4. In particular, there are no exceptional infinite dimensional algebras[3].

This classification already suggests one generalization of quantum mechanics, i.e. that in which the complex numbers are replaced by the real numbers, quaternions or octonions. It appears that real quantum mechanics is essentially equivalent to complex quantum mechanics, and quaternion quantum mechanics suffers from a surplus of imaginary units and does not yield much that is new[4]. Octonionic quantum mechanics, based on the exceptional Jordan algebra[5], is the most interesting case because exceptionality implies that no Hilbert space formulation is possible. This is therefore a genuine and radical generalization of quantum mechanics, although one wihout as yet any application.

But my purpose here is not to review the generalizations of quantum mechanics afforded by the Jordan formulation but to explain how conventional quantum mechanics is contained in it. So far we have discussed the algebra of observables but we have not discussed the representation of states. In the usual formulation, states are identified with rays in a Hilbert space. But if $(|n\rangle)$ is a basis of vectors in a Hilbert space, n=1,2,...,the Hermitian operators

$$E_n = |n\rangle\langle n| \quad ,n=1,2,... \tag{7}$$

contain the same information. These operators satisfy

$$\mathrm{tr}E = 1 \tag{8}$$

$$E^2 = E \tag{9}$$

the first relation following from the normalization of the states.

To carry this over to the Jordan formulation we have first to introduce the trace form of a Jordan algebra, which is defined to satisfy

$$\mathrm{tr}1 = \nu \qquad \mathrm{tr}e_i = 0 \tag{10}$$

with ν an integer to be specified shortly. This is equivalent to the introduction of the positive definite inner product

$$(e_i,e_j) = \frac{1}{\nu}\mathrm{tr}(e_i \circ e_j) = \delta_{ij} \qquad (11)$$

Now an element P of an algebra satisfying $P^2 = P$ is called an idempotent. For idempotents P_1, P_2, orthogonality with respect to the inner product (11) implies the seemingly stronger relation

$$P_1 \circ P_2 = 0 \qquad (12)$$

which we may take as the definition of orthogonality for idempotents. An idempotent E that cannot be expressed as the sum of two orthogonal idempotents is said to be primitive. The maximal number of primitive orthogonal idempotents is the degree of the algebra, and this complete set constitutes a decomposition of unity:

$$\sum_{i=1}^{\mathrm{deg.}} E_i = 1 \qquad (13)$$

If the trace form is normalized such that trE=1 as in (8), then the degree of the algebra is the integer ν appearing in (10). Thus a Jordan algebra of degree ν describes a quantum system of ν linearly independent states. According to a theorem of Jordan, von Neumann, and Wigner, every element of the algebra can be expressed as

$$a = \sum_{i=1}^{\nu} a_i E_i(a) \qquad (14)$$

i.e. as an expansion on a complete set of primitive idempotents, the set depending on the element chosen. The numbers a_i are the "eigenvalues" of the observable a in the state represented by $E_i(a)$.

The primitive idempotents E_i represent pure states. A mixed state is represented by

$$\rho = \sum_{i=1}^{\nu} \rho_i E_i, \quad \sum_{i=1}^{\nu} \rho_i = 1, \quad \rho_i \geq 0 \;, i=1,2,\ldots,\nu$$

i.e. by a positive definite element ρ satisfying $\mathrm{tr}\rho=1, \rho^2 \leqslant \rho$ with equality only for pure states. This is essentially the density matrix formalism and as in that case the expectation value of any observable a in the mixed state represented by ρ is

$$\langle a \rangle = \mathrm{tr}(a \circ \rho) \qquad (16)$$

A distiction between the Jordan formulation and the density matrix formulation of quantum mechanics arises when one considers time evolution. In the density matrix formulation one postulates the unitary evolution equation

$$\partial_t \rho = -i[H,\rho] \qquad (17)$$

where H is the positive definite and Hermitian Hamiltonian matrix. This equation has the property that it takes pure states into pure states. If one requires this property of the time evolution equation then in the Jordan formulation it must take the form

$$\partial_t \rho = D(\rho) \tag{18}$$

where D is a linear operator with the property that

$$D(e_I \circ e_J) = D(e_I) \circ e_J + e_I \circ D(e_J) \tag{19}$$

i.e. D is required to be a <u>derivation</u> of the algebra. The derivations form a Lie algebra under commutation and they generate the automorphism group of the algebra. It is a remarkable fact that the derivations of a Jordan algebra can all be expressed as

$$D_{x,y} = [L_x, L_y] \tag{20}$$

where L_x is the operation of multiplication by a <u>traceless</u> element. The Jacobi identity for $D_{x,y}$ is satisfied as a consequence of the Jordan identity for x and y. The representation (20) of the derivations D allows us to rewrite (18) as

$$\partial_t \rho = D_{x,y}\rho = x \circ (\rho \circ y) - (x \circ \rho) \circ y \tag{21}$$

Defining the <u>associator</u> of three elements a.b.c. as

$$[a,b,c] = a \circ (b \circ c) - (a \circ b) \circ c \tag{22}$$

we see that (21) is equivalent to

$$\partial_t \rho = [x, \rho, y] \tag{23}$$

The associator plays a role in the Jordan formulation similar to that of the commutator in the standard formulation. To make contact with the latter we work out the r.h.s. of (23) for complex Hermitian matrices x,y and ρ and the Jordan product given by (1). One finds

$$\partial_t \rho = [\ [x,y]\ , \rho] \tag{24}$$

so that we have equivalence with (17) if

$$H = i[x,y] + \lambda 1 \tag{25}$$

with λ an arbitrary real constant. Thus, in a sense the Jordan formulation is the "square root" of the standard formulation. One wonders whether there is a connection to supersymmetry here. The interpretation of the elements x and y in the case of real, quaternionic, or the octonionic Jordan algebras is problematic.

Density matrices belong to the cone of positive definite Hermitian matrices. This has a generalization to Jordan algebras and the cone C is called a domain of positivity. The group of linear transformations of coordinates of a cone is its automorphism group and if this group acts transitively the cone is said to be homogeneous. A domain of positivity is a homogeneous cone and its automorphism group is generated by the derivations of the associated algebra together with the operation L of multiplication by an element of the algebra. The derivations form a subalgebra that generate the stability group of the cone. and the linear operators L generate translations in the cone.

If one considers the possibility of a more general time evolution for which pure states do not necessarily evolve into pure states [6)] then one might think to generalize (18) to include a term $L(\rho)$ in $\partial_t \rho$. But this violates the unit trace condition on ρ. One can construct an operation which preserves the trace by rescaling ρ but as the rescaling parameter is ρ-dependent this is not a linear operation. Thus one cannot use the automorphism group of the cone to evolve density matrices into density matrices via a linear evolution equation (except of course for the stability subgroup that evolves pure states into pure states). Conversely, those linear transformations that take pure states into mixed states cannot form a group. This is in accord with the second law of thermodynamics which allows such transformations to form a semi-group but not a group, but the precise formulation of such generalized evolution equations in Jordan algebraic language has yet to be worked out.

Acknowledgements I am grateful to Drs. Gary Gibbons, German Sierra, and Murat Gunaydin for helpful conversations and correspondence.

References

1. P. Jordan, Gottingen Nachr. (1933), 209
2. P. Jordan, J von Neumann, and E. P. Wigner, Ann. Math. 56, (1934), 29.
3. K. McCrimmon, Algebras, Groups and Geometries 1 (1984), 1
4. D. Finkelstein, J. M. Jauch, S. Schiminovitch, and D. Speiser, J. Math. Phys. 3, (1962), 207.
5. M. Gunaydin, C. Piron, and H. Ruegg, Comm. Math. Phys. 61, (1978), 69.
6. S. W. Hawking, Phys. Rev. D14, (1976), 2460.

LIST OF PARTICIPANTS

Abad, J.	Univ. de Zaragoza
Achucarro, A.	Univ. del País Vasco, Bilbao
Aguila, F. del	Univ. Autónoma de Barcelona
Aldaya, V.	Univ. de Valencia
Alonso, F.	Univ. Autónoma de Madrid
Alonso, J.L.	Univ. de Zaragoza
Alvarez,E.	Univ. Autónoma de Madrid
Ametller, Ll.	Univ. Autónoma de Barcelona
Asorey, M.	Univ. de Zaragoza
Azcárraga, J.A.de	Univ. de Valencia
Ayala, C.	Univ. Autónoma de Barcelona
Bernabéu, J.	Univ. de Valencia
Bona, C.	Univ. de Palma de Mallorca
Bordes, J.M.	Univ. de Valencia
Botella, F.J.	Univ. de Valencia
Boya, L.J.	Univ. de Zaragoza
Cariñena, J.F.	Univ. de Zaragoza
Carot, J.	Univ. de Palma de Mallorca
Cerveró, J.M.	Univ. de Salamanca
Cornet, F.	Univ. Autónoma de Barcelona
Chamorro,A.	Univ. del País Vasco, Bilbao
Duff, M.J. (lecturer)	Imperial College, London
Elizalde,E.	Univ. de Barcelona
Espriu, D.	Univ. of Oxford
Freedman, D.Z. (lecturer)	M.I.T. Cambridge
Frydryszak, A.	Univ. of Wrocław
Fustero, X.	Univ. Autónoma de Barcelona
Futamase, T.	University College, Cardiff
Gaite, J.	Univ. de Salamanca
García-Canal	Univ. Nac. de La Plata
Gato, B.	I.E.M. (CSIC), Madrid

Gibbons, G.W. (lecturer)	DAMTP, Cambridge
Ginsparg, P.	Harvard Univ.
Gomis, J.	Univ. de Barcelona
González, J.	Univ. Autónoma de Madrid
Goñi, M.A.	Univ. del País Vasco, Bilbao
Grau, A.	Univ. Autónoma de Barcelona
Grifols, J.A.	Univ. Autónoma de Barcelona
Gutiérrez, M.	Univ. Complutense, Madrid
Herrero, M.J.	Univ. Autónoma de Madrid
Ibáñez, L.E.	Univ. Autónoma de Madrid
Ibort, A.	Univ. de Zaragoza
Jordán de Urríes, F.	Univ. de Alcalá de Henares
Koh, I.G.	Imperial College, London
Kowalski-Glikman, J.	Univ. of Warsaw
Lara, J.	Univ. Autónoma de Madrid
Latorre, J.I.	Univ. de Barcelona
León, J.J.	I.E.M. (CSIC), Madrid
López Almorox, A.	Univ. de Salamanca
López C.	Univ. Autónoma de Madrid
Lucha, W.	Univ. of Wien
Lukierski, J.	Univ. of Wroclaw
Macedo, P.	Univ. of Coimbra
Mas, Ll.	Univ. de Palma de Mallorca
Mató, P.	Univ. Autónoma de Barcelona
Méndez, A.	Univ. Autónoma de Barcelona
Milewski, B.	Max-Planck-Institut, München
Minnaert, P.	Univ. of Bordeaux
Miramontes, L.	Univ. de Santiago de Compostela
Muñoz, C.	Univ. Autónoma de Madrid
Narganes, F.J.	Univ. del País Vasco, Bilbao
Nelson, J.E.	Univ. of Torino
Neufeld, H.	Univ. of Wien
Olmo, M. del	Univ. de Valladolid
Orteu, F.X.	Univ. Autónoma de Barcelona

Palanques-Mestre, A.	Univ. de Barcelona
Pascual, R.	Univ. Autónoma de Barcelona
Pérez, C.	Univ. Autónoma de Madrid
Pérez-Canyellas, A.	Universidad de Valencia
Pérez-Mercader, J.A.	I.E.M.(CSIC), Madrid
Pich, A.	Univ. de Valencia
Quirós, J.M.	I.E.M.(CSIC), Madrid
Raby, S. (lecturer)	Los Alamos Nat.Lab. N.M.
Ramírez, A.	Univ. de Valencia
Ramírez, J.J.	Univ. Complutense, Madrid
Ramón-Medrano, M.	Univ. Complutense, Madrid
Ramond, P. (lecturer)	Univ. of Florida
Rañada, M.F.	Univ. de Zaragoza
Regge, T. (lecturer)	Univ. of Torino
Rodríguez, M.A.	Univ. Complutense, Madrid
Rodríguez, M.J.	Univ. Complutense, Madrid
Rodríguez, R.J.	Univ. de Zaragoza
Román-Roy, N.	Univ. de Barcelona
Ross, G.G. (lecturer)	Univ. of Oxford
Sánchez-Guillén, J.J.	Univ. de Santiago de Compostela
San José, F.	Univ. Complutense, Madrid
Santamaría, A.	Univ. de Valencia
Sanz, J.L.	Univ. de Santander
Segui, A.	Univ. de Zaragoza
Sierra, G.	Univ. Complutense, Madrid
Solá, J.	Univ. Autónoma de Barcelona
Stelle, K.S. (lecturer)	Imperial College, London
Tarancón, A.	Univ. de Zaragoza
Townsend, P. (lecturer)	DAMTP, Cambridge
Valle, J.W.F.	Rutherford Lab., Didcot
Varias, A.	Univ. de Santiago de Compostela
Verdaguer, E.	Univ. Autónoma de Barcelona
Vindel, P.	Univ. de Valencia
Zanelli, J.	I.C.T.P., Trieste